Lecture Notes in Computer Science   10879

Commenced Publication in 1973
Founding and Former Series Editors:
Gerhard Goos, Juris Hartmanis, and Jan van Leeuwen

More information about this series at http://www.springer.com/series/7410

Itai Dinur · Shlomi Dolev
Sachin Lodha (Eds.)

Cyber Security Cryptography and Machine Learning

Second International Symposium, CSCML 2018
Beer Sheva, Israel, June 21–22, 2018
Proceedings

 Springer

Editors
Itai Dinur
Ben-Gurion University of the Negev
Beer Sheva
Israel

Sachin Lodha
Tata Consultancy Services (India)
Chennai, Tamil Nadu
India

Shlomi Dolev
Ben-Gurion University of the Negev
Beer Sheva
Israel

ISSN 0302-9743   ISSN 1611-3349   (electronic)
Lecture Notes in Computer Science
ISBN 978-3-319-94146-2   ISBN 978-3-319-94147-9   (eBook)
https://doi.org/10.1007/978-3-319-94147-9

Library of Congress Control Number: 2018947568

LNCS Sublibrary: SL4 – Security and Cryptology

Printed on acid-free paper

This Springer imprint is published by the registered company Springer International Publishing AG
part of Springer Nature
The registered company address is: Gewerbestrasse 11, 6330 Cham, Switzerland

Preface

CSCML, the International Symposium on Cyber Security, Cryptography, and Machine Learning, is an international forum for researchers, entrepreneurs, and practitioners working in the theory, design, analysis, implementation, or application of cyber security, cryptography, and machine learning systems and networks, and, in particular, of conceptually innovative topics in the scope.

Information technology became crucial to our everyday life in indispensable infrastructures of our society and therefore a target for attacks by malicious parties. Cyber security is one of the most important fields of research today because of this phenomenon. The two, sometimes competing, fields of research, cryptography and machine learning, are the most important building blocks of cyber security, as cryptography hides information by avoiding the possibility to extract any useful information pattern while machine learning searches for meaningful information patterns.

The subjects include cyber security design; secure software development methodologies; formal methods, semantics and verification of secure systems; fault tolerance, reliability, availability of distributed secure systems; game-theoretic approaches to secure computing; automatic recovery, self-stabilizing and self-organizing systems; communication, authentication and identification security; cyber security for mobile and Internet of Things; cyber security of corporations; security and privacy for cloud, edge and fog computing; cryptocurrency; blockchain; cryptography; cryptographic implementation analysis and construction; secure multi-party computation; privacy-enhancing technologies and anonymity; post-quantum cryptography and security; machine learning and big data; anomaly detection and malware identification; business intelligence and security; digital forensics, digital rights management; trust management and reputation systems; and information retrieval, risk analysis, DoS.

The second edition of CSCML took place during June 21–22, 2018, in Beer-Sheva, Israel. This year the symposium was held in cooperation with the International Association for Cryptologic Research (IACR), and there was a dedicated special issue of selected papers in the *Information and Computation* journal.

This volume contains one invited paper, 16 contributions selected by the Program Committee, and six brief announcements. All submitted papers were read and evaluated by the Program Committee members, assisted by external reviewers. We are grateful to the EasyChair system for assisting in the reviewing process.

The support of Ben-Gurion University of the Negev (BGU), in particular the BGU Lynne and William Frankel Center for Computer Science, the BGU Cyber Security Research Center, ATSMA, the Department of Computer Science, Tata Consultancy Services, IBM and BaseCamp, is also gratefully acknowledged.

April 2018
Sitaram Chamarty
Itai Dinur
Shlomi Dolev
Sachin Lodha

Organization

CSCML, the International Symposium on Cyber Security Cryptography and Machine Learning, is an international forum for researchers, entrepreneurs, and practitioners in the theory, design, analysis, implementation, or application of cyber security, cryptography, and machine learning systems and networks, and, in particular, of conceptually innovative topics in the scope.

Founding Steering Committee

Orna Berry	DELLEMC, Israel
Shlomi Dolev (Chair)	Ben-Gurion University, Israel
Yuval Elovici	Ben-Gurion University, Israel
Ehud Gudes	Ben-Gurion University, Israel
Jonathan Katz	University of Maryland, USA
Rafail Ostrovsky	UCLA, USA
Jeffrey D. Ullman	Stanford University, USA
Kalyan Veeramachaneni	MIT, USA
Yaron Wolfsthal	IBM, Israel
Moti Yung	Columbia University and Snapchat, USA

Organizing Committee

General Chairs

Shlomi Dolev	Ben-Gurion University of the Negev, Israel
Sachin Lodha	Tata Consultancy Services, India

Program Chairs

Sitaram Chamarty	Tata Consultancy Services, India
Itai Dinur	Ben-Gurion University of the Negev, Israel

Organizing Chair

Timi Budai	Ben-Gurion University of the Negev, Israel

Program Committee

Yehuda Afek	Tel Aviv University, Israel
Adi Akavia	Tel Aviv Yaffo Academic College, Israel
Amir Averbuch	Tel Aviv University, Israel
Roberto Baldoni	Università di Roma La Sapienza, Italy
Michael Ben-Or	Hebrew University, Israel
Anat Bremler-Barr	IDC Herzliya, Israel

Additional Reviewers

Eran Amar
Mai Ben Adar-Bessos
Rotem Hemo
Yu Zhang

Sponsors

Contents

Optical Cryptography for Cyber Secured and Stealthy Fiber-Optic Communication Transmission

Invited Paper

Tomer Yeminy[1], Eyal Wohlgemuth[1], Dan Sadot[1],
and Zeev Zalevsky[2(✉)]

[1] Department of Electrical and Computer Engineering,
Ben-Gurion University of the Negev, 84105 Beer Sheva, Israel
[2] Faculty of Engineering, Bar Ilan University, 52900 Ramat Gan, Israel
Zeev.Zalevsky@biu.ac.il

Abstract. We propose a method for stealthy, covert, fiber-optic communication. In this method, the power of the transmitted signal is spread and lowered below the noise level both in time as well as in frequency domains which makes the signal "invisible". The method is also efficient in jamming avoidance.

1 Short Description

Short "laymen" description of the proposed approach can be as follows: The concept described in this paper aims to encode the optical information by doing it in the "photon" level before it is being sampled and converted into a "digital electron" of information. Since the encryption is done in the analog domain it can be added in parallel to all the existing digital encryption techniques which can add strength to the proposed concept and which are completely orthogonal to what we are doing. Since in optics conventional sensors capture only intensity and not phase of the arriving wavefront, then the process of capturing the information and converting the analog photon into a digital electron that can be processed and decoded already destroys large portion of the information that is needed for the decoding process. The analog encoding that we do in the photon level properly plays with the phase and the sampling scheme both in the time domain as well as in the Fourier domain such that in both domains the signal is lowered below the noise level. Thus, the information signal is below the noise existing in the system and thus it is unseen by the intruder and when the intruder attempts to capture the analog "photon" and to convert it into a digital one in order to try to see if there is an encrypted information, then the sampling process which destroys the phase (captures only intensity) and adds quantization noise completely erases the information that is being hidden below the noise. A more elaborated description including both theoretical mathematical background as well as numerical and experimental validation can be found in Refs. [1–4].

If we now adopt a more mathematical description methodology, then the proposed stealthy communications system is illustrated in the attached Fig. 1. While in Fig. 1(a)

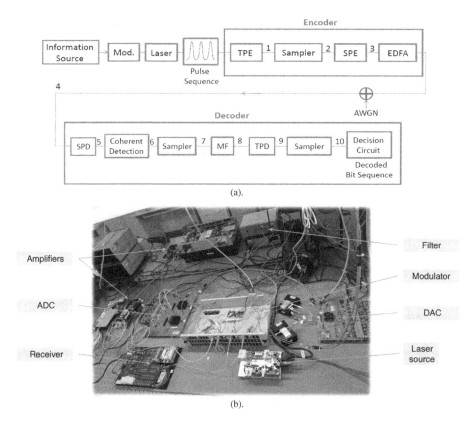

Fig. 1. (a). Proposed covert communication system. Mod.-Modulator, TPE-Temporal Phase Encoder, SPE-Spectral Phase Encoder, EDFA-Erbium Doped Fiber Amplifier, AWGN-Additive White Gaussian Noise, SPD-Spectral Phase Decoder, MF-Matched Filter, TPD-Temporal Phase Decoder. (b). The established experimental setup. ADC and DAC are analog to digital converter and vice versa respectively.

one may see the schematic system configuration and in Fig. 1(b) one can see the constructed experimental setup. The temporal phase of the pulses at the transmitted pulse sequence is encrypted in order to reduce its power spectral density (PSD) below the noise level. The PSD reduction occurs since each pulse in the sequence has different phase, hence, the pulses are added incoherently in frequency domain. Then, the spectral amplitude of the signal is deliberately spread wide, essentially enabling to transmit a signal with low PSD (lowering the signal below the noise level in the frequency domain).

The spectral spreading is achieved by optically sampling the signal (this can be implemented by an optical amplitude modulator), which generates replicas of the signal in frequency domain. The spectral phase of the sampled signal is subsequently encrypted. The spectral phase encryption has two goals. The first is signal spreading below the noise level in time domain. The second goal is to prevent signal detection in frequency domain by coherent addition of the various spectral replicas of the signal.

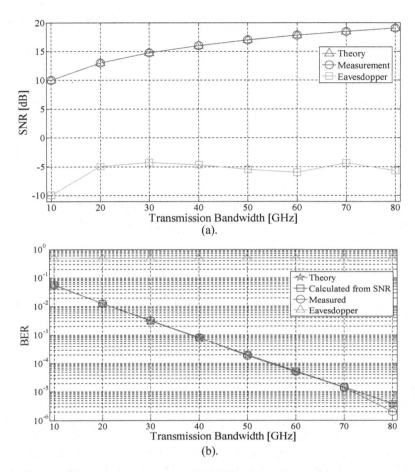

Fig. 2. Estimated performance. (a) SNR after decoding. (b) BER after decoding.

Only an authorized user having the temporal and spectral encrypting phases in hand can raise the signal above the noise level and detect it. The encrypted signal is then amplified and sent to the receiver.

At the receiver, the spectral phase of the signal is decrypted and the signal is subsequently coherently detected. Then, all the spectral replicas of the signal are folded to the baseband by means of electrical sampling, therefore the PSD of the signal is reconstructed and in turn, the signal to noise ratio (SNR) is improved. This is achieved by coherently adding all the signal's spectral replicas at the baseband (hence the signal is reinforced) whereas the spectral replicas of the noise are added incoherently (consequently they are averaged to a low value). Then, a matched filter is applied and the temporal phase of the signal is decrypted. Thus, the signal is raised above the noise level in both time and frequency domains. A decision circuit is subsequently used to recover the original transmitted symbol sequence.

It should be noted that an eavesdropper cannot decrypt the transmitted signal since using wrong decrypting temporal and spectral phases does not raise the signal above

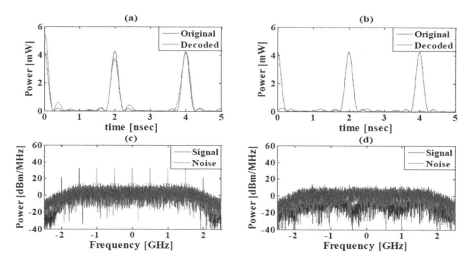

Fig. 3. Decoded signal, authorized user and eavesdropper. (a) Original noiseless pulse sequence and authorized user noisy decoded pulse sequence. (b) Original noiseless pulse sequence and eavesdropper noisy decoded pulse sequence. (c) Authorized user pulse sequence and noise power spectral density. (d) Eavesdropper pulse sequence and noise power spectral density.

the noise level. Hence, very low SNR and high bit error rate (BER) are experienced by the eavesdropper, being unable to reveal the transmission's existence.

The SNR and BER performance of an authorized user and an eavesdropper are shown in Figs. 2(a) and (b), respectively, as a function of the communication systems transmission bandwidth (higher transmission bandwidth requires higher transmission power but also results in better SNR and BER performance since more spectral replicas of the signal are added coherently at the baseband while the noise is added incoherently).

Figure 3 shows the original and decrypted signal in time and frequency domains for an authorized user and an eavesdropper. It is clearly seen that while the authorized user properly decrypts the received signal, the eavesdropper does not succeed to reveal the signal as it remains below the noise level in both time and frequency domains due to the usage of incorrect temporal and spectral decrypting phases. The parameters used for the simulations presented in the figures are common in optical communication systems.

Our stealthy communications fiber optic system can also efficiently cope with jamming. When an opponent tries to jam our secure transmitted signal by occupying its temporal and spectral domains with a high power signal, the jamming signal will be lowered by the decryption module below the noise level due to its spreading in time and frequency domains. However, our secure transmitted signal will be properly decrypted and raised above the noise level since it has the right encrypting temporal and spectral phase.

The required resources needed for our proposed encryption method are standard fiber optical communication system, optical amplitude modulators (to implement the optical sampling), temporal phase modulators and spectral phase modulators. All of the above can easily be integrated in a given photonic communication link by using currently available optics communication hardware. Therefore, we also do not require any special external interfaces and we are fully compatible with existing photonic hardware and protocols [3, 4].

2 Conclusions

This short paper gives an insight on a new way of analog photonic encryption that can strengthen the existing encryption concepts working on top of the digital electrons of information. The analog photon of information is lowered below the noise level both in the time as well as in the Fourier domain by performing temporal and spectral phase encoding and by properly sampling the signal in the time domain such the encoding phases uniquely redistribute the energy of the signal over both time and spectral domain and lower them below the noise level to make then un-visible and un-detectable as much as possible. Due to its properties the proposed scheme has high applicability in commercial cyber based configurations [5].

References

1. Yeminy, T., Sadot, D., Zalevsky, Z.: Spectral and temporal stealthy fiber-optic communication using sampling and phase encoding. Opt. Exp. **19**, 20182–20198 (2011)
2. Yeminy, T., Sadot, D., Zalevsky, Z.: Sampling impairments influence over stealthy fiber-optic signal decryption. Opt. Commun. **291**, 193–201 (2013)
3. Wohlgemuth, E., Yoffe, Y., Yeminy, T., Zalevsky, Z., Sadot, D.: Demonstration of coherent stealthy and encrypted transmission for data center interconnection. Opt. Exp. **26**, 7638–7645 (2018)
4. Wohlgemuth, E., Yeminy, T., Zalevsky, Z., Sadot, D.: Experimental demonstration of encryption and steganography in optical fiber communications. In: Proceedings of the European Conference on Optical Communication (ECOC 2017), Gothenburg, Sweden, 17–21 September 2017
5. Sadot, D., Zalevsky, Z., Yeminy, T.: Spectral and temporal stealthy fiber optic communication using sampling and phase encoding detection systems. US patent No. 9288045; EP2735111A1

Efficient Construction of the Kite Generator Revisited

Orr Dunkelman[1]($\boxtimes$)(iD) and Ariel Weizman[2](iD)

[1] Computer Science Department, University of Haifa, Haifa, Israel
orrd@cs.haifa.ac.il
[2] Department of Mathematics, Bar-Ilan University,
Ramat Gan, Israel

Abstract. The *kite generator*, first introduced by Andreeva et al. [1], is a strongly connected directed graph that allows creating a message of almost any desired length, connecting two chaining values covered by the *kite generator*. The *kite generator* can be used in second pre-image attacks against (dithered) Merkle-Damgård hash functions.

In this work we discuss the complexity of constructing the *kite generator*. We show that the analysis of the construction of the *kite generator* first described by Andreeva et al. is somewhat inaccurate and discuss its actual complexity. We follow with presenting a new method for a more efficient construction of the *kite generator*, cutting the running time of the preprocessing by half (compared with the original claims of Andreeva et al. or by a linear factor compared to corrected analysis). Finally, we adapt the new method to the dithered Merkle-Damgård structure.

1 Introduction

One of the important fundamental primitives in cryptography is cryptographic hash functions. They are widely used in digital signatures, hashing passwords, message authentication code (MAC), etc. Hence, their security has a large impact on the security of many protocols.

Up until the SHA3 competition, the most widely used hash function construction was the Merkle-Damgård one [5,11]. The Merkle-Damgård structure extends a compression function $f : \{0,1\}^n \times \{0,1\}^m \to \{0,1\}^n$ into a hash function $\mathcal{MDH}^f : \{0,1\}^* \to \{0,1\}^n$. Indeed, the Merkle-Damgård structure is still widespread, as can be seen from the wide use of the SHA2 family [12]. However, in the last fifteen years, a series of works pointed out several structural weaknesses in the Merkle-Damgård construction and its dithered variant [1,6–9].

One way for comparing between different structures of cryptographic hash functions is considering generic attacks. Naturally, generic attacks use complex algorithms and data structures, and often become used as subroutines in other attacks. In such cases, the accurate analysis of these algorithms and data structures becomes very important. For example, Kelsey and Kohno suggest a special data structure, called the *diamond structure*, which is a complete binary tree

© Springer International Publishing AG, part of Springer Nature 2018
I. Dinur et al. (Eds.): CSCML 2018, LNCS 10879, pp. 6–19, 2018.
https://doi.org/10.1007/978-3-319-94147-9_2

with 2^ℓ leaves, to support the *herding* attack [8]. Blackburn et al. [4] point out an inaccuracy in Kelsey-Kohno's analysis and fix it, resulting in an increased time complexity. Later work presented new algorithms for more efficient constructions of the diamond structure [10,14].

A different second pre-image attack is based on the *kite generator*. This is a long message (with 2^k blocks) second pre-image attack due to Andreeva et al. [1] on the Merkle-Damgård structure and its dithered variant [13]. The *kite generator* is a strongly connected directed graph of 2^{n-k} chaining values that for each two chaining values a_1, a_2 covered by the *kite generator*, there exist a sequence of message blocks of almost any desired length that connects a_1 to a_2. Their analysis claims that the *kite generator*'s construction takes about 2^{n+1} compression function calls.

We start this paper by pointing out an inaccuracy in their construction based on some theorems from the Galton-Watson branching process field: We show that the resulting graph, using the original construction, is not strongly connected and therefore is unusable in the online phase. We proceed by offering corrected analysis that shows that the construction of the kite generator takes about $(n-k) \cdot 2^n$ compression function calls.

We then show a completely different method to build the kite generator. This new method allows constructing the kite generator in time of 2^n compression function calls, i.e., it takes half the time of the originally inaccurate claim. Finally, we adapt all these issues to the dithered variant of the Merkle-Damgård structure.

This paper is organized as follows: Sect. 2 gives notations and definitions used in this paper. In Sect. 3 we quickly recall Andreeva et al.'s second preimage attack, and most importantly, the construction of the *kite generator*. We identify and analyze the real complexity of constructing a usable *kite generator* in Sect. 4. We introduce a new method for constructing kite generators in Sect. 5. We treat the analysis of the *kite generator* and the new construction in the case of dithered Merkle-Damgård in Sect. 6. Finally, we conclude the paper in Sect. 7.

2   Notations and Definitions

Definition 1. *A cryptographic hash function is a function $H : \{0,1\}^* \to \{0,1\}^n$, that takes an arbitrary length input and transforms it to an n-bit output such that $H(x)$ can be computed efficiently, while the function has three main security properties:*

1. Collisions resistance: It is hard to find (with high probability) an adversary that could produce two different messages M, M' such that $H(M) = H(M')$ in less than $\mathcal{O}(2^{n/2})$ calls to $H(\cdot)$.
2. Second pre-image resistance: Given M such that $H(M) = h$, an adversary cannot produce (with high probability) an additional message $M' \neq M$ such that $H(M') = h$ in less than $\mathcal{O}(2^n)$ calls to $H(\cdot)$.

3. Pre-image resistance: Given a hash value h, an adversary cannot produce (with high probability) any message M such that $H(M) = h$ in less than $\mathcal{O}(2^n)$ calls to $H(\cdot)$.

Definition 2 (Merkle-Damgård structure ($\mathcal{MDH}$)). *The Merkle-Damgård structure [5, 11] is a structure of an iterative hash function. Given a compression function $f : \{0,1\}^n \times \{0,1\}^m \rightarrow \{0,1\}^n$ that takes an n-bit chaining value and an m-bit message block and transforms them into a new n-bit chaining value, $\mathcal{MDH}^f$ is defined as follow: For an input message M:*

1. Padding step[1]
   (a) Concatenate '1' at the end of the message.
   (b) Let b be the number of bits in the message, and ℓ be the number of bits used to encode the message length in bits.[2] Pad a sequence of $0 \leq k < m$ zeros, such that $b + 1 + k + \ell \equiv 0 \pmod{m}$.
   (c) Append the message with the original message length in bits, encoded in ℓ bits.
2. Divide the message to blocks of m bits, so if the length of padded message is $L \cdot m$ then
$$M = m_0 || m_1 || \ldots || m_{L-1}.$$
3. The iterative process starts with a constant IV, denoted by h_{-1}, and it updated in every iteration, according to the appropriate message block m_i (for $0 \leq i \leq L - 1$), to new chaining value:
$$h_i = f(h_{i-1}, m_i).$$
4. The output of this process is:
$$\mathcal{MDH}^f(M) = h_{L-1}$$

The process is depicted in Fig. 1.

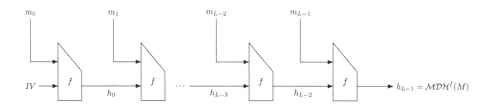

Fig. 1. The Merkle-Damgård Structure

---

[1] We describe here the standard padding step done in many real hash functions such as MD5 and SHA1. Other variants of this step exist, all aiming to achieve prefix-freeness.

[2] It is common to set $2^\ell - 1$ as the maximal length of a message.

Merkle [11] and Damgård [5] proved that if the compression function is collision-resistant then the whole structure (when the padded message includes the original message length) is also collision-resistant. Although the Merkle-Damgård structure is believed to be secure also from second pre-image attacks, in practice it is not [1,2,6,9].

Definition 3. *Let $G = (V, E)$ be a directed graph. A directed edge from v to u is denoted by (v, u). For each $v \in V$ we define the in-degree of v, denoted by $d_{in}(v)$, to be the number of edges that ingoing to v, and the out-degree of v, denoted by $d_{out}(v)$, to be the number of edges that outgoing from v.*

Definition 4 (Galton-Watson Branching Process). *A Galton-Watson branching process is a stochastic process that illustrates a population increasing, which starts from one individual in the first state S_0, and for each $t \in \mathbb{N} \cup \{0\}$ each individual from S_t produces $i \in \mathbb{N} \cup \{0\}$ offsprings for the next state S_{t+1} according to a fixed probability distribution. Formally, a Galton-Watson branching process is defined as a Markov chain $\{Z_t\}_{t \in \mathbb{N} \cup \{0\}}$:*

1. *Let $Z_0 := 1$.*
2. *For each $(i, t) \in \mathbb{N} \times \mathbb{N}$ let X_i^t be a random variable follows a fixed probability distribution $P : \mathbb{N} \cup \{0\} \to [0, 1]$ with expected value $\mu < \infty$.*
3. *Define inductively:*

$$\forall t \in \mathbb{N} : Z_t := \sum_{1 \leq i \leq Z_{t-1}} X_i^t.$$

The random variable X_i^t represents the number of offspring produced by the i'th element (if there is one) of the Z_{t-1} elements from the time $t - 1$.

A central issue in the theory of branching processes is ultimate extinction, i.e., the event of some $Z_t = 0$. One can see that $E(Z_t) = \mu^t$. Still, even for $\mu \geq 1$ as long as $Pr[X_i = 0] > 0$ the ultimate extinction is an event with a positive probability. To study such events of ultimate extinction we need to study their probability, given by

$$\lim_{t \to \infty} Pr[Z_t = 0] = Pr[\exists t \in \mathbb{N} : Z_t = 0].$$

In [3] Athreya and Ney show that the probability of ultimate extinction is the smallest fixed point $x \in [0, 1]$ of the P's moment-generating function $f_P(x)$. For example, if $X_i^t \sim Poi(\lambda)$, then the probability of ultimate extinction is the smallest solution $x \in [0, 1]$ of $e^{\lambda(x-1)} = x$.

3 The Kite Generator

In [1] Andreeva et al. suggest a method to generate second pre-images for long messages of 2^k blocks. Using an expensive precomputation of 2^{n+1} compression function calls, the online complexity of their attack is $\max(\mathcal{O}(2^k), \mathcal{O}(2^{\frac{n-k}{2}}))$ time and $\mathcal{O}(2^{n-k})$ memory.

3.1   The Attack's Steps

The Precomputation. In the precomputation the adversary constructs a data structure called the *kite generator*, which is a strongly connected directed graph with 2^{n-k} vertices. The vertices are labeled by chaining values and the directed edges by message blocks which lead one chaining value to another. Given two chaining values a_1, a_2 covered by the kite generator, this structure allows to create a message of almost any desired length that connects a_1 to a_2.

To construct the kite generator, the adversary picks a set A of 2^{n-k} different chaining values, containing the IV. For each chaining value $a \in A$ he finds two message blocks $m_{a,1}, m_{a,2}$ such that $f(a, m_{a,1}), f(a, m_{a,2}) \in A$. We note that for each chaining value $a \in A$, $d_{out}(a) = 2$, and therefore $\forall a \in A : E[d_{in}(a)] = 2$.

The Online Phase. In the online phase, given a long message M, the adversary computes $H(M)$ and finds, with high probability, a chaining value h_i, for $n-k < i < 2^k$, such that $h_i \in A$. Now the adversary creates, using the kite generator, a sequence of i message blocks, starting from the IV, that leads to h_i. It is done in the following steps:

1. The adversary performs a random walk in the kite generator of $i - (n - k)$ message blocks, from the IV. To do so, the adversary starts from the IV and chooses an arbitrary message block $m \in \{m_{IV,1}, m_{IV,2}\}$ and traverse to the next chaining value $h_1 = f(IV, m)$. The adversary continues in the same manner $i - (n - k) - 1$ times, until $h_{i-(n-k)}$ is reached. Denote the concatenation of the chosen message blocks by s_1.
2. The adversary computes all the $2^{\frac{n-k}{2}}$ chaining values reachable from $h_{i-(n-k)}$ by walking $\frac{n-k}{2}$ steps in the kite generator.
3. The adversary computes all the expected $2^{\frac{n-k}{2}}$ chaining values that may lead to h_i by walking back in the kite generator $\frac{n-k}{2}$ steps from h_i.
4. The adversary looks for a collision between these two lists (due to the birthday paradox, such a collision is expected with high probability). Denote the concatenation of the message blocks yielding from $h_{i-(n-k)}$ to the common chaining value by s_2, and the concatenation of the message blocks yielding from the common chaining value to h_i by s_3.

The concatenation $s_1||s_2||s_3$ is a sequence of i message blocks that leads from the IV to h_i, as desired. Now, the adversary creates a second pre-image:

$$M' = s_1||s_2||s_3||m_{i+1}||\cdots||m_k.$$

Figure 2 illustrates the attack.

3.2   The Attack Complexity

The Precomputation Complexity. As described in Sect. 3.1, to construct the kite generator, the adversary has to find, for each chaining value $a \in A$, two message

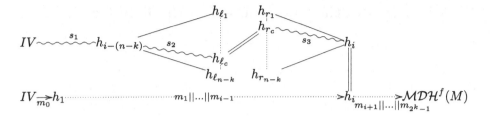

Fig. 2. A second pre-image using a kite generator.

blocks $m_{a,1}, m_{a,2}$ such that $f(a, m_{a,1}), f(a, m_{a,2}) \in A$. To do so, he generates $2 \cdot 2^k$ message blocks, each leads to one of the 2^{n-k} chaining values of A with probability of 2^{-k}. Hence, is expected to find two such message blocks, and the total complexity is about $2 \cdot 2^k \cdot 2^{n-k} = 2^{n+1}$ compression function calls.[3]

The online Complexity. First of all, the memory used to store the kite generator is $\mathcal{O}(2^{n-k})$. Second, the online phase has two steps:

1. The adversary should compute the M's chaining values to find the common chaining value with the kite generator's chaining values. This step requires $\mathcal{O}(2^k)$ compression function calls.
2. The adversary should find a collision between the two lists described in Sect. 3.1. Since each list contains about $2^{\frac{n-k}{2}}$ chaining values, this step requires $\mathcal{O}(2^{\frac{n-k}{2}})$ time and memory.[4]

Thus, the online time complexity is

$$\max(\mathcal{O}(2^k), \mathcal{O}(2^{\frac{n-k}{2}})),$$

and the online memory complexity is

$$\mathcal{O}(2^{n-k}).$$

---

[3] Note that using this method $d_{out}(a)$ follows a $Poi(2)$ distribution, and about 13% of the chaining values are expected to have $d_{out}(a) = 0$. To solve this issue, it is possible to generate for each chaining value as many message blocks as needed to find two out-edges. Now, the average time complexity needed for a chaining value a is 2^{k+1}. The actual running time for a given chaining value is the sum of two geometric random variables with mean 2^k each. Hence, the total running time is the sum of 2^{n-k+1} geometric random variables $X_i \sim Geo(2^{-k})$. Since $\sum_{i=1}^{2^{n-k+1}} (X_i - 1) \sim NB(2^{n-k+1}, 1 - 2^{-k})$, then $\sum_{i=1}^{2^{n-k+1}} X_i \sim 2^{n-k+1} + NB(2^{n-k+1}, 1 - 2^{-k})$. Therefore, $E[\sum_{i=1}^{2^{n-k+1}} X_i] = 2^{n-k+1} + \frac{(1-2^{-k})2^{n-k+1}}{2^{-k}} = 2^{n+1}$ with a standard deviation of $\frac{\sqrt{2^{n-k+1}(1-2^{-k})}}{2^{-k}} \le 2^{\frac{n+k+1}{2}}$.

[4] Andreeva et al. [1] note that it is possible to find the common chaining value by a more sophisticated algorithm which requires the same time but negligible additional memory, using memoryless collision finding. Our findings affect these variants as well.

4   A Problem in the Construction of the Kite Generator

4.1   On the Inaccuracy of Andreeva et al.'s Analysis

As described in Sect. 3.1, for constructing the kite generator Andreeva et al. [1] suggest to find from each chaining value of A two message blocks, each of them leads to a chaining value of A. They claim that as $\forall a \in A : E[d_{in}(a)] = 2$, the resulting graph is strongly connected.

We agree with their claim that due to the fact that $d_{out}(a) = 2$ then $\forall a \in A : E[d_{in}(a)] = 2$, but we claim that their conclusion, that the resulting graph is strongly connected, is wrong. The actual distribution of $d_{in}(a)$ can be approximated by $d_{in}(a) \sim Poi(2)$, as the number of entering edges follows a Poisson distribution with a mean value of 2. Hence, for each chaining value $a \in A$:

$$Pr[d_{in}(a) = 0] = \frac{e^{-2} \cdot 2^0}{0!} = e^{-2}.$$

Thus, about $2^{n-k} \cdot e^{-2} \approx 13\%$ of the chaining values in the kite generator are expected to have $d_{in}(a) = 0$. Obviously, the resulting graph is not strongly connected.

Moreover, there are more h_i's in the kite generator for which the attack fails. The construction of the "backwards" tree from h_i is a branching process with $Poi(2)$ offspring (see Definition 4). Therefore, an ultimate extinction of the branching process suggests that h_i cannot be connected to, and the online phase fails. According to the branching process theorems [3], the probability of ultimate extinction in a branching process with offspring distribute according a distribution P is the smallest fixed point $x \in [0, 1]$ of the moment-generating function of P. In our case the distribution is $Poi(2)$, and the moment-generating function is $f(x) = e^{2(x-1)}$. Hence, the probability of ultimate extinction is the smallest solution $x \in [0, 1]$ of $e^{2(x-1)} = x$. Using numerical computation we get that $x \approx 0.2032$. It means that in about 20% of the cases the "backwards" tree is limited. We note that usually this extinction happens very quickly. For example, about 85% of the "extinct" h_i do so in one or zero steps (i.e., their backwards tree is of depth of at most 1).

To fix this problem we need that $E[d_{in}(a)] = n - k$, and then the expected number of chaining values $a \in A$ with $d_{in}(a) = 0$ is $2^{n-k} \cdot e^{-(n-k)} \ll 1$. The naive approach to do so, is to generate from each chaining value $a \in A$, $n - k$ message blocks, $m_{a,1}, m_{a,2}, \ldots, m_{a,n-k}$, for which $f(a, m_{a,i}) \in A \setminus \{a\}$. Using this approach, the complexity of the precomputation increases to $(n - k) \cdot 2^n$ compression function calls.

A different approach for fixing the problem is to increase the kite generator by adding vertices such that the intersection between a message of length 2^k and the kite generator is sufficiently large (i.e., that there are several joint chaining values). Hence, even if some of the pairs of the joint chaining values fail to connect through the kite generator, there is a sufficient number of pairs that do connect. This approach does not increase the precomputation time beyond 2^{n+1} (as the additional vertices in the kite generator reduce the "cost" of connecting any vertex). At the same time, it increases the memory complexity of the attack. We do not provide a full analysis of this approach given the improved attack of Sect. 5 which does not require additional memory, and enjoys a smaller time complexity.

5   A New Method for Constructing Kite Generators

In the construction described in Sect. 3.1, the set A of the chaining values is chosen arbitrarily. We now suggest a new method for choosing the chaining values in order to optimize the complexity of constructing a kite generator. Our main idea is to define the set A iteratively in a manner that ensures that $d_{in}(a) \geq 1$ (for all but one chaining value, the IV).

Construction. For the reading convenience we consider two different message blocks m_1, m_2.[5] The following steps are required:

1. Let $L_0 := \{h_0 = IV\}$.
2. Set the second layer $L_1 = \{h_1 = f(IV, m_1), h_2 = f(IV, m_2)\}$.
3. Continue by the same method to set

$$L_i = \{f(h, m) \mid h \in L_{i-1}, m \in \{m_1, m_2\}\}, \forall 1 \leq i \leq n - k - 1$$

until L_{n-k-1} is generated.
4. Set

$$A = \bigcup_{i=0}^{n-k-1} L_i.$$

Note that[6] $|A| = \sum_{i=0}^{n-k-1} 2^i = 2^{n-k} - 1$.
5. Finally, for each chaining value a reached in the last layer L_{n-k-1}, look for two message blocks $m_{a,1}, m_{a,2}$ (probably different than m_1, m_2) that lead to some chaining value $b \in A$.

Figure 3 illustrates the construction of A.

The advantage of this method is that for each chaining value $IV \neq a \in A$, there exists another chaining value $b \in A$ and a message block m_b such that

$$f(b, m_b) = a$$

---

[5] It is not necessary to use only two different message blocks in the setting, but it is possible since they are used for different chaining values.
[6] With high probability we expect some collisions in A. This can be easily solved during the construction: If a chaining value $f(h_i, m_j)$ is already generated, replace the message block m_j one by one until a new chaining value is reached. It is easy to see that the additional time complexity is negligible.

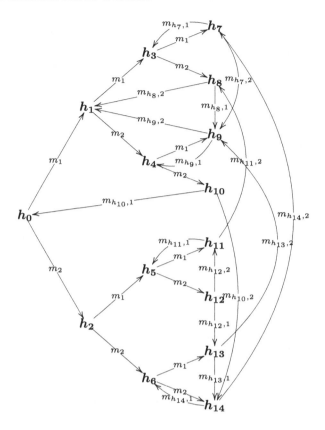

Fig. 3. An example for an iterative construction of A

i.e.,

$$d_{in}(a) \geq 1.$$

The case of $d_{in}(IV) = 0$ is not problematic, since we need the IV in the kite generator only as the source of the random walking, and it is done only with the out-edges. In addition, In this method of constructing A, each chaining value $a \neq IV$ follows $d_{in}(a) \sim 1 + Poi(1)$. It implies that $\forall a \neq IV : Pr[d_{in}(a) = 0] = 0$, and therefore the probability of ultimate extinction in the branching process defined by the backwards tree is 0.

Analysis. Steps 1–4 generate arbitrary message blocks for each reached chaining value until about 2^{n-k} chaining values are reached. They require about 2^{n-k} compression function calls. Step 5, of finding two out-edges from each chaining value that reached in the last layer L_{n-k-1}, requires about $2 \cdot 2^{n-k-1} \cdot 2^k = 2^n$ compression function calls.[7] Thus, the precomputation complexity is about

$$2^n + 2^{n-k} \approx 2^n$$

compression function calls. It means that our method not only ensures that the resulting graph is strongly connected, but is also more efficient than the original method.

Improvement I. As additional improvement, we can reduce the complexity of Step 5 described above by finding only one out-edge from each chaining value in the last layer L_{n-k-1}. Now, each chaining value $a \neq IV$ follows $d_{in}(a) \sim 1 + Poi(0.5)$, and the branching process does not extinct. To use this improvement, we need to slightly change the online phase: We need to increase the length of the sequences s_2, s_3 mentioned in Sect. 3.1 to about $\log_{\frac{3}{2}}(2) \cdot \frac{n-k}{2}$ (instead of $\frac{n-k}{2}$) to find, with high probability, a common chaining value. Using this improvement, the complexity of the precomputation is reduced by a factor of 2 to about 2^{n-1} compression function calls. There is no change in the online time complexity.

6 Adapting Our New Method to Dithered Merkle-DamgåRd

6.1 Dithered Merkle-DamgåRd

The main idea of the dithered Merkle-Damgård structure [13] is to perturb the hashing process by using an additional input to the compression function. This additional input is formed by taking elements of a fixed dithering sequence. Using this additional input, the compression of a message block depends on its position in the whole message. Thus, it decreases the adversary's control on the input of the compression function. Using the dithered Merkle-Damgård structure, some attacks such as the Dean's attack [6] and the Kelsey-Schneier's expandable-messages attack [9] are mitigated.

In order to use the dithered sequence for any message with the maximal number of message blocks in the hash function, it is reasonable to consider an infinite sequence. Let $\mathcal{B}$ be a finite alphabet, and let z be an infinite sequence over $\mathcal{B}$ and let z_i be the i'th symbol of z. The dithered Merkle-Damgård construction is obtained by replacing the iterative chaining value defined in the original Merkle-Damgård structure (see Definition 2) with

$$h_i = f(h_{i-1}, m_i, z_i).$$

---

[7] Again, in this step we actually need to generate for each chaining value as many message blocks as needed to find two out-edges. Now, the average time complexity needed for a chaining value a is 2^{k+1}. The actual running time for a given chaining value is the sum of two geometric random variables with mean 2^k each. Hence, the total running time is the sum of 2^{n-k} geometric random variables $X_i \sim Geo(2^{-k})$. Since $\sum_{i=1}^{2^{n-k}} (X_i - 1) \sim NB(2^{n-k}, 1 - 2^{-k})$, then $\sum_{i=1}^{2^{n-k}} X_i \sim 2^{n-k} + NB(2^{n-k}, 1 - 2^{-k})$. Therefore, $E[\sum_{i=1}^{2^{n-k}} X_i] = 2^{n-k} + \frac{(1-2^{-k})2^{n-k}}{2^{-k}} = 2^n$ with a standard deviation of $\frac{\sqrt{2^{n-k}(1-2^{-k})}}{2^{-k}} \leq 2^{\frac{n+k}{2}}$.

6.2   Adapting the Kite Generator to Dithering Sequence

As described in Sect. 3.1, the adversary could not know in advance the position of the message blocks to be used in the second pre-image. Thus, in order to allow the use of the kite generator at each position of the message, the adversary should consider any factor of z. To do so, Andreeva et al. [1] adapt the precomputation phase as follows: For each chaining value $a \in A$ and for each symbol $\alpha \in B$ the adversary looks for two message blocks $m_{a,\alpha,1}, m_{a,\alpha,2}$ s.t. $f(a, m_{a,\alpha,1}, \alpha), f(a, m_{a,\alpha,2}, \alpha) \in A$. Hence, The complexity of the precomputation using the original method is about

$$2 \cdot |B| \cdot 2^n$$

compression function calls.

The same problem mentioned in Sect. 4.1 carries over to this case as well. As described in Sect. 4.1, in order to fix the inaccuracy that the resulting graph is not strongly connected, the adversary should generate about $n - k$ such message blocks for each chaining value and for each dithered symbol. Hence, the complexity of the precomputation is increased to about

$$(n - k) \cdot |B| \cdot 2^n$$

compression function calls.

6.3   Adapting Our Method

Construction. As described in Sect. 5, the main idea of our new method is to choose the chaining values of A by generating two message blocks for each reached chaining value, starting from the IV. In order to adapt it to the dithered Merkle-Damgård structure, the following steps are required:

1. Choose an arbitrary symbol $\alpha \in B$, and construct the kite generator using this symbol only, according to our new method.
2. Use the original method to complete the kite generation for the remaining symbols of B, i.e., for each chaining value $a \in A$ and for each symbol $\alpha \neq \beta \in B$, look for $n - k$ message blocks that lead to another chaining value of A.

Analysis. The complexity of the first step, of constructing the kite generator using one symbol, is similar to the one in Sect. 5, i.e., about 2^{n-1} compression function calls (using the improved method). The complexity of the second step, of completing the kite generator for the remaining symbols, is about $(n - k) \cdot (|B| - 1) \cdot 2^n$ compression function calls. Thus, the total complexity of the precomputation is about

$$((n - k) \cdot (|B| - 1) + 0.5) \cdot 2^n$$

compression function calls.

6.4   Improvement II

In Sect. 6.3 we adapted our new method while considering the probability of ultimate extinction in the construction of the "backwards" tree tends to zero. We now show that by allowing a small probability of ultimate extinction, denoted by p, we can reduce the complexity as follows. We look for a $\lambda(p)$ such that the probability of ultimate extinction in a branching process with $Poi(\lambda(p))$ offspring is p. According to the branching process theorems [3] we need that

$$e^{\lambda(p)(1-p)} = p$$

that implies

$$\lambda(p) = \frac{\ln(p)}{1-p}.$$

For examples, $\lambda(0.01) \approx 4.65$, and $\lambda(0.001) \approx 6.91$. It means that in the construction described in Sect. 6.3, we can replace the second step of looking for $n-k$ message blocks per symbol for each chaining value, by looking for such a constant number. Thus, consider $p = 0.001$, the complexity of the precomputation is reduced to about

$$(6.91 \cdot (|\mathcal{B}| - 1) + 0.5) \cdot 2^n$$

compression function calls.

7   Summary

As a concluding discussion, we note that when the kite generator has 2^{n-k} chaining values and the message is of length 2^k blocks, one should expect the kite generator to contain one of the message's chaining values with probability 63%, which translates to about 50% success rate. The way to fix this issue is trivial — increase the size of the kite generator. Multiplying the number of nodes in the kite generator by a factor of 2, reduces the probability of disjoint sets of chaining values from $1/e \approx 37\%$ to merely $1/e^2 \approx 13.5\%$, and this rate can be further reduced to as small probability as the adversary wishes.[8] However, when the kite generator is not strongly connected, as our analysis shows, the success probability of the original is upper bounded by 80%, no matter how large the kite generator is taken to be.

To conclude, in this work we pointed out an inaccuracy in the analysis of the construction of the *kite generator* suggested by Andreeva et al. [1]: The *kite generator* is not strongly connected, and thus the online phase fails in probability of at least 20%. We showed that to fix the inaccuracy, we need to increase the complexity of the construction phase by a factor of $\frac{n-k}{2}$. We then suggested a

---

[8] This issue happens also in the online phase, when the adversary looks for common chaining values between the two lists described in Sect. 3.1. The fixing is similarly – increase the size of these lists accordingly.

Table 1. Comparing the complexities of the different methods.

Method	Complexity	
	Merkle-Damgård	Dithered Merkle-Damgård
Andreeva et al. [1] [a]	2^{n+1}	$|\mathcal{B}| \cdot 2^{n+1}$
Our fixed analysis	$(n-k) \cdot 2^n$	$(n-k) \cdot |\mathcal{B}| \cdot 2^n$
Our new method	2^n	$((n-k) \cdot (|\mathcal{B}|-1)+1) \cdot 2^n$
Improvement I	2^{n-1}	$((n-k) \cdot (|\mathcal{B}|-1)+0.5) \cdot 2^n$
Improvement II	Not relevant	$(6.91 \cdot (|\mathcal{B}|-1)+0.5) \cdot 2^n$

[a] — Andreeva et al.'s analysis is inaccurate.

new method to optimize the construction of the *kite generator* that is both correct and more efficient than the original method. Finally, we adapted the fixing analysis and our new method to the dithered Merkle-Damgård construction.

For comparison, we present in Table 1 the number of required compression function calls to construct a *kite generator*, using the different methods, for the Merkle-Damgård structure and its dithered variant.

Acknowledgements. The research of Ariel Weizman was supported by the European Research Council under the ERC starting grant agreement n. 757731 (LightCrypt) and by the BIU Center for Research in Applied Cryptography and Cyber Security in conjunction with the Israel National Cyber Bureau in the Prime Minister's Office.

References

1. Andreeva, E., Bouillaguet, C., Dunkelman, O., Fouque, P.-A., Hoch, J., Kelsey, J., Shamir, A., Zimmer, S.: New second-preimage attacks on hash functions. J. Cryptol. **29**(4), 657–696 (2016)
2. Andreeva, E., Bouillaguet, C., Fouque, P.-A., Hoch, J.J., Kelsey, J., Shamir, A., Zimmer, S.: Second preimage attacks on dithered hash functions. In: Smart, N. (ed.) EUROCRYPT 2008. LNCS, vol. 4965, pp. 270–288. Springer, Heidelberg (2008). https://doi.org/10.1007/978-3-540-78967-3_16
3. Athreya, K.B., Ney, P.E.: Dover books on mathematics. In: Branching Processes, pp. 1–8. Dover Publications, New York (2004). Chap. 1
4. Blackburn, S.R., Stinson, D.R., Upadhyay, J.: On the complexity of the herding attack and some related attacks on hash functions. Des. Codes Crypt. **64**(1–2), 171–193 (2012)
5. Damgård, I.B.: A design principle for hash functions. In: Brassard, G. (ed.) CRYPTO 1989. LNCS, vol. 435, pp. 416–427. Springer, New York (1990). https://doi.org/10.1007/0-387-34805-0_39
6. Dean, R.D.: Formal aspects of mobile code security. Ph.D. thesis, Princeton University, Princeton (1999)
7. Joux, A.: Multicollisions in iterated hash functions. application to cascaded constructions. In: Franklin, M. (ed.) CRYPTO 2004. LNCS, vol. 3152, pp. 306–316. Springer, Heidelberg (2004). https://doi.org/10.1007/978-3-540-28628-8_19

8. Kelsey, J., Kohno, T.: Herding hash functions and the nostradamus attack. In: Vaudenay, S. (ed.) EUROCRYPT 2006. LNCS, vol. 4004, pp. 183–200. Springer, Heidelberg (2006). https://doi.org/10.1007/11761679_12

9. Kelsey, J., Schneier, B.: Second preimages on n-bit hash functions for much less than 2^n work. In: Cramer, R. (ed.) EUROCRYPT 2005. LNCS, vol. 3494, pp. 474–490. Springer, Heidelberg (2005). https://doi.org/10.1007/11426639_28

10. Kortelainen, T., Kortelainen, J.: On diamond structures and trojan message attacks. In: Sako, K., Sarkar, P. (eds.) ASIACRYPT 2013. LNCS, vol. 8270, pp. 524–539. Springer, Heidelberg (2013). https://doi.org/10.1007/978-3-642-42045-0_27

11. Merkle, R.C.: One way hash functions and DES. In: Brassard, G. (ed.) CRYPTO 1989. LNCS, vol. 435, pp. 428–446. Springer, New York (1990). https://doi.org/10.1007/0-387-34805-0_40

12. National Institute of Standards and Technology: Secure hash standard. FIPS, PUB **17**, 3–180 (1995)

13. Rivest, R.L.: Abelian square-free dithering for iterated hash functions. In: ECrypt Hash Function Workshop, vol. 21, June 2005

14. Weizmann, A., Dunkelman, O., Haber, S.: Efficient construction of diamond structures. In: Patra, A., Smart, N.P. (eds.) INDOCRYPT 2017. LNCS, vol. 10698, pp. 166–185. Springer, Cham (2017). https://doi.org/10.1007/978-3-319-71667-1_9

Using Noisy Binary Search
for Differentially Private Anomaly
Detection

Daniel M. Bittner[1]($\boxtimes$), Anand D. Sarwate[2], and Rebecca N. Wright[3]

[1] Department of Computer Science, Rutgers University, Piscataway, NJ, USA
dbittner@cs.rutgers.edu
[2] Department of Electrical and Computer Engineering, Rutgers University,
Piscataway, NJ, USA
anand.sarwate@rutgers.edu
[3] Department of Computer Science and DIMACS, Rutgers University,
Piscataway, NJ, USA
rebecca.wright@rutgers.edu

Abstract. In this paper, we study differential privacy in noisy search. This problem is connected to noisy group testing: the goal is to find a defective or anomalous item within a group using only aggregate group queries, not individual queries. Differentially private noisy group testing has the potential to be used for anomaly detection in a way that provides differential privacy to the non-anomalous individuals while still helping to allow the anomalous individuals to be located. To do this, we introduce the notion of anomaly-restricted differential privacy. We then show that noisy group testing can be used to satisfy anomaly-restricted differential privacy while still narrowing down the location of the anomalous samples, and evaluate our approach experimentally.

1 Introduction

We consider the problem of privacy-sensitive anomaly detection—screening to detect individuals, behaviors, areas, or data samples of high interest. What defines an anomaly is context-specific: examples include a spoofed rather than genuine user attempting to log in to a web site, a fraudulent credit card transaction, or a suspicious traveler in an airport. The unifying assumption is that the number of truly anomalous points is quite small with respect to the population, so that deep screening of all individual data points would potentially be time-intensive, costly, and unnecessarily invasive of privacy. Anomaly detection is well studied (see the survey of Chandola et al. [11]), but methods to provide anomaly detection along with privacy are less well studied. In this paper we provide a framework for identifying anomalous data while guaranteeing quantifiable privacy in a rigorous sense. Once identified, such anomalies could warrant further data collection and investigation, depending on the context and relevant policies.

© Springer International Publishing AG, part of Springer Nature 2018
I. Dinur et al. (Eds.): CSCML 2018, LNCS 10879, pp. 20–37, 2018.
https://doi.org/10.1007/978-3-319-94147-9_3

While anomaly detection is important for many applications, it can also raise privacy concerns when the underlying data is sensitive. Search algorithms on private data can violate data use agreements and can make people uncomfortable with potential anomaly detection methods. In this paper, we focus on guaranteeing privacy during the deployment of anomaly detection. To achieve this, we take as our starting point the notion of *group testing* [14], which was most famously proposed for screening US military draftees for syphilis during World War II. In group testing, individuals are tested in groups to limit the number of tests. Using multiple rounds of screenings, a small number of positive individuals can be detected very efficiently. Group testing has the added benefit of providing privacy to individuals through plausible deniability—since the group tests use aggregate data, individual contributions to the test are masked by the group.

Our work takes the first steps toward strengthening and formalizing these privacy guarantees to achieve differential privacy. Differential privacy is a statistical measure of disclosure risk that was introduced in 2006 [18] and captures the intuition that an individual's privacy is protected if the results of a computation have at most a very small and quantifiable dependence on that individual's data. In the last decade, there has been an explosion of research in differential privacy, with many techniques and algorithms poised for practical application [20,27,31] and adoption underway by high-profile companies such as Apple [21] and Google [20].

Potential anomaly detection applications for group testing would rely on existing or new sensing technologies that can perform (reasonably accurate) queries in aggregate to reveal and isolate anomalous outliers. Applications might include privacy-sensitive methods for searching for outlying cell phone activity patterns or Internet activity patterns in a geographic location. These techniques are also in line with the US Department of Homeland Security's visionary goal of "screening at speed" [13]—unobtrusive screening of people, baggage, or cargo.

Our main contribution is a differentially private access mechanism for narrowing down the location of anomalies in a set of samples using noisy group testing. Our goal is to guarantee privacy for non-anomalous individuals while identifying anomalous samples. To formalize this we introduce the notion of *anomaly-restricted differential privacy*. By adding noise to group query results, we can guarantee differential privacy while allowing efficient and accurate detection of non-anomalous individuals. The adaptive sequential query design is an active learning algorithm for noisy binary search that is connected to information-theoretic models of communication with feedback.

A summary of our contributions is as follows:

- We introduce a new notion of anomaly-restriction differential privacy, which may be of independent interest.
- We provide a noisy group-based search algorithm that satisfies the anomaly-restricted differential privacy definition.
- We provide both theoretical and empirical analysis of our noisy search algorithm, showing that it performs well in some cases and exhibits the usual privacy/accuracy tradeoff of differentially private mechanisms.

2   Related Work

Machine learning methods have found widespread use in anomaly detection due to their ability to analyze and extract patterns from large amounts of data. Several surveys cover the wide variety of anomaly detection techniques and applications. For example, Hodge and Austin [23] and Agyemang et al. [2] survey anomaly detection techniques in the context of outlier detection via proximity and statistical approaches. Chandola et al. [11] provide a comprehensive survey addressing techniques in these categories as well as covering information theoretic and spectral approaches and techniques used in range of applicable fields including popular applications such as intrusion detection and fraud detection, as well as medical, industrial, image, and text anomalies.

Group testing describes a set of techniques for detection of anomalies from sets primarily containing non-anomalous items by performing testing on groups rather than querying individual items. Group testing was initially conceived during World War II as a cost-efficient method to test for syphilis by grouping multiple individuals' blood into a single sample [14]. A negative result for the single sample would imply all the individuals were negative, while a less-common positive result would require further follow up. The technique was not put into practice due to the limited number of individuals that could be tested at any one time, and group testing languished for several years before eventually being revived for industrial testing purposes [15].

Group testing has received more recent interest in the statistics and information theory communities. In particular, classical connections between group testing and error control coding have led to relaxations of the group testing problem, as surveyed in a recent paper by Mazumdar [29]. Group testing has also been used for multiaccess communications [5,37], data mining [28], molecular biology [12], and DNA screening [32]. Related concepts have been explored in constructing compressed sensing matrices [9,30].

Introduced by Dwork et al. in 2006 [18], differential privacy has become a widely studied framework for providing privacy-sensitive results from data analyses. Differential privacy for anomaly detection has been studied previously in the context of training classifiers using machine learning [22]. In contrast, our work addresses differential privacy during the deployment of an anomaly search algorithm by using differentially private group testing.

Our method of differentially private group testing makes use of noisy group testing [3,8,10], which provides methods that successfully identify anomalies using group queries among a set of items even if the answers to the group queries are not completely accurate. Specifically, we use a probabilistic binary search [4, 25,33–35], which is intimately connected to the problem of communication over noisy channels with feedback. The classical scheme by Horstein [24] uses what we would now call a Bayesian active learning approach to learn a threshold with noisy labels. In our case, the noise is used (and may even be deliberately introduced) to provide differential privacy.

3   Problem Formulation

The main idea behind our approach is to query individuals in groups and use noise to provide differential privacy. For this to work, we must have a group query which can detect the presence of an anomalous sample. As in active learning algorithms, we use multiple adaptive queries to locate the anomalies. In particular, we use a Bayesian formulation in which the algorithm maintains a probability distribution, or posterior belief, over the point representing its belief about where the anomaly lies. The number of queries can be controlled by either a stopping rule based on the belief or limits on overall privacy risk.

Notation: We generally use calligraphic script to denote sets. For any positive integer K, we denote the set $\{1, 2, \ldots, K\}$ by $[K]$.

3.1   Data Model

In this paper, we analyze a simplified version of the full problem with a single anomaly: for this setting, we can characterize the performance theoretically.

The data is a vector $\mathcal{X} = (\mathbf{x}_1, \mathbf{x}_2, \ldots, \mathbf{x}_{n+1})$ of $n + 1$ individuals, where $\mathbf{x}_i \in \mathbb{R}^+$. With some abuse of notation we write this as an ordered multiset $\{\mathbf{x}_i : i \in [n + 1]\}$. The i-th element $\mathbf{x}_i$ represents the output of some anomaly score function applied to individual i: larger $\mathbf{x}$ denotes a higher anomaly level. One of the data points is an anomaly $\mathbf{x}^*$. Let i^* be the index of the anomaly, so that $\mathbf{x}_{i^*} = \mathbf{x}^*$. Two thresholds t_ℓ and t_h separate the anomaly value of the anomalous points from the other points such that

$$\mathbf{x} \in \begin{cases} [0, t_\ell] & \mathbf{x} \neq \mathbf{x}^* \\ [t_h, \infty) & \mathbf{x} = \mathbf{x}^* \end{cases} \tag{1}$$

for a set of two thresholds $t_\ell, t_h \in \mathbb{R}^+$ where $t_\ell < t_h$. This corresponds to a scenario where there is some measurement that can distinguish the anomaly from the non-anomalous values.

The data is held by an oracle that has access to $\mathcal{X}$ and can answer queries about $\mathcal{X}$. The search algorithm knows the number of points $n + 1$ and the index set $[n+1]$, the levels t_ℓ and t_h separating anomalous from non-anomalous values, and that $\mathcal{X}$ contains a single anomalous point. However, it does not know the actual values $\{\mathbf{x}_1, \ldots, \mathbf{x}_{n+1}\}$. We wish to model a situation in which the oracle can only query groups of points. This could correspond to a situation where there is a measurement or sensor which can access aggregates (for example, all items in a given area) but not individual records.

3.2   Differential Privacy

The search algorithm queries the oracle, which provides *differentially private* responses. Traditional differential privacy protects privacy for every individual in the database [18]. The key difference in our model is that we only require that the oracle provide differential privacy for the non-anomalous points: we define a new notion of *anomaly-restricted* neighbors.

Definition 1 (Anomaly-Restricted Neighbors). *We say that two data sets* $\mathcal{D}$ *and* $\mathcal{D}'$ *are anomaly-restricted neighbors (and write* $\mathcal{D} \sim \mathcal{D}'$*) if* $\mathbf{x}^* \in \mathcal{D} \cap \mathcal{D}'$ *and* $|\mathcal{D} \cap \mathcal{D}'| = n$.

Definition 2 (Differential Privacy [18]). *A randomized mechanism* $\mathcal{A}(\cdot)$ *is* ϵ*-differentially private if for any set of measurable outputs* $\mathcal{Y}$ *and any two databases* $\mathcal{D}$ *and* $\mathcal{D}'$ *with* $\mathcal{D} \sim \mathcal{D}'$,

$$\Pr\left[\mathcal{A}(\mathcal{D}) \in \mathcal{Y}\right] \le e^{\epsilon} \Pr\left[\mathcal{A}\left(\mathcal{D}'\right) \in \mathcal{Y}\right]. \tag{2}$$

A differentially private algorithm $\mathcal{A}(\cdot)$ guarantees that neighboring databases create similar outputs: for anomaly-restricted neighbors this means that adding or removing a single non-anomalous individual does not significantly alter the output of the mechanism. The privacy parameter ϵ is the privacy risk: larger values of ϵ allow larger differences between the distributions of $\mathcal{A}(\mathcal{D})$ and $\mathcal{A}(\mathcal{D}')$ [16–18]. Differential privacy controls the error probabilities in the hypothesis test between $\mathcal{D}$ and $\mathcal{D}'$ given the output of the mechanism [26, 36].

The Laplace mechanism [18] is a common approach to making differentially private approximation to scalar functions $H(\cdot)$. This approach adds Laplace noise with a parameter that is a function of the privacy risk ϵ and the global sensitivity Δ_g of $H(\cdot)$. Corresponding to our new neighbor definition, we also need a model for anomaly-restricted global sensitivity.

Definition 3 (Anomaly-Restricted Global Sensitivity). *Let* $H(\cdot)$ *be a scalar-valued function. The anomaly-restricted global sensitivity of* $H(\cdot)$ *is*

$$\Delta_g = \max_{\mathcal{D}, \mathcal{D}' : \mathcal{D} \sim \mathcal{D}'} |H(\mathcal{D}) - H(\mathcal{D}')|. \tag{3}$$

Given ϵ and $H(\cdot)$, the Laplace mechanism computes $\mathcal{A}(\mathcal{D}) = H(\mathcal{D}) + Z$ where $Z \sim \mathsf{Lap}(\Delta_g/\epsilon)$ where the Laplace distribution $\mathsf{Lap}(\lambda)$ has density

$$p(z; \lambda) = \frac{1}{2\lambda} \exp\left(-\frac{z}{\lambda}\right). \tag{4}$$

Differential privacy satisfies several *composition properties*.

Definition 4 (Simple Sequential Composition [18]). *Given a series of* n *independent differentially private mechanisms* $\mathcal{A}_1, \mathcal{A}_2, \ldots, \mathcal{A}_n$ *with privacy parameters* $\epsilon_1, \epsilon_2, \ldots, \epsilon_n$ *computed on* $\mathcal{D}$*, the resulting function is differentially private with privacy parameter* $\sum_{i=1}^{n} \epsilon_i$.

Definition 5 (Parallel Composition [18]). *Given a series of* n *independent differentially-private mechanisms* $\mathcal{A}_1, \mathcal{A}_2, \ldots, \mathcal{A}_n$ *with privacy parameters* $\epsilon_1, \epsilon_2, \ldots, \epsilon_n$ *computed on disjoint subsets of* $\mathcal{D}$*, then the resulting function is differentially private with privacy parameter* $\max_i \epsilon_i$.

In this paper, we restrict our attention to ϵ-differentially private methods. For approximate (ϵ, δ)-differential privacy there are stronger composition results in which the total privacy risk for sequential composition grows sublinearly with the number of terms [6, 19, 26], including the so-called "moments accountant" [1].

4   Algorithms

At each time t the search algorithm issues a query $\mathcal{Q}_t \subset [n+1]$ to the oracle that depends on the responses to past queries. A search algorithm consists of rules for sequentially selecting sets $\mathcal{Q}_1, \mathcal{Q}_2, \ldots$ with privacy risks $\epsilon_1, \epsilon_2, \ldots$ where $\mathcal{Q}_t \subset [n+1]$. A standard (noiseless) bisection search algorithm receives accurate queries and can then discard non-anomalous data points with certainty. When the oracle responses are noisy, we cannot fully discard any data points. We use a discretized version [4,7] of a probabilistic bisection algorithm [24] to adaptively determine the location of the anomaly. In particular, the algorithm uses a Bayesian inference step to update a probability mass function on $[n + 1]$ that represents the *belief* about i^*.

4.1   Warmup: Randomized Response

A baseline algorithm for privacy binary search is noisy binary search using randomized response. At each time t the algorithm chooses a query $\mathcal{Q}_t$ and sends it to the oracle, which responds with

$$\mathcal{Y}_t = \mathbf{1}\left(i^* \in \mathcal{Q}_t\right) \oplus z_t \tag{5}$$

where $\oplus$ is addition modulo 2 and $z_t \sim \mathsf{Bernoulli}(p)$.

Proposition 1. *The response in* (5) *guarantees* $\log \frac{1-p}{p}$*-differential privacy.*

Given a response $\mathcal{Y}_t$ and noise parameter p, the algorithm can compute a posterior distribution on the location of the anomaly. Given $\bar{\mathcal{Q}}_t = [n + 1] \setminus \mathcal{Q}_t$, let $\mathcal{R}_t = \mathcal{Q}_t$ if $\mathcal{Y}_t = 1$ and $\mathcal{R}_t = \bar{\mathcal{Q}}_t$ if $\mathcal{Y}_t = 0$. Given an initial estimate $\mathbf{f}_{t-1}$ on $[n + 1]$, the Bayesian update is given by

$$\mathbf{f}_t(i) = \begin{cases} \dfrac{\mathbf{f}_{t-1}(i)(1-p)}{\sum_{j \in \mathcal{R}_t} \mathbf{f}_{t-1}(j)(1-p) + \sum_{k \notin \mathcal{R}_t} \mathbf{f}_{t-1}(k)p} & i \in \mathcal{R}_t \\[4mm] \dfrac{\mathbf{f}_{t-1}(i)p}{\sum_{j \in \mathcal{R}_t} \mathbf{f}_{t-1}(j)(1-p) + \sum_{k \notin \mathcal{R}_t} \mathbf{f}_{t-1}(k)p} & i \notin \mathcal{R}_t \end{cases} \tag{6}$$

Because $p < \frac{1}{2}$, this rule increases $f_{t-1}(i)$ for $i \in \mathcal{R}_t$ and decreases $f_{t-1}(i)$ for $i \notin \mathcal{R}_t$ and eventually concentrates the posterior on i^*. If at each iteration the algorithm chooses a query $\mathcal{Q}_t$ with posterior probability close to $1/2$ (i.e. a median split) this is a classic algorithm first analyzed by Burnashev and Zigangirov [7] (see also Horstein [24]) for i^* chosen uniformly in $[n + 1]$; we can initialize by uniformly permuting the indices to use their result.

4.2   Proposed Algorithm: Differentially Private Binary Search

Before presenting the search algorithm, we introduce a modified oracle. Randomized response forces the oracle to determine whether $i^* \in \mathcal{Q}_t$ or $i^* \in \bar{\mathcal{Q}}_t$ and then obfuscates that value. In some cases, the oracle may simply be a noisy privacy-preserving sensor that instead returns noisy estimates $\mathcal{A}(\mathcal{Q}_t; \mathcal{X})$ of some function $H(\mathcal{Q}_t; \mathcal{X})$. Consider an oracle that computes

$$\mathcal{Y}_t = \mathcal{A}\left(\mathcal{Q}_t; \mathcal{X}\right) \qquad \bar{\mathcal{Y}}_t = \mathcal{A}\left(\bar{\mathcal{Q}}_t; \mathcal{X}\right), \tag{7}$$

where the oracle splits the data set into $\mathcal{Q}_t$ and $\bar{\mathcal{Q}}_t = [n+1] \setminus \mathcal{Q}_t$ and returns anomaly-restricted differentially private approximation to both components. Notationally, we suppress $\mathcal{X}$ from $H(\mathcal{Q}_t; \mathcal{X})$ when it is clear from context.

There are many choices for the aggregation function $H(\cdot)$ used to calculate $\mathcal{A}$. For example, we could take the average $H(\mathcal{Q}) = \frac{1}{|\mathcal{Q}|} \sum_{i \in \mathcal{Q}} \mathbf{x}_i$. The anomaly-restricted global sensitivity is $\Delta_g = \frac{t_\ell}{|\mathcal{Q}|}$, so we can hypothetically add Laplace noise $Z \sim \mathsf{Lap}(\frac{t_\ell}{|\mathcal{Q}|\epsilon})$, $\bar{Z} \sim \mathsf{Lap}(\frac{t_\ell}{|\bar{\mathcal{Q}}|\epsilon})$ to form $\mathcal{Y} = H(\mathcal{Q}) + Z$ and $\bar{\mathcal{Y}} = H(\bar{\mathcal{Q}}) + \bar{Z}$, respectively.

In this work, we consider instead the max function:

$$H\left(\mathcal{Q}\right) = \max\{\mathbf{x}_i : i \in \mathcal{Q}\}. \tag{8}$$

Due to our definition of anomaly-restricted sensitivity, averages that include the anomaly can "dilute" the effect of the anomaly level. The max function can show the difference between $\mathcal{Y}$ and $\bar{\mathcal{Y}}$ in a way the depends less strongly on the distribution of the non-anomalous population. It has a higher sensitivity than the average function but we demonstrate its effectiveness empirically.

Lemma 1. *The anomaly-restricted global sensitivity of the aggregation function* $H(\mathcal{Q}; \mathcal{X}) = \max\{\mathbf{x}_i : i \in \mathcal{Q}\}$ *in* (8) *is* $\Delta_g(H) = t_\ell$.

Proof. Let $\mathcal{Q}$ be any query. Consider two anomaly-restricted neighboring data sets $\mathcal{X}$ and $\mathcal{X}'$ and let i^* be the index of the anomalous point. If $i^* \in \mathcal{Q}$ then $|H(\mathcal{Q}; \mathcal{X}) - H(\mathcal{Q}; \mathcal{X}')| = 0$ and $|H(\bar{\mathcal{Q}}; \mathcal{X}) - H(\bar{\mathcal{Q}}; \mathcal{X}')| \leq t_\ell$. If $i^* \in \bar{\mathcal{Q}}$ then $|H(\mathcal{Q}; \mathcal{X}) - H(\mathcal{Q}; \mathcal{X}')| \leq t_\ell$ and $|H(\bar{\mathcal{Q}}; \mathcal{X}) - H(\bar{\mathcal{Q}}; \mathcal{X}')| = 0$. Thus $\max |H(\mathcal{Q}; \mathcal{X}) - H(\mathcal{Q}; \mathcal{X}')| = t_\ell$. □

The oracle can then provide a differentially private query mechanism $\mathcal{A}$ for $H(\mathcal{Q}) = \max\{\mathbf{x}_i : i \in \mathcal{Q}\}$ by generating

$$\mathcal{A}\left(\mathcal{Q}\right) = \max\{\mathbf{x}_i : i \in \mathcal{Q}\} + Z \qquad \text{and} \qquad \mathcal{A}\left(\bar{\mathcal{Q}}\right) = \max\{\mathbf{x}_i : j \notin \mathcal{Q}\} + \bar{Z}, \tag{9}$$

where Z and $\bar{Z}$ are independent random variables with distribution $\mathsf{Lap}(t_\ell/\epsilon)$.

Given this revised oracle, we can turn to the search algorithm. The search is greedy: the searcher picks a query set which yields the most information (measured with respect to its belief) about the location of the anomaly. To represent our relative certainty about whether a given point is the anomaly, our search procedure updates a probability mass function $\mathbf{f}_t$ on $[n+1]$ where $\mathbf{f}_t(i) = \Pr(i^* = i)$. At each iteration we treat the previous posterior as a new prior and use $\mathbf{f}_{t-1}$ to determine the new query $\mathcal{Q}_t$. Since we do not have any prior knowledge about what element of $\mathcal{X}$ is the anomaly, at $t = 0$, we assume that each point is equally likely to be the anomaly: the initial prior distribution $\mathbf{f}_0$ is uniformly distributed on $[n+1]$, so $\mathbf{f}_0(i) = \frac{1}{n+1}$.

The algorithm uses the probability mass function $\mathbf{f}_{t-1}$ in order to select a query at each iteration $\mathcal{Q}_t$. First, the algorithm chooses a uniformly chosen random permutation σ on $[n+1]$. The corresponding permutation of the prior distribution is $\tilde{\mathbf{f}}_{t-1}(\sigma(i)) = \mathbf{f}_{t-1}(i)$. For a probability mass function on $[n+1]$ define the median $\mathcal{M}(\mathbf{f}) = \max\{m : \sum_{i=1}^{m} \mathbf{f}(i) < \sum_{i=m+1}^{n+1} \mathbf{f}(i)\}$.

The algorithm selects a query that maximizes information gain by dividing each query along the median of the permuted probability mass function.

At each iteration t the algorithm queries the oracle with

$$\mathcal{Q}_t = \left\{ i : \sigma(i) \leq \mathcal{M}\left(\tilde{\mathbf{f}}_{t-1}\right) \right\}. \tag{10}$$

Let $q_{t-1} = \sum_{i=0}^{\mathcal{M}(\tilde{\mathbf{f}}_{t-1})} \tilde{\mathbf{f}}_{t-1}(i)$ be the probability mass of the query set $\mathcal{Q}_t$. Note that $q \leq \frac{1}{2}$. Correspondingly, randomly choosing σ prevents reductions in information gain when q deviates significantly from $\frac{1}{2}$.

The oracle returns noisy values $\mathcal{Y}_t$ and $\bar{\mathcal{Y}}_t$ using (7) and (9) and the algorithm updates using a Bayesian update step similar to the case of randomized response. Given a prior belief $\mathbf{f}_{t-1}(i)$ that $i^* = i$, the likelihood of observing $(\mathcal{Y}_t, \bar{\mathcal{Y}}_t)$ is approximated by

$$\phi\left(\mathcal{Y}_t, \bar{\mathcal{Y}}_t \mid i^* = i\right) = \begin{cases} \frac{\epsilon^2}{4t_\ell^2} \exp\left(-\frac{\epsilon}{t_\ell}|\mathcal{Y}_t - t_h|\right) \exp\left(-\frac{\epsilon}{t_\ell}|\bar{\mathcal{Y}}_t - t_\ell|\right) & i \in \mathcal{Q}_t \\ \frac{\epsilon^2}{4t_\ell^2} \exp\left(-\frac{\epsilon}{t_\ell}|\mathcal{Y}_t - t_\ell|\right) \exp\left(-\frac{\epsilon}{t_\ell}|\bar{\mathcal{Y}}_t - t_h|\right) & i \in \bar{\mathcal{Q}}_t \end{cases}. \tag{11}$$

We can use this approximation in the Bayes update:

$$\mathbf{f}_t(i) = \frac{\mathbf{f}_{t-1}(i)\,\phi\left(\mathcal{Y}_t, \bar{\mathcal{Y}}_t \mid i^* = i\right)}{\sum_{j \in [n+1]} \mathbf{f}_{t-1}(j)\,\phi\left(\mathcal{Y}_t, \bar{\mathcal{Y}}_t \mid i^* = j\right)}. \tag{12}$$

There are two ways in which this procedure can halt. The first is if the algorithm expends the *privacy budget*. From the composition results, after T queries with ϵ-differentially private responses, the algorithm has incurred privacy risk $T\epsilon$. Given a total privacy budget b, we therefore halt the algorithm when $(T+1)\epsilon > b$.

The second halting condition is on the estimated posterior distribution $\mathbf{f}_t$. If the posterior has concentrated around a single point or small interval, we can halt the procedure and output the posterior distribution. This is characterized by computing some stopping time $\tau(\mathbf{f}_t)$. For example, Ben-Or and Hassidim [4] proposed a multi-epoch recursive search strategy and suggest taking $\tau(\mathbf{f}) = 1(\max_i \mathbf{f}_t(i) > \epsilon_{par})$ for $\epsilon_{par} = (24 \log n)^{-1/2}$ to prune the initial set $[n+1]$ into a smaller set of indices with larger posterior probability. In the approach studied by Burnashev and Zigangirov [7], the algorithm terminates when $\max_i \log \frac{\mathbf{f}_t(i)}{1 - \mathbf{f}_t(i)} > \log(1/\delta)$ for a target error probability δ. In this case, the goal is to guarantee that the largest posterior probability is $\mathbf{f}_t(i^*)$ with probability $1 - \delta$.

Pseudocode for the algorithm is shown in Algorithm 1.

Algorithm 1. PrivateBinarySearch($\mathcal{X}, \epsilon, b, t_\ell, t_h, \epsilon_{par}$)

1: $\mathbf{f}_0 \leftarrow \frac{1}{|\mathcal{X}|}$ for $i = 1, 2, \ldots, |\mathcal{X}|$, $t = 1$
2: **while** $\tau(\mathbf{f}_{t-1}) \neq 1$ and $t\epsilon < b$ **do**
3:      Draw σ uniformly at random from permutations on $[n + 1]$.
4:      $\mathcal{Q}_t \leftarrow \{i : \sigma(i) \leq \mathcal{M}(\hat{\mathbf{f}}_{t-1})\}$
5:      $\mathcal{Y}_{\mathcal{Q}_t} \leftarrow \mathcal{A}(\mathcal{Q}_t)$ and $\mathcal{Y}_{\bar{\mathcal{Q}}_t} \leftarrow \mathcal{A}(\bar{\mathcal{Q}}_t)$ from (9)
6:      Update $\mathbf{f}_t$ using (12)
7:      $t \leftarrow t + 1$
8: **end while**
9: **return** $\mathbf{f}_{t-1}$

4.3 Finding the Output

The search algorithm uses a halting condition based on $\mathbf{f}_{t-1}$ and then outputs $\mathbf{f}_{t-1}$, leaving open the question of how to determine the location of the anomaly i^*. If the algorithm waits for $\mathbf{f}_{t-1}$ to concentrate significantly, then with high probability the largest value in $\mathbf{f}_{t-1}$ corresponds to i^*. If instead it prioritizes the privacy budget, then it could pass a list of the largest entries of $\mathbf{f}_{t-1}$ for further processing. More issues regarding practical deployment of this algorithm are discussed in Sect. 7.

5 Analysis

The sensitivity of the max query in Lemma 1 immediately implies that each iteration guarantees ϵ-differential privacy.

Proposition 2. *Each query in Algorithm 1 is ϵ-differentially private. After t iterations of the loop, the overall privacy risk is $t\epsilon$.*

Proof. The result follows from the fact that the noisy computation in (9) guarantees ϵ-differential privacy for $Z, \bar{Z} \sim \mathsf{Lap}(t_\ell/\epsilon)$. Fix neighboring anomaly-restricted datasets $\mathcal{X}$ and $\mathcal{X}'$ and queries $\mathcal{Q} \subset [|\mathcal{X}|]$ and $\mathcal{Q}' \subset [|\mathcal{X}'|]$. Since each iteration of the algorithm splits the dataset into disjoint subsets and applies $\mathcal{A}$ to each independently, by demonstrating that each $\mathcal{A}$ is ϵ-differentially private, we can apply the parallel composition theorem of differential privacy in Definition 5. If $\mathcal{Q} = \mathcal{Q}'$, then clearly

$$\Pr\left[\mathcal{A}\left(\mathcal{Q}\right) = \mathcal{Y}\right] = \Pr\left[\mathcal{A}\left(\mathcal{Q}'\right) = \mathcal{Y}\right], \tag{13}$$

so the application of $\mathcal{A}$ is ϵ-differentially private. We are therefore left with the case where $\mathcal{Q}$ and $\mathcal{Q}'$ differ in a single non-anomalous point. By the post-processing invariance of differential privacy [18], it is sufficient to show that $\mathcal{Y} = \mathcal{A}(\mathcal{Q})$ is ϵ-differentially private. This follows from Lemma 1 and the differential privacy of the Laplace mechanism.   □

Analyzing the convergence of Algorithm 1 is challenging because using Laplace noise means the amount of "progress" made by the algorithm using (12) varies from iteration to iteration. Furthermore, because we only know bounds on the non-anomalous and anomalous values, the update rule is performing an approximation to a Bayes update.

To understand the convergence of the method, we show that a modified version of the update reduces the problem to a noisy binary search. There are two changes: firstly, we do away with the random permutation and secondly, we compute a binary response from $(\mathcal{Y}_t, \bar{\mathcal{Y}}_t)$ and then apply the same Bayes update as randomized response update in (6). More specifically, the algorithm computes $\mathcal{Z}_t = \mathbf{1}\left(\mathcal{Y}_t > \bar{\mathcal{Y}}_t\right)$ and performs a Bayesian update of the prior distribution $\mathbf{f}_{t-1}$ to form the posterior $\mathbf{f}_t$. Because the determination of the subset containing the anomaly $\mathcal{Z}_t$ may be inaccurate, in order to perform the update, we must determine $p = \Pr(i^* \in \mathcal{Z}_t)$.

Lemma 2.

$$\Pr\left(i^* \in \mathcal{Z}_t\right) \geq 1 - \left(\frac{1}{2} + \frac{t_h - t_\ell}{t_\ell} \cdot \frac{\epsilon}{4}\right) \exp\left(-\epsilon \frac{t_h - t_\ell}{t_\ell}\right). \tag{14}$$

Proof. Without loss of generality, let us assume $i^* \in \mathcal{Q}$. We want to find the probability that the following difference is positive:

$$\mathcal{Y} - \bar{\mathcal{Y}} = \max\{\mathbf{x}_i : i \in \mathcal{Q}\} + Z - \max\{\mathbf{x}_i : i \in \bar{\mathcal{Q}}\} - Z'. \tag{15}$$

By assumption, $H(\mathcal{Q}) \geq t_h$ and $H(\bar{\mathcal{Q})\} \leq t_\ell$, thus $H(\mathcal{Q}) - H(\bar{\mathcal{Q}}) \geq t_h - t_\ell$. Therefore $\Pr(Z' - Z) > t_h - t_\ell$ serves as a lower bound on the probability of that the query will return an erroneous result due to noise.

Since the Z and Z' both have zero mean, the distribution of $W = Z' - Z$ is the same as that of $Z + Z'$, which can be found by convolving the two Laplace densities given by (4) with parameter $\lambda = t_\ell/\epsilon$. By assumption, $t_h - t_\ell > 0$, so the probability density function for $w > 0$ is

$$\mathbf{f}(w) = \int_{-\infty}^{\infty} \frac{1}{2\lambda} \exp\left(-|z|/\lambda\right) \frac{1}{2\lambda} \exp\left(-|z - w|/\lambda\right) dz \tag{16}$$

$$= \int_{-\infty}^{0} \frac{1}{4\lambda^2} \exp\left((2z - w)/\lambda\right) dz + \int_{0}^{w} \frac{1}{4\lambda^2} \exp\left(-w/\lambda\right) dz$$

$$+ \int_{w}^{\infty} \frac{1}{4\lambda^2} \exp\left(-(2z - w)/\lambda\right) dz \tag{17}$$

$$= \frac{1}{8\lambda} \exp\left(-w/\lambda\right) + \frac{w}{4\lambda^2} \exp\left(-w/\lambda\right) + \frac{1}{8\lambda} \exp\left(-w/\lambda\right) \tag{18}$$

$$= \frac{\lambda + w}{4\lambda^2} \exp\left(-w/\lambda\right). \tag{19}$$

The cumulative distribution function for $w > 0$ is

$$\mathcal{F}(W \leq w) = \frac{1}{2} + \int_0^w \frac{\lambda + u}{4\lambda^2} \exp\left(-u/\lambda\right) \, du \tag{20}$$

$$= \frac{1}{2} + \left[-\frac{1}{4} \exp\left(-u/\lambda\right)\right]_{u=0}^{w} + \left[-\frac{u}{4\lambda} \exp\left(-u/\lambda\right)\right]_{u=0}^{w}$$

$$- \int_0^w -\frac{1}{4\lambda} \exp\left(-u/\lambda\right) \, du \tag{21}$$

$$= \frac{1}{2} - \frac{1}{4} \exp\left(-w/\lambda\right) + \frac{1}{4} - \frac{w}{4\lambda} \exp\left(-w/\lambda\right) - \frac{1}{4} \exp\left(-w/\lambda\right) + \frac{1}{4} \tag{22}$$

$$= 1 - \left(\frac{1}{2} + \frac{w}{4\lambda}\right) \exp\left(-w/\lambda\right). \tag{23}$$

Now, plugging in $\lambda = \frac{t_\ell}{\epsilon}$ and $w = t_h - t_\ell$ we have (14). $\qquad \square$

Thus, we define

$$p = \left(\frac{1}{2} + \frac{t_h - t_\ell}{t_\ell} \cdot \frac{\epsilon}{4}\right) \exp\left(-\epsilon \frac{t_h - t_\ell}{t_\ell}\right). \tag{24}$$

from (14) and apply the Bayes update in (6).

Proposition 3. *Suppose the anomaly i^* is uniformly distributed in $[n+1]$. For any $\delta \in (0, 1)$, let*

$$T = \min\left\{t : \max_i \log \frac{\mathbf{f}_t(i)}{1 - \mathbf{f}_t(i)} > \log \frac{1}{\delta}\right\}. \tag{25}$$

Set the stopping time $\tau(\mathbf{f}_{t-1}) = \mathbf{1}(t = T)$. Then the modified version of Algorithm 1 using $\mathcal{Z}_t = \mathbf{1}\left(\mathcal{Y}_t > \bar{\mathcal{Y}}_t\right)$ and (24) with update (6) satisfies

$$\mathbb{E}[T] \leq \frac{\log(n+1) + \log(1/\delta) + \epsilon}{1 - h_{\mathrm{b}}(p)} \tag{26}$$

where $h_{\mathrm{b}}(p) = -p \log p - (1-p) \log(1-p)$ is the binary entropy function.

Proof. The result follows by mapping the algorithm to the interval estimation problem studied by Burnashev and Zigangirov [7]. The main difference is that when using $\mathcal{Z}_t$, (24) is only an upper bound on the error probability of the oracle for randomized response. However, this means that the oracle is only potentially *less* noisy than the randomized response oracle. Using the stopping rule in (25), we get the upper bound on the expected number of queries [7, Theorem 3]. $\quad \square$

6   Experimental Results

We demonstrate the practical performance of our approach through experiments on a data set for anomaly detection. The experiments investigate how different configurations of input parameters and constraints on the datasets can affect accuracy and total privacy risk. Specifically, we are interested in the impact of the thresholds t_h and t_ℓ, the oracle response configuration, and the halting conditions τ and privacy budget b.

6.1   Dataset

The experiments use the A1 Benchmark from the Yahoo Labeled Anomaly Detection Dataset, part of the Yahoo Webscope reference library [38]. Each dataset in the benchmark is preprocessed down to single anomaly by selecting the largest anomalous point in each dataset and selecting thresholds by letting $\mathbf{x}_j = \max\{\mathbf{x}_i : i \neq i^*\}$ and setting $t_h = \mathbf{x}_{i^*} - .1(\mathbf{x}_{i^*} - \mathbf{x}_j)$ and $t_\ell = \mathbf{x}_j + .1(\mathbf{x}_{i^*} - \mathbf{x}_j)$. Some experiments are run specifically on datasets 6 and 8 in order to explore the effects of the non-anomalous point distribution on the algorithm performance. These two datasets exemplify the two primary distributions for sets contained in benchmark: datasets that are a mixture of normal distributions, and datasets where points are heavily skewed toward 0.

6.2   Procedure

Because we are interested in approximate detection of the anomaly, we declare that the algorithm succeeds if it halts and can output a small set S of indices such that $i^* \in S$. In particular, we choose $|S| = 4$ and set S to be the indices with the 4 largest posterior probabilities. This selection is to capture the difference between $\mathbf{f}(i^*)$ being the close to the largest posterior probability and being much smaller. Cases where $\mathbf{f}(i) = \mathbf{f}(j)$ for $i \neq j$ are prevented in practice by the randomized permutation of the probability mass function after each iteration. For these experiments, $\tau = \mathbf{1}(\max\{\mathbf{f}(i) : i \in [n+1]\} > 0.5)$ is used as a halting condition when not otherwise specified. Each configuration of the algorithm parameters are run for a set number of cycles c. The approximate average error rate for the configuration is $(1 - \frac{\sum_{i=1}^{c} \mathbf{1}(i^* \in S_i)}{c})$ and the average total privacy risk is $\frac{\sum_{i=1}^{c}(t\epsilon)_i}{c}$. For these experiments, we take $c = 100$.

6.3   Results

We demonstrate the algorithm's performance as a function of the privacy parameter ϵ. Smaller ϵ values result in noisier responses from the oracle which require more iterations to reach the halting condition. Correspondingly, larger values of ϵ decrease noise which requires fewer total iterations, but at greater privacy cost per iteration. The tradeoff between error rate and total privacy risk forms a concave upward curve. Lower values of the privacy parameter are more costly in total privacy risk as the noise at each iteration strongly decreases $\Pr(i^* \in \mathcal{R})$. Increasing the privacy parameter increases $\Pr(i^* \in \mathcal{R})$ at a greater rate than the privacy cost per iteration increases, thus decreasing total privacy risk. However, these improvements have diminishing returns. Eventually, increasing the privacy parameter no longer improves the error rate as $\Pr(i^* \in \mathcal{R}) \to 1$. At this point, increasing the privacy parameter doesn't improve the error rate, but continues to increase total privacy risk.

Threshold Ratios. Figure 1 demonstrates the effect of the thresholds on the algorithm's performance. Each point in the figure depicts the error rate as a function of that dataset's threshold ratio $\frac{t_h - t_\ell}{t_\ell}$ with privacy parameter set to $\epsilon = 1$. A dataset with a higher threshold ratio tends to perform better than an equivalent dataset with a lower threshold ratio for a given value of the privacy parameter. This is due to $\Delta_g = t_\ell$, which causes smaller differences between thresholds $t_h - t_\ell$ to be more likely to be overcome by noise. The steep improvement in error rate for small changes in the threshold ratio highlight the importance of tuning the privacy parameter to the thresholds of the dataset. (Note that datasets 6 and 8 were selected to have similar threshold ratios at 0.647 and 0.701 respectively).

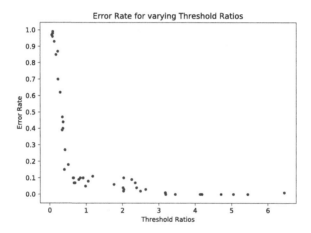

Fig. 1. Error rate for each dataset as a function of the threshold ratio $\frac{t_h - t_\ell}{t_\ell}$.

Oracle Response Constructions. Figure 2 demonstrates how different constructions of the oracle response and Bayesian update methods affect the error rate. The proposed oracle response approaches include the randomized response oracle (5), the binarized noisy response oracle (14) and the direct noisy result oracle (11). Despite all constructions achieving $t\epsilon$-differential privacy, there is a strong difference in effect on the error rate and total privacy risk.

Randomized response has the worst error rate because the oracle error probability is fixed. This contrasts with the oracle mechanisms that use the noisy aggregations: the actual noisy response depend on the values $(\mathcal{Y}, \bar{\mathcal{Y}})$, which can be more informative depending on the noise. For example, when the actual difference between $\mathcal{Y}$ or $\bar{\mathcal{Y}}$ exceeds the difference between t_h and t_ℓ, added noise is less likely to cause incorrect responses than in randomized response. Similarly, the oracle that directly uses the noisy response performs better than the

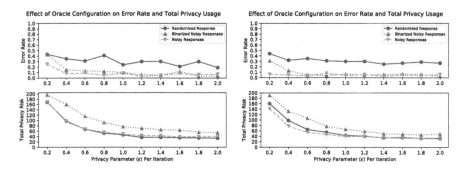

Fig. 2. The error rate and total privacy risk as a function of the privacy parameter ϵ for different oracle response constructions on data sets 6 and 8.

binary oracle construction as the likelihood for the binary oracle at each iteration is a lower bound given by (14) which gives up some information gain on each iteration. Because the binarized construction is a lower bound on the actual likelihood, more updates become required to achieve the same effect as the other constructions and thus ends up having greater total privacy risk.

Algorithm Halting Conditions. The algorithm's two termination conditions, τ and total privacy risk exceeding budget b, are explored in Figs. 3 and 4. Figure 3 depicts the algorithm's error rate with varying budget constraints where the halting constraint τ has been removed. When the total privacy risk passes pre-assigned budget checkpoints, S is checked for the presence of the anomaly and the algorithm continues. Similarly Fig. 4 depicts various halting constraints where the budget constraint has been removed and again checks S at pre-assigned halting checkpoints. When the algorithm is forced to preemptively halt because total privacy risk exceeds the budget, errors are excessively high. This is due to the increased chance that not enough iterations have been run to allow the algorithm to overcome noisy oracle responses. When the privacy parameter ϵ is larger, the

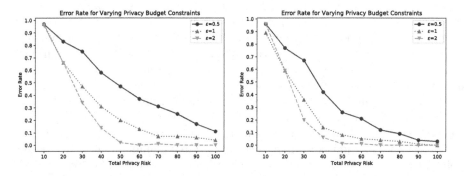

Fig. 3. Error rates for varying inputs of the privacy parameter ϵ with differing maximum budget constraints b for datasets 6 and 8.

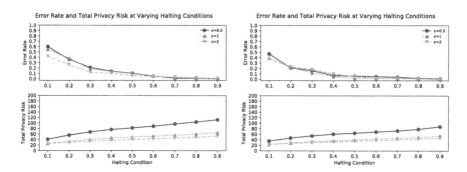

Fig. 4. Error rates and total privacy risk across varying halting constraints τ for data sets 6 and 8.

algorithm is more likely to suffer errors from the algorithm terminating early. Correspondingly, when total privacy risk does not prevent early termination due to budget b, larger values of ϵ result in fewer errors. Thus, a proper privacy budget should be allocated to perform enough iterations to prevent errors due to halting early.

Figure 4 demonstrates how different halting conditions τ affect the error rate with unlimited privacy budget. Specifically, the figure depicts the effect of altering α for $\tau = \mathbf{1}(\max\{\mathbf{f}(i) : i \in [n+1]\} > \alpha)$. As the halting condition serves as a requirement of convergence of the probability mass toward a single point, the algorithm can steadily improve the error rates by increasing α. This requires correspondingly more iterations to achieve, incurring greater total privacy risk for any run of the algorithm.

7   Discussion

We have described a differentially private search algorithm using noisy binary search with applications to anomaly detection. For this application, we defined a new notion of anomaly-restricted neighboring databases to capture the idea that anomalous points (which potentially merit scrutiny even if it is privacy-invasive) are not given privacy guarantees. The noise in the algorithm provides quantifiable privacy during the search. We showed theoretically and empirically that the greedy Bayesian search strategy can quickly narrow down a small set of samples that contain the anomaly.

There are a number of practical considerations that must be further addressed for our work to be useful in particular applications. For example, in most cases, it will be necessary to handle multiple anomalies rather than only a single anomaly. If a good upper bound is known on the expected maximum number of anomalous points, then one approach for using our method would be to first divide the set into disjoint subsets that with high probability contain only a single anomaly, and then proceeding to apply our method to each of those subsets individually.

In any particular application, it is also necessary to specify what points the algorithm should return. This depends on various factors, including what will be done with those points. We envision a scenario in which the points returned undergo some further screening, presumably after appropriate policies are followed. However, this creates a tradeoff between false positives and false negatives. To provide the most privacy, it would be desirable for the returned set to be as small as possible. However, narrowing down too far increases the chance of returning a set that misses the anomalous point. In very large search spaces or problems with many anomalies, one option would be to recursively prune out non-anomalous points: while this should work well in practice, theoretically analyzing the corresponding privacy-utility tradeoffs may be quite complex.

Our method uses a fixed privacy loss ϵ_t per iteration, not without loss of generality. Varying ϵ_t across iterations in a decaying manner could correspond to active learning or noisy search under the Tsybakov noise condition. Results from active learning can yield bounds on convergence to interpret the error/privacy tradeoff. A key difference between our search model and standard noisy search is that we can design the noise to optimize the privacy-utility tradeoff.

In order to provide privacy without relying on a trusted party, our method relies on the existence of a sensor or other measurement device that carries out the noisy aggregate queries directly, without carrying out individual queries and computing a noisy aggregate result from them. Practical use of our techniques therefore depends on the practical creation and deployment of such sensors.

Acknowledgements. This work was partially supported by NSF under award CCF-1453432, DARPA and SSC Pacific under contract N66001-15-C-4070, and DHS under award 2009-ST-061-CCI002 and contract HSHQDC-16-A-B0005/HSHQDC-16-J-00371.

References

1. Abadi, M., Chu, A., Goodfello, I., McMahan, H.B., Mironov, I., Talwar, K., Zhang, L.: Deep learning with differential privacy. In: Proceedings of the 2016 ACM SIGSAC Conference on Computer and Communication Security (CCS 2016), Vienna, Austria, 24–28 October 2016, pp. 303–318. ACM (2016). https://doi.org/10.1145/2976749.2978318
2. Agyemang, M., Barker, K., Alhajj, R.: A comprehensive survey of numeric and symbolic outlier mining techniques. Intell. Data Anal. **10**(6), 521–538 (2006)
3. Atia, G.K., Saligrama, V.: Boolean compressed sensing and noisy group testing. IEEE Trans. Inf. Theory **58**(3), 1880–1901 (2012). https://doi.org/10.1109/TIT.2011.2178156
4. Ben-Or, M., Hassidim, A.: The Bayesian learner is optimal for noisy binary search (and pretty good for quantum as well). In: 49th Annual IEEE Symposium on Foundations of Computer Science (FOCS 2008), pp. 221–230 (2008). https://doi.org/10.1109/FOCS.2008.58
5. Berger, T., Mehravari, N., Towsley, D., Wolf, J.: Random multiple-access communication and group testing. IEEE Trans. Commun. **32**(7), 769–779 (1984). https://doi.org/10.1109/TCOM.1984.1096146

6. Bun, M., Steinke, T.: Concentrated differential privacy: simplifications, extensions, and lower bounds. In: Hirt, M., Smith, A. (eds.) TCC 2016. LNCS, vol. 9985, pp. 635–658. Springer, Heidelberg (2016). https://doi.org/10.1007/978-3-662-53641-4_24

7. Burnashev, M.V., Zigangirov, K.S.: An interval estimation problem for controlled observations. Probl. Inf. Transm. **10**, 223–231 (1974)

8. Cai, S., Jahangoshahi, M., Bakshi, M., Jaggi, S.: GROTESQUE: noisy group testing (quick and efficient). Technical report arXiv:1307.2811 [cs.IT], ArXiV, July 2013. http://arxiv.org/abs/1307.2811

9. Calderbank, R., Howard, S., Jafarpour, S.: Construction of a large class of deterministic sensing matrices that satisfy a statistical isometry property. IEEE J. Sel. Topics Sig. Process. **4**(2), 358–743 (2010). https://doi.org/10.1109/JSTSP.2010.2043161

10. Chan, C.L., Jaggi, S., Saligrama, V., Agnihotri, S.: Non-adaptive group testing: explicit bounds and novel algorithms. IEEE Trans. Inf. Theory **60**(5), 3019–3035 (2014). https://doi.org/10.1109/TIT.2014.2310477

11. Chandola, V., Banerjee, A., Kumar, V.: Anomaly detection: a survey. ACM Comput. Surv. (CSUR) **41**(3), 15 (2009)

12. Chen, H.B., Hwang, F.K.: A survey on nonadaptive group testing algorithms through the angle of decoding. J. Comb. Optim. **15**(1), 49–59 (2008). https://doi.org/10.1007/s10878-007-9083-3

13. Department of Homeland Security: Screening at speed (2017). https://www.dhs.gov/science-and-technology/apex-screening-speed. Accessed 3 Aug 2017

14. Dorfman, R.: The detection of defective members of large populations. Ann. Math. Stat. **14**(4), 436–440 (1943). http://www.jstor.org/stable/2235930

15. Du, D.Z., Hwang, F.K.: Combinatorial group testing and its applications, vol. 12, 2nd edn. World Scientific (1999). https://doi.org/10.1142/4252

16. Dwork, C.: Differential privacy. In: Bugliesi, M., Preneel, B., Sassone, V., Wegener, I. (eds.) ICALP 2006. LNCS, vol. 4052, pp. 1–12. Springer, Heidelberg (2006). https://doi.org/10.1007/11787006_1. https://www.microsoft.com/en-us/research/publication/differential-privacy/

17. Dwork, C.: A firm foundation for private data analysis. Commun. ACM **54**(1), 86–95 (2011)

18. Dwork, C., McSherry, F., Nissim, K., Smith, A.: Calibrating noise to sensitivity in private data analysis. In: Halevi, S., Rabin, T. (eds.) TCC 2006. LNCS, vol. 3876, pp. 265–284. Springer, Heidelberg (2006). https://doi.org/10.1007/11681878_14

19. Dwork, C., Rothblum, G., Vadhan, S.: Boosting and differential privacy. In: 2010 51st Annual IEEE Symposium on Foundations of Computer Science (FOCS), Las Vegas, NV, pp. 51–60, October 2010. https://doi.org/10.1109/FOCS.2010.12

20. Erlingsson, Ú., Pihur, V., Korolova, A.: RAPPOR: randomized aggregatable privacy-preserving ordinal response. In: Proceedings of the 2014 ACM SIGSAC Conference on Computer and Communications Security (CCS 2014), pp. 1054–1067 (2014). https://doi.org/10.1145/2660267.2660348

21. Evans, J.: What Apple users need to know about differential privacy. IComputerWorld, June 2016. http://www.computerworld.com/article/3088179/apple-mac/what-apple-users-need-to-know-about-differential-privacy.html

22. Ghassemi, M., Sarwate, A.D., Wright, R.N.: Differentially private online active learning with applications to anomaly detection. In: Proceedings of the 9th ACM Workshop on Artificial Intelligence and Security (AISec), Vienna, Austria, pp. 117–128, October 2016

23. Hodge, V.J., Austin, J.: A survey of outlier detection methodologies. Artif. Intell. Rev. **22**(2), 85–126 (2004)

24. Horstein, M.: Sequential transmission using noiseless feedback. IEEE Trans. Inf. Theory **9**(3), 136–143 (1963). https://doi.org/10.1109/TIT.1963.1057832

25. Jedynak, B., Frazier, P.I., Sznitman, R.: Twenty questions with noise: Bayes optimal policies for entropy loss. J. Appl. Probab. **49**(1), 114–136 (2012). https://doi.org/10.1239/jap/1331216837

26. Kairouz, P., Oh, S., Viswanath, P.: The composition theorem for differential privacy. IEEE Trans. Inf. Theory **63**(6) (2017). https://doi.org/10.1109/TIT.2017.2685505

27. Machanavajjhala, A., Kifer, D., Abowd, J.M., Gehrke, J., Vilhuber, L.: Privacy: theory meets practice on the map. In: IEEE 24th International Conference on Data Engineering (ICDE), pp. 277–286 (2008). https://doi.org/10.1109/ICDE.2008.4497436

28. Macula, A.J., Popyack, L.J.: A group testing method for finding patterns in data. Discrete Appl. Math. **144**(1–2), 149–157 (2004). https://doi.org/10.1016/j.dam.2003.07.009

29. Mazumdar, A.: Nonadaptive group testing with random set of defectives. IEEE Trans. Inf. Theory **62**(12), 7522–7531 (2016). http://ieeexplore.ieee.org/document/7577749/

30. Mazumdar, A., Barg, A.: Sparse-recovery properties of statistical RIP matrices. In: Proceedings of the 49th Allerton Conference on Communication, Control and Computing, pp. 9–12, September 2011. https://doi.org/10.1109/Allerton.2011.6120142

31. Mir, D.J., Isaacman, S., Cáceres, R., Martonosi, M., Wright, R.N.: DP-WHERE: differentially private modeling of human mobility. In: Proceedings of the 2013 IEEE International Conference on Big Data, October 2013. https://doi.org/10.1109/BigData.2013.6691626

32. Ngo, H.Q., Du, D.Z.: A survey on combinatorial group testing algorithms with applications to DNA library screening. In: Du, D.Z., Pardalos, P.M., Wang, J. (eds.) Discrete Mathematical Problems with Medical Applications. DIMACS Series in Discrete Mathematics and Theoretical Computer Science, vol. 55. AMS (2000)

33. Nowak, R.: Generalized binary search. In: 2008 46th Annual Allerton Conference on Communication, Control, and Computing, pp. 568–574. IEEE (2008)

34. Nowak, R.: Noisy generalized binary search. In: Advances in Neural Information Processing Systems, pp. 1366–1374 (2009)

35. Waeber, R., Frazier, P.I., Henderson, S.G.: Bisection search with noisy responses. SIAM J. Control Optim. **51**(3), 2261–2279 (2013)

36. Wasserman, L., Zhou, S.: A statistical framework for differential privacy. J. Am. Stat. Assoc. **105**(489), 375–389 (2010). https://doi.org/10.1198/jasa.2009.tm08651

37. Wolf, J.: Born again group testing: multiaccess communications. IEEE Trans. Inf. Theory **31**(2), 185–191 (1985). https://doi.org/10.1109/TIT.1985.1057026

38. Yahoo Labs: S5 - a labeled anomaly detection dataset, version 1.0 (2016). https://webscope.sandbox.yahoo.com/catalog.php?datatype=s&did=70

Distributed Web Mining of Ethereum

Trishita Tiwari$^{(\boxtimes)}$, David Starobinski, and Ari Trachtenberg

Department of Electrical and Computer Engineering,
Boston University, Boston, MA 02215, USA
{trtiwari,staro,trachten}@bu.edu

Abstract. We consider the problem of mining crytocurrencies by harnessing the inherent distribution capabilities of the World Wide Web. More specifically, we propose, analyze, and implement `WebEth`, a browser-based distributed miner of the Ethereum cryptocurrency. `WebEth` handles Proof-of-Work (PoW) calculations through individualized code that runs on the client browsers, and thereafter collates them at a web server to complete the mining operation. `WebEth` is based on a lazy evaluation technique designed to function within the expected limitations of the clients, including bounds on memory, computation and communication bandwidth to the server. We provide proofs-of-concept of `WebEth` based on `JavaScript` and `WebAssembly` implementations, with the latter reaching hash rates up to roughly 40 kiloHashes per second, which is only 30% slower than the corresponding native `C++`-based implementation. Finally, we explore several applications of `WebEth`, including monetization of web content, rate limitation to server access, and private Ethereum networks. Though several distributed web-based cryptominers have appeared in the wild (for other currencies), either in malware or in commercial trials, we believe that `WebEth` is the first open-source cryptominer of this type.

Keywords: Crypto-currency · Ethereum · Distributed computing
Web-browser computing · Mining

1 Introduction

Cryptocurrencies are increasingly gaining traction as a viable form of currency. This has been accompanied by a correspondingly increasing interest in the efficient validation of cryptocurrency transactions. Whereas initial efforts in this domain have focused on creating dedicated hardware for this task [30], more recent approaches have examined repurposing existing infrastructure. Indeed, one such class of efforts has focused on the use of client web browsers as a platform for distributed computing [11]. The growing popularity of CoinHive [13] is a case in point of the potential success of distributed in-browser cryptocurrency mining as a commercial (if malicious) enterprise.

In this work, we propose `WebEth`, a browser-based distributed miner for the popular Ethereum block chain [28]. `WebEth` tackles the challenge of achieving a profitable hash rate within a distributed ensemble of browsers under constrained

© Springer International Publishing AG, part of Springer Nature 2018
I. Dinur et al. (Eds.): CSCML 2018, LNCS 10879, pp. 38–54, 2018.
https://doi.org/10.1007/978-3-319-94147-9_4

memory, computation and network usage. Indeed, every browser needs to store a data structure of at least 1 GB in memory in order to mine Ethereum. Clearly, it is unfeasible to transfer this entire data structure every time a browser loads a web page. Instead, WebEth employs a lazy approach to generate this data structure while mining. Through our experiments, we show that this approach takes at most five minutes to reach a steady state hash rate – making it ideal for web applications where users spend time, such as gaming and video streaming. Our experiments also show that WebEth yields a hashing rate of up to 40 kilo-Hashes/s, which, despite the overhead from running the algorithm in a browser, is only 30% smaller than the performance of a corresponding miner running natively.

The main contributions of this work are as follows:

- We propose WebEth [1], an open-source implementation of a distributed web-based Ethereum cryptominer in both JavaScript and WebAssembly that can operate under relatively resource constrained environments. Though miners for other currencies exist in the wild (e.g. CoinHive [13]), they are all proprietary and closed-source.
- We provide theoretical analysis and experimental evidence of the potential efficacy of the lazy approach adopted by WebEth to achieve high hashing rates.
- We propose a number of potential applications built upon WebEth, including rate-limiting server access, usage tracking, and content monetization.

Related Work: Distribution of a common task is hardly a new concept [7–9], but the growing popularity and efficiency of dynamic web content and client-side scripting languages like JavaScript and WebAssembly have made web browsers an enticing implementation option [10,11,14]. Coinhive [13] has built into this environment a proprietary method for mining the Monero cryptocurrency, but this is often done on the browsers of unsuspecting users. A more ethical, open-sourced alternative, the Basic Attention Token [32], is an Ethereum-based ERC20 token [35] currently in development to be used in conjunction with the Brave browser [4] to generate ad revenue for website publishers by *measuring a user's attention* on an advertisement. This platform promises to balance the ties between website users, publishers, and advertisers to ensure that users get only ads they would accept, advertisers pay for actual users (instead of click bots), and publishers actually get revenue instead of begging users to turn off their ad blockers.

Roadmap: The remainder of the paper is organized as follows. In Sect. 2 we cover the relevant background and related literature for our work. In particular, we provide a self-contained description of Ethash, the Ethereum mining Proof-of-Work (PoW), and WebAssembly, the language in which we implement an efficient miner. Section 3 describes WebEth, including its lazy approach to distributing the Ethash PoW over numerous, resource-constrained browsers. This section also includes a performance analysis. We present the experimental results of our implementations in Sect. 4. In Sect. 5 we discuss several potential applications of our mining platform. We conclude in Sect. 6.

2 Background

2.1 Cryptocurrencies: A General Overview

Most cryptocurrencies like Ethereum involve storing transactions in blocks, and the entire history of transactions is collated in a data structure known as a block chain. The block chain is managed by dedicated machines called *client nodes*; where each client node typically stores the entire block chain. However, because each client node operates on its own copy of the block chain, the block chains on different nodes may go out of sync. Hence, there needs to be an accepted mechanism to decide the order in which new transaction blocks are appended to this block chain. For this, every time a new transaction takes place, the transaction is pooled together with all other transactions that have been broadcasted to the network, but haven't been added to the block chain yet. For most currencies, the data structure that stores these unconfirmed transactions is known as the *mem pool*. A miner picks valid transactions from his mem pool and creates a new block out of these transactions. Once this happens, the goal of the miner is to have his/her block appended to the block chain, which is achieved through a process called *mining*. Mining involves a race amongst miners to solve a *Proof-of-Work* (PoW) puzzle, which is usually an energy intensive computation. The winner of this race gets to have his block appended to the block chain. The winner also receives a payout, which acts as an incentive to mine.

Ethereum uses Ethash as its PoW algorithm, which is explained in detail below.

2.2 Ethereum Proof of Work

Ethereum is a crypto-currency that was released in July 2015 by Vitalik Buterin, Gavin Wood and Jeffrey Wilcke. Ethereum uses the Ethash algorithm (derived from the Dagger and Hashimoto algorithms [6]) for its PoW for mining blocks. Before we discuss how mining works with Ethash, we first establish basic terminology about the data structures involved in the PoW.

A *block header* contains meta-data related to the transactions of the corresponding block, and it is provided as an input to the Ethash algorithm together with an integer *nonce*. The nonce is chosen in a brute-force fashion in order to hash, together with the block header, into a value that matches a specific pattern (based on a predefined *difficulty threshold*). The process of finding an appropriate nonce for a given block is known as *mining*.

Once a block has been mined, it is propagated to other client nodes so that they can update their copies of the block chain. However, before each client node does so, it must validate whether the miner is submitting a legitimate block – i.e., check whether the miner genuinely solved the hash as claimed. This is easily done by putting the block header and nonce associated with the block through the Ethash algorithm and checking whether the output follows the pattern prescribed by the difficulty threshold. *Light weight* client nodes do not mine new blocks, but rather only verify whether any new block submitted by a miner is valid or not. *Full* client nodes, on the other hand, both mine and verify new blocks.

An *epoch* is a unit of "time" that spans 30,000 blocks. All of the data structures in Ethash (mentioned below) need a 256 bit *seed*, which differs for every epoch. For the first epoch the seed is a Keccak-256 [29] hash of a series of 32 bytes of zeros. For every other epoch it is always the Keccak-256 hash of the previous seed hash. The seed for a particular epoch is used to compute a data structure known as the *Cache*, which is an array of 4 byte integers [16]. The Cache production process involves using the seed hash to first sequentially fill up the Cache, then performing two passes of the RandMemoHash algorithm [12] on top of it to get the final Cache value.

Light weight clients may use this Cache for verifying hashes, while full node clients can use it to calculate a Directed Acyclic Graph (DAG) dataset, as described below. A Ethereum *Directed Acyclic Graph* is stored as a large byte array (around 1 GB in size on the private Ethereum Network used for our experiments) and has the following two attributes:

1. *Node*: Each DAG node in this byte array spans 64 bytes, and node indices are therefore aligned at a 64 byte boundary.
2. *Page*: Each DAG page spans 2 nodes, however, page accesses are not aligned at a 2 node boundary. The mining process involves accessing some DAG pages and hashing them together with the block header and nonce.

Each node in the DAG is generated by combining data from 256 pseudo-randomly selected Cache nodes and hashing them together. This process is repeated to generate all nodes in the DAG. Finally, intermediate byte arrays used to store temporary results in the Ethash algorithm are known as *Mixes*.

We must point out that as time goes on, mining Ethereum becomes more and more difficult, as the size of the Cache and DAG increases with every epoch.

Mining is performed by starting with the current block header hash and nonce, which are combined to get a 128-byte wide Mix, as seen in step 1 of Fig. 1. The Mix is then used to fetch a specific page of the DAG from memory. After this, the Mix is updated with the fetched page of the DAG (step 2). Then, this updated Mix is used to fetch a new page of the DAG (step 3). This process of sequentially fetching parts of the DAG is repeated 64 times (step 4), and the final value of the Mix is put through a transformation to obtain a 32 byte digest (step 5). This digest is then compared to the threshold (step 6). If it is smaller than the threshold, the nonce is valid and the block is successfully mined and can be broadcast to the network. However, if the digest is greater than the threshold, the nonce is unsuccessful, and the entire process must be repeated with a new nonce [16].

It is important to note that the pages of the DAG that are used to compute the hash for a particular block depend on the nonce used, hence there is no way to pre-determine which pages will be useful to have in memory. This therefore forces miners to store entire DAG in memory, making Ethereum mining "Memory Hard".

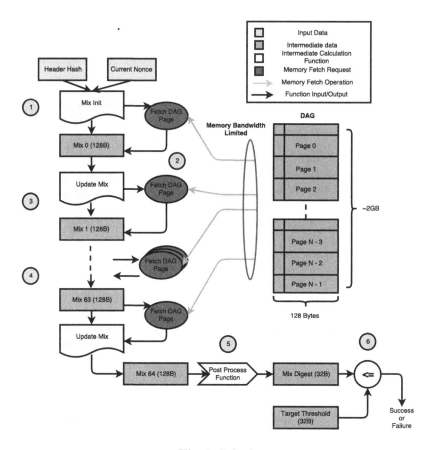

Fig. 1. Ethash

Whereas mining is memory intensive, verification is relatively lightweight. This is because of the property that each node in the DAG depends on a set of pseudo-randomly selected items from the Cache. Hence, the Cache is used to regenerate only the specific pages of the DAG that are needed to recalculate the hash for the particular nonce. And so, only the Cache needs to be stored by light weight clients that only perform verification. In fact, as we will see in the next section, we use this property of being able to generate parts of the DAG as needed to our advantage in order to alleviate some of the memory and network bandwidth restrictions that browsers typically face.

2.3   JavaScript and WebAssembly

Introduced in 1995 by Netscape Communications Corporation [31], JavaScript was a meant to be a light scripting language in order to make web content dynamic. Over the span of 23 years, it has grown to become one of the most popular client-side web development languages used to make dynamic user interfaces.

In fact, up to recently, `JavaScript` has been the only language available to make dynamic client-side web content. However, the situation has changed since the advent of `WebAssembly` in 2016. As per its creators, `WebAssembly` is a "binary instruction format for a stack-based virtual machine" [18]. WASM is designed to be compiled from high-level languages like `C/C++/Rust`, and is supported by 4 major browser platforms – Firefox, Safari, IE, and Chrome [18]. The `WebAssembly` stack machine is designed to be encoded in a "size-and load-time-efficient binary format" [18], and aims to execute near native speed by utilizing common hardware capabilities present on a wide variety of platforms [18]. The language is meant to improve performance for computationally intensive use cases such as image/video editing, games, streaming, etc. [18]. This makes it the language of choice to implement a miner within a browser.

3   WebEth

In this Section, we present our Web mining architecture for Ethereum. A diagram of the `WebEth` architecture is depicted in Fig. 2. The architecture of the miner itself involves the browser connecting to a central node as soon as the web page loads (Steps 1 and 2 in Fig. 2). On connecting, the browser then receives the current block header hash and Cache (Step 3), using which it begins mining using the lazy evaluation algorithm discussed below (Step 4). Then, the browser could take one of two paths – it could either have solved the block (Step 5a), in which case it sends the solution to the central node and asks for the next block to solve. Or it could timeout (Step 5b), in which case it polls the central node again for the current block header hash and Cache and then resumes mining. This architecture, including the lazy evaluation mining algorithm, is described in detail below.

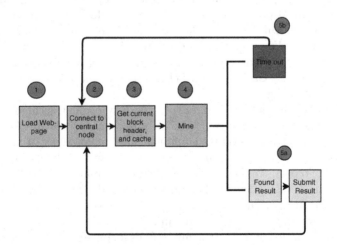

Fig. 2. WebEth architecture

3.1   Lazy Evaluation

The mining itself (Step 3) in WebEth is based on a lazy evaluation to alleviate the network and memory requirements for mining Ethereum in a distributed scenario on browsers. Specifically, as soon as each browser connects to the webserver, the server sends to the browser the current block header hash and the Cache. Once the browser receives the Cache, it allocates an array buffer to store the DAG nodes.

   Once the buffer is allocated, the browser can start iterating over nonces to compute hashes. Since, to begin with, the browser does not store any nodes of the DAG, it must compute each node on the fly using the Cache. However, every node that the browser computes is stored in the buffer, for quicker access in the future. Hence, as time passes, the buffer starts filling up, such that more and more nodes are quickly accessed from the buffer rather than being computed from the ground-up, which makes hash computations faster with time. This has the effect that the longer the user remains on the web-page, the better the hash rate gets for that user.

3.2   Implementation

The WebEth architecture is centered around a central node and client-side Ethereum miners.

   We have two implementation for the client-side miners: in JavaScript and in WebAssembly. For our miner, we model the JavaScript implementation after the node.js implementation of Ethash [3]. The WebAssembly version is the JavaScript version transpiled to C++, which in turn is compiled to WebAssembly using the Emscripten compiler [19].

   The central node itself coordinates all workers (browsers). Its implementation is based on a modified version of geth [17], a real world Ethereum miner written in Go. geth typically runs as a standalone miner that mines on the machine on which it is running. We modify the code so that instead of doing all the mining all by itself, the node sends over the necessary data (namely, the hash of the Block Header and the Cache) to any client that connects to it on port 9000 (Steps 1, 2 and 3 in Fig. 2).

   After receiving the necessary data, per the lazy evaluation algorithm, each browser allocates a buffer for the DAG in order to store future DAG nodes. (Note that the buffer for the DAG is implemented as an array of ints, so as to make each lookup in the buffer constant in time.) Now, the client-side miner can begin to mine (Step 4). At the beginning, the miner creates a random nonce and computes the hash (using the Cache and the buffered DAG) as discussed in the previous section. It continues to perform this action on new nonces until one of two following scenarios occur (Steps 5a and 5b).

   In the first scenario (Step 5a), the miner finds a nonce such that the computed hash is below the given threshold. In that case, the browser submits the result back to the central node and then asks the central node for the new block header hash and the Cache. It then uses these new inputs and continues to mine.

In the second scenario (Step 5b), the algorithm times out without finding a result. In that case, the miner polls the central node for the current versions of the block header hash and Cache and continues to mine using the new inputs. This process continue until the user moves away from the website or closes the browser. This time out is necessary since the browser should work with the most recent block header and Cache. The block header can become stale if that particular block has already been mined, and the Cache can become stale if the Ethereum network transitions into a new epoch (this happens once every 30,000 blocks).

Finally, we must point out that both our current implementations in `JavaScript` and `WebAssembly` require no external dependencies, and therefore can be directly embedded into any website. Furthermore, the fact that the central server does not have to keep track of each client makes the system quite scalable.

3.3   Performance Analysis

In this section, we perform a back-of-the envelope calculation for the number of hashes needed till `WebEth` fills up almost all the buffer. This is important because the hash rate reaches its maximum steady-state only once the buffer in the browser is almost full. Specifically, we show that filling a buffer the size of the DAG till only $5 * 10^{-7}\%$ of it is empty should take on average about 1.85 million hashes, while filling it up till $5 * 10^{-1}\%$ of this buffer is empty takes much lesser – about 700 thousand hashes. The approach we take is based on the Coupon Collector problem [24].

For simplicity, we assume that 128 nodes in the DAG are *randomly* sampled in order to compute each hash, whereas, in reality, this is not entirely true. This is because Ethash samples 128 *pages* per hash (rather than 128 *nodes*). Since each page is two nodes wide, two *neighboring* DAG nodes are sampled for each page computation. Hence, the DAG nodes are not accessed completely randomly. Nonetheless, this estimation still provides us a good approximation.

For our analysis, we introduce the following notation. We denote by N the total number of nodes in the DAG, by a the number of DAG nodes needed to compute a hash, by δ the failure probability of finding a specific node in the buffer (i.e., the buffer miss rate), by ω the failure probability of computing a hash using nodes already stored in the buffer, by $E(X)$ the expected number of hashes to fill the buffer with a failure probability δ, and by H_n the n-th Harmonic number.

Claim. For $\delta \ll 1$, $E(X) \approx \frac{N}{a}(H_N - H_{N\delta})$.

Proof: The number of nodes needed in the buffer to achieve a failure probability δ is $\lceil N(1 - \delta) \rceil$. This means that even though we are allocating a buffer that can hold N nodes, we are willing to forgo $\lfloor N\delta \rfloor$ nodes (to simplify notation, from now and on, we assume $N\delta$ is an integer).

Using results from the Coupon Collector's problem [24], we know that the expected number of trials for obtaining the i-th new node after having buffered

$i-1$ nodes is $N/(N-(i-1))$. Thus, the expected number of trials in order to fill up the buffer with $N(1-\delta)$ nodes is given by

$$E(t) = N \sum_{i=1}^{N(1-\delta)} \frac{1}{N-i+1}. \tag{1}$$

Splitting this expression into two sums, we get

$$E(t) = N \left(\sum_{i=1}^{N} \frac{1}{N-i+1} - \sum_{i=N(1-\delta)+1}^{N} \frac{1}{N-i+1} \right), \tag{2}$$

or

$$E(t) = N(H_N - H_{N\delta}). \tag{3}$$

We now relate the failure probability of calculating a hash using nodes already stored in the buffer ω with the failure probability of having a specific node in the buffer δ. By the independence assumption,

$$1 - \omega = (1 - \delta)^a$$

For $\delta \ll 1$, we have $(1-\delta)^a \approx 1 - a\delta$. Hence,

$$\delta \approx \frac{\omega}{a}. \tag{4}$$

Hence, it follows from Eq. (4) that

$$E(X) \approx \frac{E(t)}{a}. \tag{5}$$

Finally, from Eqs. (3) and (5) and, we obtain

$$E(X) \approx \frac{N}{a}(H_N - H_{N\delta}). \tag{6}$$

∎

We use the following approximation on the Harmonic Numbers to compute Eq. (6):

$$H_n \approx \ln n + \gamma + o(1) \tag{7}$$

where $\gamma = 0.57721566...$ is the Euler-Mascheroni constant.

Specifically, using Eqs. (6) and (7), and setting $N = 16777186$ (the number of nodes in the DAG for our experiments), $a = 128$ and $\delta = 5 * 10^{-9}$, we get $E(X) \approx 1.85$ million hashes. However, if we increase δ to $5 * 10^{-3}$, we get $E(X) \approx 700$ thousand hashes. Hence, we see that even computing merely 700 thousand hashes fills a DAG buffer as large as the entire DAG within a margin of $5 * 10^{-1}\%$, as opposed to calculating 1.85 million to fill it within a margin of $5 * 10^{-7}\%$.

In fact, as we will see in the results, filling up 99.5% of a buffer as large as the DAG already starts giving us good hash rates for a browser – 35 kH/s for the WebAssembly miner. This shows that we need to have the buffer almost – but not completely – full in order to do well in terms of hash rates. And as we have seen, making the buffer almost full is not nearly as hard as filling it up entirely. This means that reaching a reasonably steady state is not as hard as it seems at face value, making WebEth viable for web settings where users might not stay on websites for long. However, given this, we would also like to point out that in reality, while it takes a lot more hashes to fill up the buffer, it does not take a lot more *time* to fill it up. This is because as discussed in the next section, when the buffer gets closer and closer to being full, the hash rate spikes up and so computing the remaining number of hashes to fill the buffer becomes quite fast. Hence, it does not take too long to completely fill up the buffer (around five minutes in our experimental setup), which is practical for many web applications, such as streaming.

4 Results

4.1 Experimental Set up

Our experimental set up consisted of a machine with an Intel i7-7700HQ processor with 8 cores and 16 GB ram. These results were obtained from a private Ethereum test network at epoch 0. The DAG size was 16777186 nodes (1.074 GB). The cache size was 1.677 MB. We ran the implementations in JavaScript and WebAssembly in the browser, and a native miner written in C++ that employs the same lazy evaluation approach outside of the browser for control results. Each miner was run till 800 kHashes were computed and the hash rate and buffer hit rate were sampled every 10k hashes. (Note that both the hash rate and buffer hit rate sampled at a particular time reflected the values over the 10k most recent hashes).

4.2 Implementation Results

Figures 4, 5 and 6 below shows a heat-map of how the hash rate varies for each of the three implementations as a function of both the size of the buffer allocated to store the DAG (as a percentage of the size of the entire DAG) and the number of hashes computed in the browser. Further, Fig. 7 then shows a heat-map of how the DAG buffer hit-rate varies as a function of both these parameters. As expected, both the hash rate and the buffer hit rate increases with the buffer size and the number of hashes computed for all three miners. We must also note that the experimental results suggest that it takes 700 kHashes to reach a buffer hit rate of 99.5%, which agrees with our predicted value from the mathematical analysis in the previous section. In fact, we can validate that our experimental

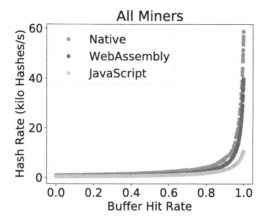

Fig. 3. All miners' hash rate

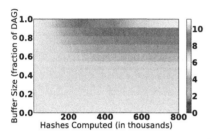

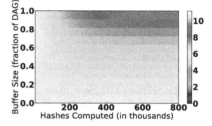

Fig. 4. WASM miner $\log(HashRate)$ Fig. 5. JS miner $\log(HashRate)$

results are typically in agreement with the predicted results for all other hit rates as well.

In order to closely examine the relationship between the hash rate and the hit rate, we show the correlations between the two parameters for all 3 miners in Fig. 3. It is interesting to see how the hash rate drastically spikes after the hit rate surpasses 95%. This suggests that accessing DAG pages from the buffer is orders of magnitude faster than computing them, so much so that even a few computations bring down the hash rate drastically. Most importantly, from the experiments, the time it takes to reach this steady state hash rate is not long (about 5 min) thereby making this approach ideal for streaming/gaming websites. Furthermore, WebEth is also ideal as a web miner as it is not very resource intensive – throughout the experiments, it did not use more than 12.5% of the CPU of our testing machine (the utilization value is normalized over 8 cores).

One might think that a way to reach the steady state hash rate faster for a given sized buffer would be to start out with a partially filled out buffer instead of an empty one. The only way this could work is by sending over part of the DAG over the network to the client. However, as it turns out, this is not feasible since, for a DAG with 16777186 nodes (1.074 GB) sending even 10% of the DAG would be sending roughly 100 MB of data. With the global average download speed for desktops and smart phones being around 5.34 MBps and 2.77 MBps respectively [20], the web page load time would be in the order of 20 s to a minute – which is too long to get a mere 10% boost in the buffer storage.

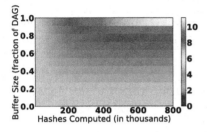
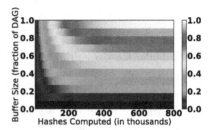

Fig. 6. Native C++ miner log($HashRate$)   **Fig. 7.** All miners' hit rate

Table 1 shows us how the performance of each of the miners compare. Interestingly, the performance variation between different implementations is not uniform across different buffer sizes and hit rates. For instance, the variation between the WebAssembly and JavaScript miners is only 35.9% when the buffer hit rate is 0, but the performance difference increases to 73.2% when the hit rate becomes greater than 0.99. We also see that for obvious reasons, the native miner outperforms both the JavaScript and the WebAssembly miner. However, the WebAssembly miner is at most 40.5% slower than the native miner – which is not very far off considering the overhead of running programs within browsers. The JavaScript miner, on the other hand, is **at least** 47.2% slower (and at most 82.0% slower), making WebAssembly the better of the two candidates for WebEth.

Table 1. Performance variations across different Implementations

	WASM/JS	Native/JS	Native/WASM
% Diff in smallest hash rates	35.9%	55.0%	38.9%
% Diff in peak hash rates	73.2%	81.3%	30.2%
Min perf % diff	24.4%	47.2%	23.9%
Max perf % diff	73.3%	82.0%	40.5%
Avg. hash rate % diff (averaged over all buffer sizes and buffer hit rates)	30.6%	55.6%	35.9%

Finally, we tabulate the most important results from our analysis in Table 2. We see that it takes all miners only a few hundred seconds to fill the buffer up, which is good considering the fact that most users don't stay on a particular website for very long. Furthermore, we see that the `WebAssembly` miner is the better of the two miners, since it gives a better terminal hash rate of 40 kH/s.

Table 2. Main results

	Native	JavaScript	WebAssembly
Median hash rate (for a buffer the size of the DAG)	15278 H/s	5290 H/s	10504 H/s
Peak hash rate	56800 H/s	10626 H/s	39651 H/s
Time taken to 99.76% buffer hit rate	163.601 s	879.857 s	257.7692 s
Avg. hash rate % diff with Native Miner (averaged over all buffer sizes)	NA	55.6%	35.9%

5   Potential Applications

We envision that `WebEth` could be used for a variety of applications.

Web Content Monetization: With the growth in global Internet usage, hosting websites has become a lucrative business. As a result, new methods of monetizing electronic content have surfaced with time. Though some are more successful than others, all of them have associated issues. For instance, selling advertisement space is now resulting in declining revenue for website owners due to the advent of new technologies such as AdBlock [2], Brave Browser [4,5]; when coupled with an increased load time, browser slow-down, and placement challenges, online ads adversely affect user experience. Thus, we envision that website content can be monetized through client-size coin mining, utilizing techniques such as those presented in this work. Note though that earning real cash requires a significant subscriber base or a large amount of time spent on the website, making this an ideal approach for video streaming/gaming websites. In fact, with a hash rate of 40 kH/s, a website would need to have around 8000 users at any given time in order to obtain around $500 per month [36].

Web Authentication Rate Limiting: Another potential application relates to rate limiting of web-authentication. Many tools are openly available for brute forcing web login pages [21,22]. Currently, the way website owners mitigate these attacks is by locking out a user for a certain amount of time after a fixed number of unsuccessful login attempts or presenting a captcha [25]. Lock out presents a Denial of Service potential by locking out legitimate users as a consequence of an attack. Captcha techniques can be used for third-party value [26] and have been successfully attacked through machine learning techniques [33] and crowdsourcing [34].

We posit a more user-friendly approach to this problem involving embedding a Proof-of-Work computation in a web page, *e.g.*, using `WebEth`, that the user's browser needs to successfully solve in order to be able to login. The PoW would amplify the computational power needed for brute force attempts, thereby selectively thwarting any attacker that attempts to brute force the login *without* significantly penalizing the legitimate user. `WebEth` is an especially good candidate for such an implementation because one could manually set the difficulty to obtain a balance between user experience and security.

Proof of Web Traffic: Another use-case of `WebEth` involves website advertisement companies. Today, website advertising sponsors decide on the remuneration for a website based on summarized server logs as a measure for site traffic. These logs can be manipulated by a website owner to generate the impression of a large amount of traffic [23] or by ad injectors [27]. As a solution to this, `WebEth` could be embedded by a website owner within the website, thereby making the site visitors compute PoW hashes. The advertiser would then ask the website owner to submit hashes that pass a certain difficulty threshold (i.e., the value of the hash being less than a certain number), and the larger the number of hashes that the website owner can provide, the more the remuneration the site receives. This would be more difficult for the website owners to fake since they would have to compute hashes themselves, an endeavor that might be more expensive than the potential ad payout.

Private Ethereum Test Networks: Finally, we would also like to note that Ethereum is an extremely flexible currency in the sense that it allows for private coin networks – *i.e.,* networks that do not mine the public Ethereum block-chain, but rather a private (and often smaller) instance of the cryptocurrency. `WebEth` can be used on any such private network to serve the network owner's specific interests.

6   Conclusion

We have designed and implemented `WebEth`, an open-source and distributed web-based Ethereum miner, with potential applications toward monetizing electronic content, rate limiting, private test networks, user tracking for advertisers, and the like. `WebEth` is standalone, implemented in both `JavaScript` and `WebAssembly`, and requires no external dependencies, meaning that both of these implementations can be readily embedded within many existing websites. We have also provided analyses and experimental data to help in engineering our proposed applications.

Future Work: Many interesting issues remain open. For one, our current implementation is still slower than traditional mining methods. One way to speed this process up is to tap into the client machine's GPU. There is a `JavaScript` library called WebCL that binds to the OpenCL library which allows `JavaScript` to interact directly with the GPU to achieve better parallel performance. Knowing that Ethereum was created for GPU mining, it should provide a substantial improvement.

Another interesting issue is whether the server should notify the clients once a new valid hash is found. One may expect that with a sufficiently large number of clients, this may lead to better performance than waiting for a timeout. In fact, experimenting with the length of the period before timeout might also be an interesting avenue to pursue.

The Ethereum Foundation is also currently developing Casper, a Proof-of-Stake algorithm, which has already been deployed in private testnets. Since Casper is open-source, it should be possible to create a Proof-of-Stake distributed browser miner implementation. However, users would most likely have to provide "stakes" in order for such an implementation to be possible [15]. Further research will be necessary to determine whether browser mining for Casper is viable or not, as the final form of Casper is still uncertain and exactly how much "stake" is required to successfully mine is unknown.

Acknowledgment. The authors would like to thank Dennis Your for his contributions during the early stages of this research. This research was supported in part by NSF under grant CCF-1563753. Any opinions, findings, and conclusions or recommendations expressed in this material are those of the authors and do not necessarily reflect the views of the NSF.

References

1. Tiwari, T., et al.: WebEth. GitHub, 1.0, GitHub, 10 April 2018. github.com/trishutiwari/web-ethereum-mining
2. Gundlach, M.: AdBlock browser extension. AdBlock. Software (2009)
3. Wampler, M., et al.: Ethash. Computer software. GitHub. Vers. 23.1. GitHub, 11 January 2015. https://github.com/ethereum/ethash. Accessed 24 Feb 2018
4. Eich, B., Bondy, B.: Brave Browser. Brave Software. Software (2015)
5. Hern, A.: Adblock Plus: the Tiny Plugin Threatening the Internet's Business Model. The Guardian, Guardian News and Media, 14 October 2013. www.theguardian.com/technology/2013/oct/14/the-tiny-german-company-threatening-the-internets-business-model
6. Buterin, V., et al.: Ethereum/Wiki. GitHub, 9 February 2014. https://github.com/ethereum/wiki/wiki/Dagger-Hashimoto
7. Ramamritham, K., Stankovic, J.A.: Dynamic task scheduling in hard real-time distributed systems. IEEE Softw. **1**(3), 65 (1984)
8. Shirazi, B.A., Kavi, K.M., Hurson, A.R.: Scheduling and Load Balancing in Parallel and Distributed Systems. IEEE Computer Society Press, Los Alamitos (1995)
9. Bal, H.E., Frans Kaashoek, M., Tanenbaum, A.S.: Orca: a language for parallel programming of distributed systems. IEEE Trans. Softw. Eng. **18**(3), 190–205 (1992)

10. Bhatia, D., Burzevski, V., Camuseva, M., Fox, G.C.: WebFlow - A Visual Programming Paradigm for Web/Java Based Coarse Grain Distributed Computing. Northeast Parallel Architecture Center (1997)
11. Cushing, R., et al.: Distributed computing on an ensemble of browsers. IEEE Internet Comput. **17**(5), 54–61 (2013). www.computer.org/csdl/mags/ic/2013/05/mic2013050054.html
12. Lerner, S.D.: Strict Memory Hard Hashing Functions (Preliminary V0. 3, 01-19-14)
13. The Coinhive Team: Coinhive browser extension. Coinhive. Software (2017)
14. Duda, J., Dłubacz, W.: Distributed evolutionary computing system based on web browsers with JavaScript. In: Manninen, P., Öster, P. (eds.) PARA 2012. LNCS, vol. 7782, pp. 183–191. Springer, Heidelberg (2013). https://doi.org/10.1007/978-3-642-36803-5_13. ACM Digital Library. dl.acm.org/citation.cfm?id=2451764.2451780
15. Dale, O.: Beginner's Guide to Ethereum Casper Hardfork: What You Need to Know. Blocknomi, 7 November 2017. (https://blockonomi.com/ethereum-casper/)
16. Wood, G.: Ethereum: a secure decentralised generalised transaction ledger. Ethereum Project Yellow Pap. **151**, 1–32 (2014)
17. Szilgyi, P., et al.: Geth. Computer software. GitHub. Vers. 1.8.1. GitHub, 22 December 2013. https://github.com/ethereum/go-ethereum. Accessed 24 Feb 2018
18. W3C Team. WebAssembly. Program documentation. WebAssembly. Vers. 1.0. WebAssembly, 17 March 2017. http://webassembly.org. Accessed 28 Mar 2018
19. Emscripten Community. Emscripten. Vers. 1.37.36. Emscripten, 11 November 2012. http://kripken.github.io/emscripten-site/docs/getting_started/Tutorial.html. Accessed 28 Mar 2018
20. Ookla. Speedtest Global Index Monthly Comparisons of Internet Speeds from around the World. Speedtest Global Index, Ookla, 25 March 2018. www.speedtest.net/global-index
21. Fogerlie, G.: Brute Force Website Login Attack Using Hydra - Hack Websites - Cyber Security. Brute Force Website Login Attack Using Hydra - Hack Websites - Cyber Security, YouTube, 24 September 2013. www.youtube.com/watch?v=ZVngjGp-oZo
22. Mahmood, O.: Brute Force Website Login Page Using Burpsuite. SecurityTraning, 5 February 2018. securitytraning.com/brute-force-website-login-page-using-burpsuite/
23. FoxBrewster, T.: 'Biggest Ad Fraud Ever': Hackers Make $5M A Day By Faking 300M Video Views. Forbes, Forbes Magazine, 20 December 2016. https://www.forbes.com/sites/thomasbrewster/2016/12/20/methbot-biggest-ad-fraud-busted/
24. Neal, P.: The Generalised coupon collector problem. J. Appl. Probab. **45**(3), 621–629 (2008). https://doi.org/10.1239/jap/1222441818
25. Google Recaptcha. https://www.google.com/recaptcha/intro/
26. Von Ahn, L., Maurer, B., McMillen, C., Abraham, D., Blum, M.: Recaptcha: human-based character recognition via web security measures. Science **321**(5895), 1465–1468 (2008)
27. Thomas, K., et al.: Ad injection at scale: assessing deceptive advertisement modifications. In: IEEE Symposium on Security and Privacy (2015)
28. Wood, G.: Ethereum: a secure decentralised generalised transaction ledger. Ethereum Project Yellow Paper. http://gavwood.com/paper.pdf
29. Bertoni, G., Daemen, J., Peeters, M., Van Assche, G.: Keccak. In: Johansson, T., Nguyen, P.Q. (eds.) EUROCRYPT 2013. LNCS, vol. 7881, pp. 313–314. Springer, Heidelberg (2013). https://doi.org/10.1007/978-3-642-38348-9_19

30. Taylor, M.B.: The evolution of bitcoin hardware. Computer **50**(9), 58–66 (2017)
31. Peyrott, S.: A brief history of JavaScript. Auth0 - Blog, Auth 0, 16 January 2017. auth0.com/blog/a-brief-history-of-javascript/
32. Brave Software. Basic Attention Token. Basic Attention Token, 1.0, Brave Software, 13 March 2018. basicattentiontoken.org/
33. Geitgey, A.: How to break a CAPTCHA system in 15 minutes with machine learning. Medium, 13 December 2017. medium.com/@ageitgey/how-to-break-a-captcha-system-in-15-minutes-with-machine-learning-dbebb035a710
34. Danchev, D.: Inside India's CAPTCHA solving economy. ZDNet, 4 December 2015. www.zdnet.com/article/inside-indias-captcha-solving-economy/
35. ERC20 Token Standard. https://theethereum.wiki/w/index.php/ERC20_Token_Standard
36. CryptoCompare. Mining Calculator Bitcoin, Ethereum, Litecoin, Dash and Monero. CryptoCompare. www.cryptocompare.com/mining/calculator/eth

An Information-Flow Control Model for Online Social Networks Based on User-Attribute Credibility and Connection-Strength Factors

Ehud Gudes and Nadav Voloch[(✉)]

Ben-Gurion University of the Negev, P.O.B. 653, 8410501 Beer-Sheva, Israel
voloch@post.bgu.ac.il

Abstract. During the last couple of years there have been many researches on Online Social Networks (OSN). The common manner of representing an OSN is by a user-based graph, where the vertices are different OSN users, and the edges are different interactions between these users, such as friendships, information-sharing instances, and other connection types. The question of whether a certain user is willing to share its information to other users, known and less known, is a question that occupies several researches in aspects of information security, sharing habits and information-flow models for OSN. While many approaches take into consideration the OSN graph edges as sharing-probability factors, here we present a novel approach, that also combines the vertices as well-defined attributed entities, that contain several properties, in which we seek a certain level of credibility based on the user's attributes, such as number of total friends, age of user account, etc. The edges in our model represent the connection-strength of two users, by taking into consideration the attributes that represent their connection, such as number of mutual friend, friendship duration, etc. and the model also recognizes resemblance factors, meaning the number of similar user attributes. This approach optimizes the evaluation of users' information-sharing willingness by deriving it from these attributes, thus creating an accurate flow-control graph that prevents information leakage from users to unwanted entities, such as adversaries or spammers. The novelty of the model is mainly its choice of integrated factors for user credibility and connection credibility, making it very useful for different OSN flow-control decisions and security permissions.

Keywords: Online Social Networks security
Information-flow networks control · Trust-based security models

1 Introduction

Handling Online Social Networks (OSN) security and information flow issues as an analytic graph problem is a well-known method, presented in different papers over the past couple of years. In [1] a sharing-habits privacy control model is established, where a community directed and weighted graph with an Ego-node is defined, and the other graph vertices are defined in three other closeness categories (close friend,

© Springer International Publishing AG, part of Springer Nature 2018
I. Dinur et al. (Eds.): CSCML 2018, LNCS 10879, pp. 55–67, 2018.
https://doi.org/10.1007/978-3-319-94147-9_5

acquaintance, and adversary). The edges are directed since they describe information-sharing instances, where a certain user gives information (outflow) to another user.

The edges' weights are defined by the probability that the Ego node user is willing to share information with the connected user node, as defined in [2]. This probability is calculated by the ratio of outflow and inflow instances between the two users.

The main problem OSN privacy models deal with is the information leakage issue, in which data shared by a certain user can be reached by an adversary, that could be a spammer, a professional foe, or just an unknown user, depending on the type of data and the preferences of the data sharing user. In [3] different types of these information leakage scenarios are described. Most of these vulnerabilities occur from discretionary privacy policies of OSN users, that create a misleading knowledge of the number and type of users exposed to this shared data. Most of the solutions suggested demand changes in these specific policies. This paper presents a novel model for information flow control in an OSN, in which the factors taken into consideration in the information sharing decision are both user credibility factors, calculated by the single-user attributes, and connection-based factors, estimating the level of trust between two users.

The rest of this paper is structured as follows: Sect. 2 discusses the background for our work, with explanations on the related papers it relies on, Sect. 3 describes and defines our model thoroughly with several examples of its operation, Sect. 4 discusses the model's implementation and connection to the work done in [1], and Sect. 5 is the conclusion of the paper, with future prospect on further research on this subject.

2   Background and Related Work

OSN and general networks flow control and access control models have been studied extensively over the past decade. [4] gives an interesting analysis of organizational networks, using threat vulnerability assessments based on the network's topology, using a clustering criterion (average clustering coefficient – denoted CC) that helps define information leaks potential. The threat defined is specifically harvesting sensitive organizational information by an attacker using social-bots. The data used in this research is specifically from Twitter and Flickr. An important addition to these researches is treating the OSN user attributes and connection attributes as significant factors for access and flow control of the OSN as presented in the following parts of this paper.

In [5] an important novel approach is presented, in which the OSN graph is not necessarily a user-relationship graph. The OSN is manifested as a multi-functional graph, in which the nodes are described as "things" such as users, photos, posts, pages etc., whilst the edges are connections between these things, such as friendships between users. Another important feature that is presented in this research is the fields of the things, meaning information attributes of users, posts, etc. The analysis done on these graphs takes into consideration the dynamic modularity of a social graph, by presenting instances of these graphs in several timestamps. This approach is one of the most close-to-real-life models of OSN, and is done specifically about Facebook, being a multi-functional OSN, serving as a professional and social OSN, and with other purposes and features as well. More data about the above-mentioned user-fields can be found in [6], that elaborates on this specific subject, and gives a 30-item list of attributes, some direct and some derived, that define the user information in an OSN. An important ranking is given to these attributes, based on information gaining from each attribute, assessing their importance in the closeness approximation between users, thus evaluating their willingness to share information.

A more exact approach of specifically detecting users that we want to deny from our information is the spammer detection approach. Basing the detection on user attributes is handled in [7] specifically for Twitter, where several important user attributes, such as age of user account, fraction of tweets with spam words and fraction of followers per followees, were checked on real data from Twitter, and with these values a spammer profile could be characterized.

In [8] this detection is also done on real data, and generalized to all OSN, though it is mainly relevant to Facebook-like networks, where it is shown that spammers usually have noticeable differences in values of certain attributes such as number of likes, hashtags and mentions. [9] shows an interesting cross-platform evaluation of user behavior, based on attributes such as name, location, date of creation of account, etc. The data is cross-referenced for three main networks: Twitter, Instagram and Foursquare, and used to detect and evaluate spammer behaviors. In the following parts of this paper, we use the evaluations and methods mentioned above to create a more accurate and elaborate model for information flow control in OSN, that combines both single-user attributes for credibility evaluation, and user-connection attributes for creating a resemblance factor that will help us evaluate whether the information should or should not be shared with a certain user. In [10] an interesting algorithm for Early Detection of Spamming accounts (ErDOS) in service provider networks is presented, in which both user (spammer) characteristics and interaction (internal/external behavior) characteristics are taken into consideration in identifying potential spammers, while [11] defines novel features of malware propagation patterns, by analyzing webmail attachments and their attributes such as number of distinct file names, number of countries, etc. These works are the basis for our model, presented in the following section.

3   OSN Information-Flow Control Model

3.1   Description of the Flow-Control Problem

For creating our information-flow model, the main issue we wish to figure is whether a certain user (our Ego-node) is willing to share information with another user (our Target-node). At this point we do not describe friends, acquaintances or adversaries, since this information is derived from the model itself, as we intend to discover whether our Target-node is an acquaintance or an adversary. Our only connection definition for adjacent vertices is the pure OSN friend definition, meaning a direct edge between two user-nodes. A friend of a friend, or any other vertex that is in a distance >1 is of an unknown definition at the time, and it is either an acquaintance or an adversary, depending on the attribute values of both the user itself, and the connections towards it from the Ego-node.

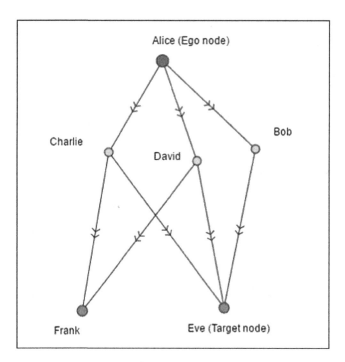

Fig. 1. Graph instance of six OSN users, where the outflow checked is from Alice to Eve. Eve (and Frank) could be either an acquaintance or an adversary, depending on the attribute values, that are still not shown.

In Fig. 1 we can see a simple six-node graph instance, in which Alice is the Ego node, Bob, Charlie and David are here friends (direct edges connect her to them), and Eve is the Target node, also connected to Bob, Charlie and David, but not directly to Alice, making her a "Friend of a Friend" to Alice. Frank is also a "Friend of a Friend", and at this point still not checked. We need to determine whether Eve is an acquaintance or an adversary, meaning whether Alice is willing to share information with her. This is, of course, only a minimal manifestation for the simplicity purpose of the example, whilst the actual problem covers much more connections and bigger distances from source to target. Here we need to formally define the problem attended in the model, and its suggested approach.

3.2   OSN Graph Definitions for the Model

Let $G = (V, E)$ be a directed graph that describes OSN activities, where V is the set of users and E is their connected social activities. $v_{src} \in V$ is the Ego source node, that holds the information to be shared, and $v_{trgt} \in V$ is the Target node, meaning the user that should or should not get the information from v_{src}. For $1 \leq i \leq n$, $PATH_i^{src \rightarrow trgt}$ is defined as the set $\{v_{src}, E^{src \rightarrow 1}, v_1, E^{1 \rightarrow 2}, \ldots, E^{k \rightarrow trgt}, v_{trgt}\}$, where the number of intertwined user-nodes is k, and the number of all paths from v_{src} to v_{trgt} is n.

For example, in Fig. 1 we can easily see $n = 3$, and $k = 1$, since there are 3 possible paths from Alice to Eve: *Alice* → *Charlie* → *Eve*, *Alice* → *David* → *Eve*, and *Alice* → *Bob* → *Eve*. In these 3 paths $k = 1$ since there is a single node between The Source (*Alice*) and Target (*Eve*).

For finding all possible paths between source and target nodes, there are several efficient algorithms, such as the Ford-Fulkerson algorithm [12], the Edmonds–Karp algorithm [13] for computing the maximum flow in a flow network in $O\ (V\ E^2)$ time, and the Dinic algorithm [14], also for a maximum flow network, that achieves a better time of $O\ (V^2\ E)$. In our model we need to find all the paths to determine whether an information instance will be passed from the Ego-node to the Target node, which is the same problem covered by the Dinic algorithm mentioned above.

3.3   The User-Credibility and Connection-Strength Attributes

The set $\{v_{src},\ E^{\ src \rightarrow 1},\ v_{1,}\ E^{\ 1 \rightarrow 2},\ .....,\ E^{\ k \rightarrow trgt},\ v_{trgt}\}$ that represents $PATH_i^{src \rightarrow trgt}$ is divided to V and E in this manner: V holds the independent user-credibility attributes, and E holds the attributes that involve the connection between two users. In Table 1 we denote the different variables by the attributes they represent.

Table 1.   User and Connection attribute variables for the model

Variable	Attribute	User/Connection
TF	Total number of friends	User (*V*)
AUA	Age of user account (OSN seniority)	User (*V*)
FFR	Followers/Followees ratio	User (*V*)
MF	Mutual friends	Connection (*E*)
FD	Friendship duration	Connection (*E*)
OIR	Outflow/Inflow ratio	Connection (*E*)
RA	Resemblance attributes	Connection (*E*)

Here we need to explain the choice of these specific variables.

Total number of Friends (TF) is an attribute handled in [6] and presented there as the most important user-credibility feature for information gain of a user profile. Spammers or fake users usually have an allotted number of friends, if any.

OSN seniority - Age of User Account (AUA) is also a very important feature shown in [8] and refers to the duration of account existence. This feature is meaningful since fake profiles and spammers are usually created and then deleted after a while, by the user itself or due to OSN security policies, detecting new accounts with suspicious activity forms.

Followers/Followees Ratio (FFR) also appears in [8], and the rational behind this attribute is that data-harvesting is usually done by the action of following, meaning a social-bot will follow users, and not be followed by them.

Mutual Friends (MF) and Friendship Duration (FD) are the two most important connection-based features shown in [6], excluding the attribute of "Amount seen together", that was not taken into consideration here since it only refers to Facebook-like OSN, whilst the general pattern of OSN does not necessarily include such features.

Outflow/Inflow Ratio (OIR) is presented in [2] and used in [1] as a good indication of sharing willingness of a source node to a target one, given the fact that more outflow than inflow instances usually portray a high probability of data sharing approval of the source node.

Resemblance Attributes (RA) is a generalization of attributes presented in [6], estimating similarity between users by their common features, assuming user friendships often have common grounds.

3.4   Values for the Variables

Setting the values for the variables is done in this model in a probability form scale of 0 to 1, since the decision of sharing information with a certain user is defined as a probability variable, 0 being no sharing willingness at all, 1 being definite sharing willingness, for clarification purposes we use c (credibility) for user attributes and p (probability) for connection attributes, though both represent the user's sharing willingness probability.

For calculating the variables' values, the following notations are given:

- c is the credibility value of a user (node) attribute. For example, c_{TF} is the credibility value for the Total Friends attribute.
- p is the probability value of a connection (edge) attribute. For example, p_{MF} is the probability value for the Mutual Friends attribute.

The values given in the following assignments are partly debatable. Some, as FFR, are based on previous researches (for FFR, it is [8]), and some are based on given estimations for the purpose of this model, making them flexible for rendering and justifications.

These values are as follows:

$$c_{TF} = \begin{cases} \frac{TF}{100} & (TF < 100), \\ 1 & (TF \geq 100). \end{cases} \quad (1)$$

c_{TF} value is based on the estimation of the TF attribute of [6], having fake profiles, social-bots, etc., with an allotted number of friends. A profile of 100 friends and above is with a high probability of being a genuine user profile.

$$c_{AUA} = \begin{cases} \frac{AUA}{365} & (AUA < 365), \\ 1 & (AUA \geq 365). \end{cases} \quad (2)$$

c_{AUA} value is calculated in days. It is based on the estimation of the AUA attribute of [8], that an active spammer profile will not remain active for more than a year, due to OSN security updating policies, usually done annually.

$$c_{FFR} = \begin{cases} FFR & (FFR < 1), \\ 1 & (FFR \geq 1). \end{cases} \tag{3}$$

c_{FFR} value is derived directly from [8]. It is a given ratio from the FFR attribute itself, and its rational is that spammers and fake profiles follow more users and are usually less followed themselves.

$$p_{MF} = \begin{cases} \frac{MF}{20} & (MF < 20), \\ 1 & (MF \geq 20). \end{cases} \tag{4}$$

p_{MF} value is based on the estimation of the MF attribute of [6], having fake profiles, social-bots, or even adversaries, with a small number of mutual friends, if any. A profile of 20 mutual friends and above is with a high probability of being an acquaintance profile, with a high probability of an actual friend potential.

$$p_{FD} = \begin{cases} \frac{FD}{365} & (FD < 365), \\ 1 & (FD \geq 365). \end{cases} \tag{5}$$

p_{FD} value is calculated in days. It is based on the estimation of the FD attribute of [6], having an acquaintance, or even a fake profile or spammer, being friends with a certain user less than a year, is of an adversary potential, like the time estimation of c_{AUA}.

$$p_{OIR} = \begin{cases} OIR & (OIR < 1), \\ 1 & (OIR \geq 1). \end{cases} \tag{6}$$

p_{OIR} value is derived directly from [1, 2]. It is a given ratio from the OIR attribute itself, and its rational is that spammers and fake profiles give the user much more inflow actions (advertisement, data-harvesting, etc.), than outflow actions from the user itself.

For the p_{RA} value a preliminary explanation is necessary. In [6] there is a list of user attributes such as gender, hometown, etc. These attributes are meaningful in the resemblance factor calculation, and only non-null attributes can be taken into consideration (if the user did not define a value for a certain attribute, it is not counted).

The independent (not derived from the connection) user attributes taken from [6] are:

- Hometown.
- Current country.
- Current city.
- Home country.
- Gender.

- Language.
- Religion.

These are the most important attributes mentioned in [6], excluding the ones that we did not take into consideration for two reasons: Facebook-like OSN attributes (such as events, movies, etc.) were omitted since we need a more general OSN definition, and the ones that are difficult to estimate (such as music, age, etc.) were also omitted.

Let us denote the following factors:

- TA_{src} is the total number of non-null attributes (from the list of 7 attributes mentioned above) of the source user. The values of these attributes must be defined by non-null values.
- $TRA_{src,\ trgt}$ is the total number of non-null resembling attributes (from the list of 7 attributes mentioned above) of the source user and the target user. The values of these attributes must be defined by non-null values.

Now we can define p_{RA}:

$$p_{RA} = \frac{TRAsrc, trgt}{TAsrc} \tag{7}$$

This value cannot be larger than 1, since the maximal number of common attributes could be the total number of source attributes at most. In [6] it is shown that a good sharing probability estimation can be done by similarity checking between users by their common features, assuming user friendships often have common grounds.

For these resemblance cases, the Pearson Correlation Coefficient [15] is often used for ratio calculation, but it defines a symmetric value for both ends of the connection, whilst our model describes an asymmetric one, since the target node is the one being checked for credibility, in relevance to the source one, and not vice versa.

Now we can define for every user the total value of credibility and for every connection the total value of sharing probability, by averaging the different factors noted above.

$$c = \frac{WcTF + WcAUA + WcFFR}{3} \tag{8}$$

$$p = \frac{WpMF + WpFD + WpOIR + WpRA}{4} \tag{9}$$

These averages indicate the same effect for every attribute-factor in c and p, assuming initially that all attribute-weights (W) are equal and can now be assigned to the OSN graph of users and their connections.

Assigning these values will be done to each path of the graph from the source vertex to the target one. The user credibility attribute c is assigned to the vertices and the connection probability p is assigned to the edges connecting the vertices.

3.5   Assigning the Values to the OSN Graph

As denoted before, for $1 \leq i \leq n$, $PATH_i^{src \rightarrow trgt}$ is defined as the set $\{v_{src}, E^{src \rightarrow 1}, v_1, E^{1 \rightarrow 2}, \ldots, E^{k \rightarrow trgt}, v_{trgt}\}$, where the number of intertwined user-nodes is k, and the number of all paths from v_{src} to v_{trgt} is n.

For every path we now need to define the total sharing-probability value, that will be the indicator for the choice of sharing the information from source to target. We take into consideration that c for the v_{src} is omitted since the Ego-user does not need to be checked for credibility. We denote this Total Sharing Probability value as TSP, and this value is calculated for a single $PATH$ from the source note to the target node.

$$TSP(PATH^{src \rightarrow trgt}) = \prod_{i=1}^{k} c_i p_i \qquad (10)$$

We can easily see that since i begins in 1, v_{src} is omitted, and the product is of every node property of c and every edge property of p in $PATH_i^{src \rightarrow trgt}$.

Table 2 shows example values assigned to the graph shown in Fig. 1. c and p are calculated by Eqs. 8 and 9 shown above.

Table 2. User and Connection attribute variables values for the graph in Fig. 2

User/Connection	c_{TF}	c_{AUA}	c_{FFR}	p_{MF}	p_{FO}	p_{OIR}	p_{RA}	c	p
Bob	0.89	0.54	0.91	–	–	–	–	0.78	–
Charlie	0.78	0.34	0.92	–	–	–	–	0.68	–
David	0.97	0.98	0.96	–	–	–	–	0.97	–
Eve	0.76	0.14	0.78	–	–	–	–	0.56	–
Frank	0.95	0.92	0.98	–	–	–	–	0.95	–
Alice → Bob	–	–	–	0.31	0.81	0.34	0.86	–	0.58
Alice → Charlie	–	–	–	0.44	0.84	0.33	0.71	–	0.58
Alice → David	–	–	–	1	0.46	1	0.86	–	0.83
Bob → Eve	–	–	–	0.28	0.26	0.19	0.43	–	0.29
David → Eve	–	–	–	0.62	0.9	0.58	0.86	–	0.74
Charlie → Eve	–	–	–	0.22	0.11	0.16	0.43	–	0.23
David → Frank	–	–	–	0.81	0.95	0.88	1	–	0.91
Charlie → Frank	–	–	–	0.5	0.64	0.72	0.86	–	0.68

Table 3 shows the TSP calculation (Eq. 10) of all the $PATH$s in the graph, using the values of Table 2.

The graph with these attributes of Tables 2 and 3 is shown in Fig. 2, having Eve as the target node that is an adversary and Frank as the target node than is an acquaintance.

Table 3. *TSP* values for all the *PATH*s of the graph

PATH	TSP
Alice → Bob → Eve	0.58 * 0.78 * 0.29 * 0.56 = **0.07**
Alice → Charlie → Eve	0.58 * 0.68 * 0.23 * 0.56 = **0.05**
Alice → David → Eve	0.83 * 0.97 * 0.74 * 0.56 = **0.33**
Alice → David → Frank	0.83 * 0.97 * 0.91 * 0.95 = **0.7**
Alice → Charlie → Frank	0.58 * 0.68 * 0.68 * 0.95 = **0.25**

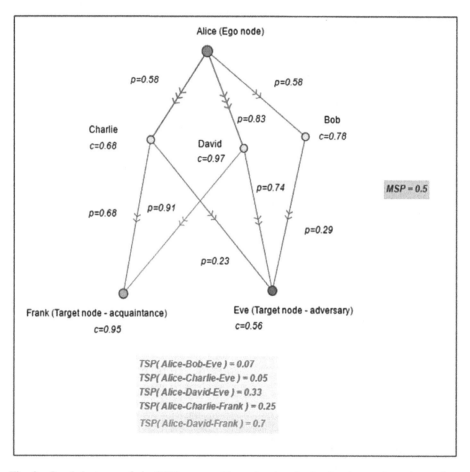

Fig. 2. Graph instance of six OSN users, with vertices' and edges' values, where the outflow checked is from Alice to Eve and Frank. Eve is detected as an adversary since none of the *PATH*s *TSP* values achieve the *MSP*. Frank is detected as an acquaintance since the *TSP* value of the *PATH* Alice-David- Frank achieves the *MSP*.

3.6   Algorithm for Determining an Acquaintance or an Adversary

For deciding whether the target node is an acquaintance or an adversary, we first need to set a threshold numeric value for *PATH*, thus including the decision of information sharing, by defining the *PATH* as safe or not safe. This threshold of Minimum Sharing Probability value is denoted here as *MSP*. The value chosen for this model is the median value of 0.5, having *TSP* $\geq$ 0.5 meaning that the *PATH* is safe, and that the target node is necessarily an acquaintance, not an adversary. This value is of course debatable and flexible for rendering and justifications.

The algorithm is as follows:

isAnAcquaintanceNotAnAdversary (Graph G, Vertex v_{src}, Vertex v_{srgt})

1. $MSP \leftarrow 0.5$

2. $\{AllPaths^{src \to trgt}\} \leftarrow Dinic\ (G)$ // $|\{AllPaths\}| = n$

3. for $1 \leq i \leq n$

    ○   *if* $TSP(PATH_i^{src \to trgt}) \geq MSP$

       ▪   return *true*

4. return *false*

The first step, as mentioned above, is setting *MSP*. The second stage is finding all possible paths between the source and target nodes. This is done, as mentioned in previous parts of this paper, by the Dinic algorithm [14] for a maximum flow network, that achieves a time of $O(V^2\ E)$. The third step is finding all the *TSP*s of the *PATH*s, and if a certain *TSP* is $\geq MSP$, returning *true*, the node is set as an acquaintance, not an adversary. If such *TSP* is not found, then we return *false* in the last step. Since the attribute values calculated in *TSP* are consisted of given OSN user-attribute and connection-attribute values, this calculation is of constant complexity, giving the algorithm a total complexity of $O(V^2\ E)$.

Figure 2 shows the examples of detecting an adversary (Eve) and an acquaintance (Frank). All the *PATH*s from Alice to Eve do not achieve the needed *MSP*, necessarily making her an adversary. The *PATH* Alice – David - Frank achieves the needed *MSP*, necessarily making him an acquaintance.

We can clearly see that if a certain *PATH* has a *TSP* value $\geq MSP$, it is then noted as safe for information flow from source to target, making the target node declared as an acquaintance, to which we are willing to share information with.

4   Discussion

Our model, described in the previous section, was initially based on the model presented in [1], that showed a novel approach of privacy control in OSN. This model was primarily based on a Min-Cut algorithm that blocked several edges in the Ego-user community graph, thus preventing information leakage to potential adversaries. This approach has some disadvantages since it has a large overhead for every information sharing action and, more importantly, cuts the information-flow between users, that are not necessarily adversaries (the intertwined non-Ego user nodes). The main problem [1] addresses is preventing information leakage from a certain OSN user to other users, adversaries or acquaintances. The method presented in [1] is cutting edges in the OSN graph, thus preventing this leakage, but without any discernment of the users being cut off. This problem can and does result in an unwanted information prevention. Our model solves this problem by deciding which user gets the data. No unnecessary edge cuts are done, and no information is being denied from non-adversary users, and the problem of [1] is handled in a manner that improves the information sharing decisions by taking into consideration many factors based both on the single-user credibility and the connection strength between two users. With these factors we optimize the assumptions of information sharing preferences and decisions done by the Ego node, thus achieving a more reliable and accurate form of privacy control in the information flow of OSN. This model gives every Ego node a graph-map for every connected user, that indicates whether it is an adversary or an acquaintance. This map is then transferred as a label with the information itself to the other approved nodes (acquaintances-having permission to this information), thus creating a clear privacy picture for every information instance, preventing leakage of this instance to potential adversaries, and by that implementing sharing-based privacy control in OSN.

5   Conclusion and Future Work

In this paper we have presented a Flow-Control model for OSN. The novelty of the model is its combination of user credibility and connection strength. The attributes of this model were carefully picked, but there could be flexibility in these choices, as well as in the values of these attributes, that are debatable. This model is now being tested by statistical data and examined on a prototype system we have built for this research, since the question of detecting an adversary is not necessarily the problem of spammer detection. An unknown user could be a regular user we do not wish to share information with. Other modeling aspects of this problem are currently a work in progress.

References

1. Levy, S., Gudes, E., Gal-Oz, N.: Sharing-habits based privacy control in social networks. In: Ranise, S., Swarup, V. (eds.) DBSec 2016. LNCS, vol. 9766, pp. 217–232. Springer, Cham (2016). https://doi.org/10.1007/978-3-319-41483-6_16
2. Ranjbar, A., Maheswaran, M.: Using community structure to control information sharing in online social networks. Comput. Commun. **41**, 11–21 (2014)
3. Li, Y., Li, Y., Yan, Q., Deng, R.H.: Privacy leakage analysis in online social networks. Comput. Secur. **49**, 239–254 (2015)
4. Bokobza, Y., Paradise, A., Rapaport, G., Puzis, R., Shapira, B., Shabtai, A.: Leak sinks: the threat of targeted social eavesdropping. In: 2015 IEEE/ACM International Conference on Advances in Social Networks Analysis and Mining (ASONAM), pp. 375–382. IEEE (2015)

5. Patil, V.T., Shyamasundar, R.K.: Undoing of privacy policies on Facebook. In: Livraga, G., Zhu, S. (eds.) DBSec 2017. LNCS, vol. 10359, pp. 239–255. Springer, Cham (2017). https://doi.org/10.1007/978-3-319-61176-1_13

6. Misra, G., Such, J.M., Balogun, H.: Improve-identifying minimal profile vectors for similarity-based access control. In: 2016 IEEE Trustcom/BigDataSE/I SPA, pp. 868–875. IEEE (2016)

7. Benevenuto, F., et al.: Detecting spammers on Twitter. In: Collaboration, Electronic Messaging, Anti-abuse and Spam Conference (CEAS), vol. 6 (2010)

8. Zheng, X., et al.: Detecting spammers on social networks. Neurocomputing **159**, 27–34 (2015)

9. Han Veiga, M., Eickhoff, C.: A cross-platform collection of social network profiles. In: Proceedings of the 39th International ACM SIGIR Conference on Research and Development in Information Retrieval. ACM (2016)

10. Cohen, Y., Gordon, D., Hendler, D.: Early detection of spamming accounts in large-scale service provider networks. Knowl. Based Syst. **142**, 241–255 (2017)

11. Cohen, Y., Hendler, D., Rubin, A.: Detection of malicious webmail attachments based on propagation patterns. Knowl. Based Syst. **141**, 67–79 (2018)

12. Ford, L.R., Fulkerson, D.R.: Maximal flow through a network. Can. J. Math. **8**, 399–404 (1956)

13. Edmonds, J., Karp, R.M.: Theoretical improvements in algorithmic efficiency for network flow problems. J. ACM. **19**(2), 248–264 (1972). Association for Computing Machinery

14. Dinic, Y.: Algorithm for solution of a problem of maximum flow in a network with power estimation. Doklady Akademii nauk SSSR **11**, 1277–1280 (1970)

15. Benesty, J., et al.: Pearson correlation coefficient. In: Cohen, I., Huang, Y., Chen, J., Benesty, J.: Noise Reduction in Speech Processing, pp. 1–4. Springer, Heidelberg (2009). https://doi.org/10.1007/978-3-642-00296-0_5

Detecting and Coloring Anomalies in Real Cellular Network Using Principle Component Analysis

Yoram Segal[✉], Dan Vilenchik, and Ofer Hadar

Communication Systems Engineering Department, Ben Gurion University of the Negev (BGU),
84105 Beer-Sheva, Israel
yoramse@post.bgu.ac.il

Abstract. Anomaly detection in a communication network is a powerful tool for predicting faults, detecting network sabotage attempts and learning user profiles for marketing purposes and quality of services improvements. In this article, we convert the unsupervised data mining learning problem into a supervised classification problem. We will propose three methods for creating an associative anomaly within a given commercial traffic data database and demonstrate how, using the Principle Component Analysis (PCA) algorithm, we can detect the network anomaly behavior and classify between a regular data stream and a data stream that deviates from a routine, at the IP network layer level. Although the PCA method was used in the past for the task of anomaly detection, there are very few examples where such tasks were performed on real traffic data that was collected and shared by a commercial company.

The article presents three interesting innovations: The first one is the use of an up-to-date database produced by the users of an international communications company. The dataset for the data mining algorithm retrieved from a data center which monitors and collects low-level network transportation log streams from all over the world. The second innovation is the ability to enable the labeling of several types of anomalies, from untagged datasets, by organizing and prearranging the database. The third innovation is the abilities, not only to detect the anomaly but also, to coloring the anomaly type. I.e., identification, classification and labeling some forms of the abnormality.

Keywords: Anomaly detection · PCA · Data mining · Machine learning

1 Introduction

Anomaly detection which is based on Network traffic analysis tools are the foundation stones for network upgrades, protecting against cyber-attacks, and are a marketing tool for analyzing user profiles. Many heuristics can serve as starting points for filtering out data that flows at extremely high speeds. Analysis of network traffic is the most effective means of reducing search within the amount of information required for further analysis.

This work was supported by the Israel Innovation Authority (Formerly the Office of the Chief Scientist and MATIMOP).

© Springer International Publishing AG, part of Springer Nature 2018
I. Dinur et al. (Eds.): CSCML 2018, LNCS 10879, pp. 68–83, 2018.
https://doi.org/10.1007/978-3-319-94147-9_6

Business companies use network traffic testing tools as the primary means of their solution architecture for intelligence and law enforcement bodies that monitor national internet services providers (ISP). It is also a significant focus on the solution concept of companies that offer optimization and advertising solutions based on network transportation.

Traffic anomaly detection has received a great deal of attention in the research literature. While there has been some work that leverages data structures to find heavy-hitters [1, 2], most papers have utilized statistical-analysis techniques to detect outliers in traffic time series. Numerous methods have been evaluated, including wavelets [3], moving average variants, Fourier transforms [4, 5], Kalman filters [6], and PCA [7]. Early work in this area often analyzed data from a single link [3], whereas more recent papers have shown promising results by examining network-wide measurements [8]. With such a large body of work, it becomes increasingly important to be able to compare presented approaches. While there have been a few papers that analyzed a subset of the statistical-analysis techniques [4, 5], researchers have only very recently begun investigating how data-reduction technologies impact the ability to detect traffic anomalies [9]. Much in the same way that early papers on traffic anomaly detectors had a limited scope, this new line of work has analyzed the impact of only one form of data-reduction [10], on only one type of traffic anomaly [11], or analyzed data from a small number of links [12].

We are focusing on unsupervised techniques for big cellular data set. Our observation vectors have 97 different parameters. In the literature, various strategies proposed for dimensionality reduction [13]. The actual dimensionality reduction methods can classify into two classes: Feature extraction and Feature selection. Feature selection aims to seek optimally or suboptimal subsets of the original features [14], by preserving the main information carried by the collected complete data, to facilitate future analysis for high-dimensional problems. Another approach is the opposite approach, instead of reducing the dimensionality, Breiman [15] suggested to increase the dimensionality by adding many functions of the predictor variables. Two outstanding examples of work in this direction are the AmitGeman method [16] and support vector machines [17]. In feature extraction model [18], the original features in the measurement space initially transformed into a new dimension-reduced space via some specified transformation. Significant characteristics determined in the new axis.

Viswanath et al. [19] used PCA to classify Facebook users as either "normal" or "anomalous" (user considered anomalous if its behavior was tagged as such by Facebook). Other papers that applied PCA successfully for anomaly detection include [20–24].

The ability to enable the labeling of several types of anomalies, from untagged datasets presented in some other works such as [25]. In [25] the validation data is split into two sets, one set that represents nominal data, and the other that represents potentially anomalous data. In some instances, benign anomalies may appear in the validation of nominally categorized data where there was no prior suspicion of them. In our case, we are adding external knowledge such as geographical location or period which allows us to classify the data without mixing between anomalies and regular sets.

Our study in this article identifies and evaluates three main challenges: (1) Identifying anomalies from logs of real network traffic. (2) Development of new statistical algorithms to identify anomalies that are adapted to the unique problem. (3) Verification of the quality of results by breaking the data into normal and the rest according to some parameter: cell congestion, time rather than statistical methods only.

2 Anomaly Detection Technics

The article deals with two main challenges: The first one is that there is no definition of what an anomaly is, no training sets for anomalies. In practice the data is unlabeled. The second challenge is handling big-data stream, off-line and certainly in an online situation is a complicated technological challenge. The techniques for identifying anomalies can be divided into two types: Techniques, which are unsupervised and assume that most of the database observations represent normal or normal cases. For example, cluster analysis techniques can be used to characterize typical representation. A representation that does not belong to any cluster defined as an anomaly. Supervised techniques in which database observations were pre-categorized for "normal" or "abnormal" observations. In this case, computational learning methods can use for categorized training, which enables the classification of new observation that we have not encountered in the learning process.

We will use the PCA method which trained on normal behavior and identifies deviations from this behavior. We are showing characteristics that best explain the normal behavior. PCA will do this by projecting on a base with a smaller or the same dimension on which we will perform statistical analyzes.

Now we are going to explain the PCA model. The first principal component (PC) is defined to be the direction (unit vector) $V_1 \in \mathbb{R}^p$ in which the variance of x is maximal. The variance of x in direction v is given by the expression $v^T \Sigma v$. Therefore $V_1 = argmax_{v \in \mathbb{R}^p} \in v^T \Sigma v$. The latter is the Rayleigh Quotient definition of the largest eigenvalue of a matrix, therefore V_1 is the leading eigenvector of Σ and $\lambda_1 = v_1^T \Sigma v_1$ is the variance explained by V_1. The remaining PCs are defined in a similar way and together they form an orthonormal basis of $\mathbb{R}^p$. The sample PCs $\hat{v}_1, \ldots, \hat{v}_p$ are the eigenvectors of the sample covariance matrix $\hat{\Sigma}$. Under various reasonable assumptions it was proven that the principal components $V_1, \ldots, V_p$ converge to the sample ones $\hat{v}_1, \ldots \hat{v}_p$ [26, 27]. We assume that this is true in our case, and we justify it by the fact that we are in the "fixed p large n" regime, where the ratio p/n tends to 0.

3 Creating an Anomaly Database

This research deals with the study of traffic of a cellular communication network to discover anomaly based on traffic data. A cellular network contains many access points to the Internet. Designated routers serve as a bridge between the Internet and the cellular data flow. These routers regularly monitored so that the traffic information through them is centralized into an information center, allowing a holistic, international view of the behavior of network traffic (Fig. 1).

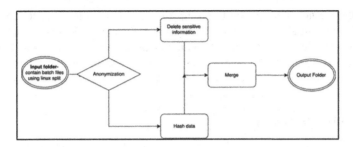

Fig. 1. Block diagram to preserve data confidentiality

Naturally, this information center (which based on the Log Center) generates significant data at the rate of tens of gigabytes per second. It should emphasize that the stream of information and information content is not constant and changes according to use. Therefore, we averaged each measured parameter, separately, in time units. Such as averaging over an hour of HTTP request size. Another problem we had to deal with was maintaining anonymity and confidentiality. The cellular networks traffic logs contain private user information. There is a need for log anonymization platform scalable for big-data. As a result, we defined a batch based anonymization tool (Fig. 2).

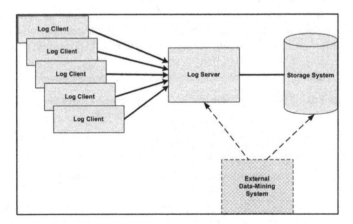

Fig. 2. Data center architecture

The database fields divided into three types: Anonymous fields- Those fields used as is; Fields that reveal user information- Those fields have been deleted; Fields that can be used but still have indirect information about the user and therefore have a low risk of user exposure. For those fields, we used at the beginning the well-known PBKDF2 anonymization algorithm. PBKDF2 is very secure and used for protecting password on almost every server. The drawback is that the algorithm is slow. It makes the anonymization process to be prolonged. Therefore, currently, we are working with SHA-256 due to resource constraints. It takes one minute to anonymize a log of 1 Gb while almost a day with PBKDF2.

Our database includes fields of the complete set of transaction log records and their formats, but the transaction log fields in any specific geographical location depend on its local configuration. HTTPS records contain data per connection, as transactions not identified. HTTPS and HTTP Tunneled transactions records include only fields that captured during their limited processing (e.g., timing, data amounts IP addresses, etc.)

4   Experiments and Results

This section describes our PCA model, methodology and software for detecting and coloring the traffic anomaly by manipulate the same database in three major ways.

4.1   Time-Period Traffic Analysis

The first method for discovering anomaly based on different time-period traffic analysis. The information divided into three-periods categories: Night\Early Morning, Morning and Evening from all geographical locations. The motivation was to examine whether traffic congestion can discover based on the assumption that each time profile has a unique pattern. Based on the observed time profile, we injected vector information belonging to other time profiles, and tried to discover them as an anomaly.

Initially, the time profiles tested naively, and elementary statistical parameters such as mean and standard deviation were measured to characterize each period by mean and standard deviation of its bytes stream volume. Sample results presented in Table 1.

Table 1. Elementary parameters from some data sets examples

Log file	Records	Date	Time	Hours reordered	Mean	Std.dev.
#1 (1 GB)	890,650	Sunday, 08.05.16	Night\Early Morning 00:00–07.59	6	35,105.823	700,800
#2 (1 GB)	20,131,028	Tuesday, 06.12.16	Morning 08:00–11.59	1	36,531.608	639,683.435
#2 (3 GB)	28,676,280	Monday, 05.12.16	Eveing, 21:00–23:59	3	40,428.878	820,315.637

Table 1 demonstrates the fact that an attempt to classify periods via first and second order statistical characteristics does not allow proper classification. The average plus the variance of each period creates an overlap that does not allow sufficient separation.

Since the naive method of detecting the anomaly of different time periods is not relevant, we used the PCA method to identify an anomaly in datasets gathered in one period and reached the system at a different time.

As a conclusion from Fig. 3, the PC effect is negligibly starting from the fourth eigenvalue. Table 2 presents some of the eigenvectors components for each relevant eigenvalue and associates them to the original dataset components.

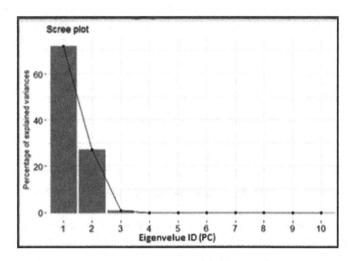

Fig. 3. Time-periods major PCs

Table 2. Eigenvectors and time-periods components association table

PC3	PC2	PC1	Original vector components
2.60E-06	−7 69E-07	7.99E-08	RESPONSE CODE
−0.50005	0 006683	−0.00453	RESPONSE_ORIG_SIZE
−0.49974	0.006682	−0.00453	DOWNLOAD DATA SIZE
−8.35E-10	−2.65E-09	7.30E-10	COMPRESSION_LVL
−1.95E-07	4.41E-08	−8.84E-09	CONTENT TYPE
−9 77E-05	−1.32E-05	8.88E-07	UPSTREAM SIZE
−0.49975	0.006689	−0.00453	DOWNSTREAM SIZE

To colored anomaly, compare to the Weekday Night Hours (WNH) dataset, we transformed all time-periods datasets into a new PCA space. We used the Normal State Transformation Matrix (NSTM), calculated by performing principal components analysis on the WNH dataset. We extracted the two independent eigenvectors and performed the projection of the datasets of all time-periods on a single shared two-dimensional graph (Fig. 4). We received a reduction of the dimension of information from a space of 97 dimensions to a 2-dimensional space that allows us to present a point of view that represents the distribution in a state without anomaly (Night hours). Similarly, we carried out the information with Weekend hours, and Evening hours and these points were marked in Yellow and Red respectively.

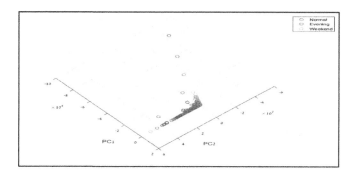

Fig. 4. Colored time-periods classes

After transforming each time-period dataset, separately, by the NSTM, we figured, per time-period dataset, the average absolute value of projections, in the direction of each eigenvector, individually. The following graph (Fig. 5) shows the average absolute values of the projection on each PC.

Significant PCs for Weekday Night hours (Blue color) in importance order: 1. PC1, 2. PC2, 3. PC3
Significant PCs for Weekend hours (Orange color) in importance order: 1. PC56, 2. PC54, 3. PC55

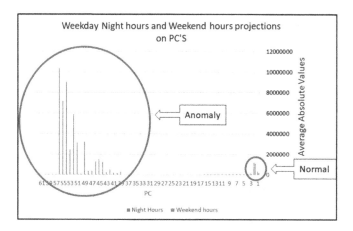

Fig. 5. Comparing average absolute values of time-periods projections on all PCs (Color figure online)

With PCA model and NSTM, we rotate the original dataset axis system so that the eigenvectors become the basis of the new axial system. The PC column in Table 3 indicates the sequence number of the most relevant eigenvectors which the dataset vectors projected on it (most relevant PC's are the PCs with the highest average absolute values after projections into the PC direction). Our dataset eigenvectors are in dimension 97, meaning that each eigenvector has 97 components that can be interpret as eigenvector weights. In Table 3, the weight column presents the X highest eigenvector weights per PC. A projection toward a PC is a linear combination (inner product) between original dataset vectors and eigenvector weights. Therefore, the eigenvector weights can interpret as the importance of the dataset vectors components (before projection). The last column in Table 3 connects between the PCA space and the real dataset log components. It allows us to interpret more efficient our log data and to characterize the most relevant features that have the highest influence on the data transportation during different time-periods.

Table 3. Eigenvectors components weights interpretation

PC	Weight	Original dataset columns description
1	0.998	The payload size only (from cache)
	0.049	The estimated connection bandwidth at session beginning
	0.004	The size of response from WEB without the headers
	0.004	Size of original response on WEB containing headers
	0.004	Size of the response data on the RAN side, including headers.
2	0.998	The estimated connection bandwidth at session beginning
	0.049	The payload size only (from cache)
3	0.500	The size of response from WEB without the headers
	0.500	Size of original response on WEB containing headers
	0.499	Size of the response data on the RAN side, including headers.
54	0.514	When a Media file or Software download file is requested in range requests, this field holds the full resource length
	0.437	The number of times the video stopped playing
	0.410	The time it took the video to start playing in milliseconds
	0.261	This field indicates the method by which the file was processed for Multi-Level Transcoding and Dynamic Rate Adaptation
55	0.659	The stalls average time in milliseconds
	0.409	When a Media file or Software download file is requested in range requests, this field holds the full resource length
	0.398	This field indicates the method by which the file was processed for Multi-Level Transcoding and Dynamic Rate Adaptation
56	0.700	The stalls average time in milliseconds
	0.591	When a Media file or Software download file is requested in range requests, this field holds the full resource length
	0.287	The time it took the video to start playing in milliseconds

When examining anomaly at different times of the week, it is easy to see that the distribution of the evening and morning hours is almost identical. But when compared to the weekend we got an extreme deviation, when in fact all significant PCs that belong to the "normal traffic" dataset are not substantial in the weekend traffic. The significant PC's for the weekend hours focused mainly on watching the video, and moreover, it was noticeable that most of the video views had been interrupted (indicating a traffic load).

4.2 Congestion Traffic Analysis

Second data structure: classification by congestion fields. The database contains some columns describing the level of transportation load. That refers to three levels of the number of Bytes per second passing through the examined network junction. 0 - low load level, 1 - medium load level and 2 - high load level.

After computing the PCA model on the low-level congestion dataset (level 0), sorting them and selecting the 3 with the highest eigenvalues, we extracted the three independent eigenvectors (the ones that belong to the three highest eigenvalues) and performed the projection of the datasets of all levels on a single shared three-dimensional graph. If to be more precise, our 97×3 matrix operator transformed each vector that belongs to level 0 into a three-dimensional vector and colored them as a blue dot in the graph. We received a reduction of the dimension of information from a space of 97 dimensions to a 3-dimensional space that allows us to present a point of view that represents the distribution in a state without anomaly. Similarly, we carried out the information with a congestion level 1, and a congestion level 2 and these points were marked in purple and red respectively. Since the base is three dimensions we could display the results in a 3D graph, and we obtained the following results:

To colored anomaly, compare to the conjunction 0 dataset principle components (PCs), we transformed all conjunctions levels (0, 1, 2) into a new PCA space. We used the Normal State Transformation Matrix (NSTM), obtained by performing principal components analysis on the level 0 conjunction dataset. After transforming each conjunction level dataset, separately, by the NSTM, we calculated, per conjunction level dataset, the average absolute value of projections, in the direction of each eigenvector, separately. The following graph shows the average conjunction of the projection on each PC (Fig. 6).

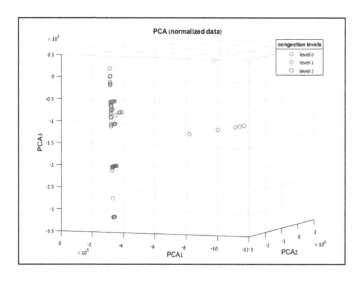

Fig. 6. Congestion levels via PCA (Color figure online)

PC-s significant for conj1 (Level 0 - Blue color) in importance order: 1. PC17, 2. PC41, 3. PC38. PC-s are significant for conj2 (Level 1 - Orange color) in importance order: 1. PC41, 2. PC38, 3. PC32 (Fig. 7).

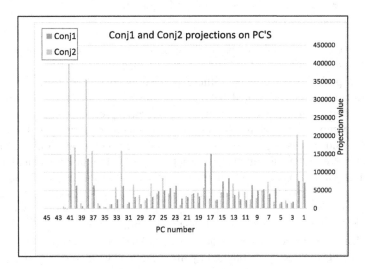

Fig. 7. Comparing average absolute values of congestion projections on all PCs. (Color figure online)

The PCA method allows us to distinguish between different conjunctions levels by performing a linear transformation of a new incoming measurement vector to the trained PC's space. If the new vector components (after transformation) will present in its component 41, 28 and 32 values which are significantly higher compare to its other components, then we know that there is an abnormal state and the reason for the anomaly is that we have moved from level 0 to congestion level 1.

It is important to emphasize that the level of congestion does not represent a single parameter whose value has exceeded a specific threshold value, that can interpret as a sole conjunction criterion. The conjunction criteria is a linear combination of 97 different measurements (components), each of which can be at its normal values range. Only the linear combination indicates an increase in the level of congestion. Therefore, a naïve and manual attempt to detected and recognize an anomaly in a vector of 97 dimensions is in the range of difficult to the point of impossible. The PCA method allows us to lower the vector dimension and also introduces interpretation that can detect and recognize congestion anomalies in low-level network transportation.

4.3   Geographical Traffic Analysis

The third data structure deals with geographical location. Routers that spread all over the world collected data stream flow from anonymous internet domains (150 different domains - one column per domain, each line is one hour aggregated bytes flow). Those datasets contain three months transportation log data. It divided into three continents groups: Africa, North America, and South America. The aim was to reveal information coming from one mainland within another mainland (for example, learning about the African continent, injecting vectors from the North American continent, and coloring such vectors as anomalies).

Remark: The original database was 97 dimensions and at a size that required analysis with big-data tools such as SPARC and HADOOP. One of the ways to reduce big-data is the use of preliminary network traffic expert knowledge. Therefore, based on the expert's guidance, which explained that hour resolution and domains transportation load is enough to detect a geographic anomaly, we performed preliminary processing on the original 97-dimensional database. Instead of doing machine learning with heavy-duty distributed cloud processing power, we conducted pre-processing utility that reduced the data into several tens of gigabytes without compromising the quality and ability to detect and classify anomalies. We extracted only columns with domain loads (The domain names converted to symbols for to preserve user confidentiality). Additionally, we reduced our dataset from 97 to 10 dimensions by selecting the top ten domains (classified by traffic average) on each of the three continents.

Figure 8 expose the common variance between African samples (X-axis) and North American samples (Y-axis) as obtained by the Canonical Component Analysis (CCA) operation. (The correlation coefficient is 0.82). It can understand that there is a great deal of commonality between the two sources of information and therefore we cannot expect to identify anomaly naively (as it presented with time-periods or conjunctions level).

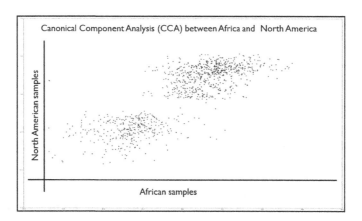

Fig. 8. Canonical Component Analysis (CCA) between Africa and North America, correlation = 0.82

The attempt to use the method used for time-period and congestion level, in a way that each class has other PCs that describe the specific type is inappropriate for the geographical case because here there is a strong correlation between the different PCs (see Figs. 8, 9 and 10). So, in the geographical situation, we look at the visual graph form, obtained after the projection. It can see that in the PC space each geographic region is placed elsewhere in the graph. And therefore, it is possible to perform separation using a linear regression line in the PC domain as a threshold between the different locations.

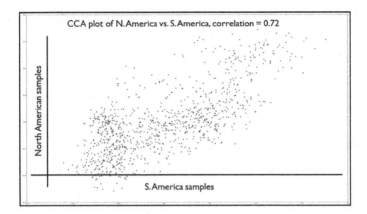

Fig. 9. CCA of N. America vs. S. America, correlation = 0.72

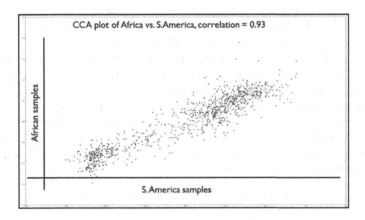

Fig. 10. CCA plot of Africa vs. S. America, correlation = 0.93

The geographical dataset is an example of analyzing different utilization mixtures with different locations and the ability to detect context (geographical) according to its pattern in the PCA space.

In Fig. 11 PCs trained on N. America data as normal dataset. We found in the training set that two eigenvalues can explain most of the variance (Explained variance for PC1 is 84% and for PC2 is 13%). Of the two eigenvectors belonging to the most explanatory eigenvalues (above eigenvalues), we extracted the eigenvectors components with the highest weights. The domains that multiplied during the PCA transformation with those most upper weights are the most dominant domain in the North American continent - Domains marked as 0, 1 and 2 were dominant in the North American continent.

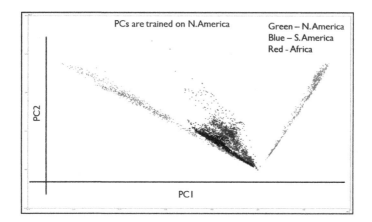

Fig. 11. PCS that trained on N. America domains Explained variance: PC1 = 84%, PC2 = 13%

In addition to extracting the 150 dominant domains, the pre-processing utility allows us to produce another query on the geographical dataset. It enabled the extraction of traffic classification by 80 different types of communications protocols (HTTP, AAC, UDP, F4V, etc.). The protocols arranged in columns. Each table row is the amount of traffic per hour. (Each table is a different continent, each column in the table is a different protocol, each line is the amount of traffic at a given time.

In Fig. 12 PCs trained on N. America protocol as normal datasets (vector of 80 dimensions). After the PCA transformation, two eigenvalues in the training set can explain most of the variance (Explained variance: 84%,13%). From the two eigenvectors belonging to the most explanatory eigenvalues, we extracted, once again, the eigenvectors components with the highest weights. The protocols that multiplied during the PCA transformation with those most upper weights are the most dominant protocol in the North American continent - the top 5 features of PC1 are HTTP.Other, Image.WebP, HTTPS.Web Messaging, and Torrent.

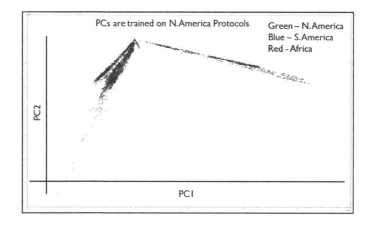

Fig. 12. PCs are trained on N. America Protocols Explained variance: PC1 = 81%, PC2 = 13%

The summary of the results of the PCA transformation of the cellular network transportation, in favor of the geographical investigation, by the cross-domain and by the cross-protocol queries, is summarized in Table 4.

Table 4. Summary of PCA geographical projections

	Trained on	Africa	N. America	S. America
Domains	N. America S. America Africa			
	Explained variance	85%, 7%	84%,13%	84%,12%
	Top Domains of PC1	0,1,2	0,1, 2	0,1
Protocols	N. America S. America Africa			
	Explained variance	76%,18%	81%,13%	76%,18%
	Top 5 features of PC1	*HTTPS.Web, Audio.AAC, Video.Facebook.CDN, Video.Google, HTTP.Other,*	*HTTP.Other, Image.WebP, HTTPS.Web, Messaging, Torrent*	*HTTPS.Web, Video.Facebook.CDN, Video.F4V, Facebook.CDN, UDP.Other*

5 Reliability and Validity

The t-tests have used for verifying the accuracies. Statistical analyses are used to conclude if the accuracies taken with the proposed approach are significantly distinct from the others (whereas both the distribution of values were normal). The test for assessing whether the data come from normal distributions with unknown, but equal, variances is the Lilliefors test. Obtaining results by comparing the results produced by 100 trials (at each trial we used a different split of the data). Obtaining a test decision for the null hypothesis that the data comes from independent random samples from normal distributions with equal means and equal but unknown variances. Results show a statistical significant effect in performance (p-value < 0.05, Lilliefors test $H = 0$).

6 Conclusions and Future Directions

In this article, we convert the unsupervised learning problem into a supervised classification problem. We proposed four methods for creating an associative anomaly within a given commercial traffic data database. We demonstrated how, using the PCA algorithm, we can detect the network anomaly behavior and classify between a regular data stream and a data stream that deviates from a routine, at the IP network layer level. The experiments we performed showed high and stable results, for example, it obtained that the detection and coloring of the time-period anomaly was PD = 90.2% and PF = 0.5%. and PD = 89.9% and PF = 1.5% for the detection of a geographical domains anomaly. Similar results obtained for the detection of anomalies in traffic congestions and for the geographical protocols anomalies.

The next direction that this study can take is the usage of advanced time series tools such as Facebook's Prophet tool. With time series tools, we expect to find trends and cycles in the dataset that will enable us to make an expectation forecast graph that any deviation from a predefined threshold around the forecasting graph will be defined as an anomaly.

References

1. Estan, C., Savage, S., Varghese, G.: Automatically inferring patterns of resource consumption in network traffic. In: ACM SIGCOMM, Karlsruhe, Germany, pp. 137–148 (2003)
2. Zhang, Y., Singh, S., Sen, S., Duffield, N., Lund, C.: Online identification of hierarchical heavy hitters: algorithms, evaluation, and applications. In: ACM Internet Measurement Conference, Taormina, Sicily, Italy, pp. 101–114 (2004)
3. Barford, P., Kline, J., Plonka, D., Ron, A.: A signal analysis of network traffic anomalies. In: ACM Internet Measurement Workshop, Marseille, France, pp. 71–82 (2002)
4. Krishnamurthy, B., Sen, S., Zhang, Y., Chen, Y.: Sketch-based change detection: methods, evaluation, and applications. In: ACM Internet Measurement Conference, Miami Beach, FL, USA, pp. 234–247 (2003)
5. Zhang, Y., Ge, Z., Greenberg, A., Roughan, M.: Network anomography. In: ACM Internet Measurement Conference, Berkeley, California, USA, October 2005
6. Soule, A., Salamatian, K., Taft, N.: Combining filtering and statistical methods for anomaly detection. In: ACM Internet Measurement Conference, Berkeley, California, USA, October 2005
7. Lakhina, A., Crovella, M., Diot, C.: Mining anomalies using traffic feature distributions. In: ACM SIGCOMM, Philadelphia, Pennsylvania, USA, pp. 217–228 (2005)
8. Lakhina, A., Crovella, M., Diot, C.: Diagnosing network-wide traffic anomalies. In: ACM SIGCOMM, Portland, Oregon, USA, pp. 219–230 (2004)
9. Soule, A., Ringberg, H., Silveira, F., Rexford, J., Diot, C.: Detectability of traffic anomalies in two adjacent networks. In: Passive and Active Measurement Conference (2007)
10. Mai, J., Chuah, C.-N., Sridharan, A., Ye, T., Zang, H.: Is sampled data sufficient for anomaly detection? In: ACM Internet measurement Conference, Rio de Janeriro, Brazil, pp. 165–176 (2006)
11. Mai, J., Sridharan, A., Chuah, C.-N., Zang, H., Ye, T.: Impact of packet sampling on portscan detection. IEEE J. Sel. Areas Commun. **24**, 2285–2298 (2006)
12. Brauckhoff, D., Tellenbach, B., Wagner, A., May, M., Lakhina, A.: Impact of packet sampling on anomaly detection metrics. In: ACM Internet Measurement Conference, Rio de Janeiro, Brazil, pp. 159–164 (2006)
13. Fodor, I.K.: A Survey of Dimension Reduction Techniques, Technical report UCRL-ID-148494, Lawrence Livermore Nat'l Laboratory, Center for Applied Scientific Computing, June 2002

14. Mao, K.Z.: Identifying critical variables of principal components for unsupervised feature selection. IEEE Trans. Syst. Man Cybern. Part B **35**, 339–344 (2005)
15. Breiman, L.: Statistical modeling: the two cultures. Stat. Sci. **16**(3), 199–215 (2001)
16. Amit, Y., Geman, D.: Shape quantization and recognition with randomized trees. Neural Comput. **9**(7), 1545–1588 (1997)
17. Guyon, I., Weston, J., Barnhill, S., Vapnik, V.: Gene selection for cancer classification using support vector machines. Mach. Learn. **46**, 389–422 (2002)
18. Webb, A.R.: Statistical Pattern Recognition, 2nd edn. Wiley, Chichester (2002)
19. Viswanath, B., Bashir, M., Crovella, M., Guha, S., Gummadi, K., Krishnamurthy, B., Mislove, A.: Towards detecting anomalous user behavior in online social networks. In: 23rd USENIX Security Symposium (USENIX Security 14), pp. 223–238 (2014)
20. Bian, L.X., Crovella, F., Diot, M., Govindan, C., Iannaccone, R., Lakhina, A.: Detection and identification of network anomalies using sketch subspaces. In: Proceedings of the 6th ACM SIGCOMM Conference on Internet Measurement, pp. 147–152 (2006)
21. Lakhina, A., Crovella, M., Diot, C.: Characterization of network-wide anomalies in traffic flows. In: Proceedings of the 4th ACM SIGCOMM Conference on Internet Measurement, pp. 201–206, (2004)
22. Lakhina, A., Crovella, M., Diot, C.: Diagnosing network-wide traffic anomalies. SIGCOMM Comput. Commun. Rev. **34**(4), 219–230 (2004)
23. Lakhina, A., Crovella, M., Diot, C.: Mining anomalies using traffic feature distributions. SIGCOMM Comput. Commun. Rev. **35**(4), 217–228 (2005)
24. Lakhina, A., Papagiannaki, K., Crovella, M., Diot, C., Kolaczyk, E., Taft, N.: Structural analysis of network traffic flows. SIGMETRICS Perform. Eval. Rev. **32**(1), 61–72 (2004)
25. Martin, R.A., Schwabacher, M., Oza, N., Srivastava, A.: Comparison Of Unsupervised Anomaly Detection Methods For Systems Health Management Using Space Shuttle Main Engine Data. Researchgate (2007)
26. Anderson, T.W.: An Introduction to Multivariate Statistical Analysis. Wiley Series in Probability and Mathematical Statistics, 2nd edn. Wiley, New York (1984)
27. Muirhead, R.J.: Aspects of Multivariate Statistical Theory. Wiley, New York (1982)

Self-stabilizing Byzantine Tolerant Replicated State Machine Based on Failure Detectors

Shlomi Dolev[1], Chryssis Georgiou[2], Ioannis Marcoullis[2(✉)],
and Elad M. Schiller[3]

[1] Department of Computer Science, Ben-Gurion University of the Negev,
Beersheba, Israel
dolev@cs.bgu.ac.il
[2] Department of Computer Science, University of Cyprus, Nicosia, Cyprus
{chryssis,imarco01}@cs.ucy.ac.cy
[3] Department of Computer Science and Engineering,
Chalmers University of Technology, Gothenburg, Sweden
elad@chalmers.se

Abstract. Byzantine Fault Tolerant (BFT) replication leverages highly available cloud services and can facilitate the implementation of distributed ledgers, e.g., the blockchain. Systems providing BFT State Machine Replication (SMR) work under severe system assumptions, for example, that less than a third of replicas may suffer a Byzantine failure. Infrequent arbitrary violations of such design assumptions, may lead the system to an unintended state, and render it unavailable thereafter, requiring human intervention. Self-stabilization is a highly desirable system property that can complement Byzantine fault tolerant systems, and allow them to both tolerate Byzantine-failures and *automatically* recovery from any unintended state that assumption violations may lead to.

This paper contributes the first self-stabilizing State Machine Replication service that is based on failure detectors. We suggest an implementable self-stabilizing failure detector to monitor both responsiveness and the replication progress. We thus encapsulate weaker synchronization guarantees than the previous self-stabilizing BFT SMR solution. We follow the seminal paper by Castro and Liskov of Practical Byzantine Fault Tolerance and focus on the self-stabilizing perspective. This work can aid towards building distributed blockchain system infrastructure enhanced with the self-stabilization design criteria.

A technical report of this work appears on https://arxiv.org/.

S. Dolev—Partially supported by the Rita Altura Trust Chair in Computer Sciences; the Lynne and William Frankel Center for Computer Science; the Ministry of Foreign Affairs, Italy; the grant from the Ministry of Science, Technology and Space, Israel, and the National Science Council (NSC) of Taiwan; the Ministry of Science, Technology and Space, Infrastructure Research in the Field of Advanced Computing and Cyber Security; and the Israel National Cyber Bureau.

I. Marcoullis—Partially supported by a Doctoral Scholarship program of the University of Cyprus.

I. Dinur et al. (Eds.): CSCML 2018, LNCS 10879, pp. 84–100, 2018.
https://doi.org/10.1007/978-3-319-94147-9_7

Keywords: Byzantine Fault-Tolerance · Self-stabilization
State Machine Replication · Fault detection

1   Introduction

Motivation and Prior Work. Modern Cloud systems offering data storage achieve high availability by employing redundancy of processors (servers or replicas) to store data and provide service. This facilitates robust services that can withstand either stop-fail (crash) failures or Byzantine failures that are more severe. The latter are usually modeled as an adversary that does not follow the protocol, and sends faulty messages that aim at hindering the system's progress, polluting the non-faulty replicas' state, or stopping service provision all together. The State Machine Replication (SMR) [13] paradigm provides fault tolerant replication and replica consistency by imposing an ordering of state transitions to every correct replica in the system.

Byzantine Fault Tolerance (BFT) [6] allows replication in the presence of Byzantine processors. It takes the form of repeated consensus to reach agreement on an ordering of state transitions. Besides providing high availability to cloud systems, BFT was also shown to enable the construction of distributed ledgers [1], e.g., blockchains. Systems providing BFT require liveness assumptions such as synchronization of stronger or weaker forms, failure detection, or randomization to overcome the FLP impossibility [12] (which states that agreement cannot be reached in asynchrony, in the face of even one crash fault).

BFT requires a series of assumptions. A standard safety assumption is that only less than one third of the system's processors may experience Byzantine faults [14]. Most existing systems also assume a consistent initial replica state and system variables. Temporal violation of system assumptions can drive the system to an illegal state, rendering it indefinitely unavailable. Systems that do not facilitate recovery mechanisms from such transient faults, depend on human intervention to return the system back to its intended behavior, i.e., to a legal state. Self-stabilizing systems [7] are designed to *automatically* drive the system back to a legal state. As such, self-stabilization is a desired system property that adds value and enforces fault tolerance.

The only work available on self-stabilizing Byzantine Fault Tolerant replication is by Binun et al. [4] that assumes a semi-synchronous setting and employs a self-stabilizing byzantine-tolerant clock synchronization algorithm [10] to enforce a new instance of Byzantine agreement upon every clock pulse.

Our Approach. In this work, we follow the classic BFT protocol of Castro and Liskov [6], but enhanced and modified to handle transient faults along with Byzantine behavior. Replicas operate within a *view*, identified by an integer number, in which a *primary* replica acts as the coordinator. In particular, each replica in view i considers p_i as the primary. The primary must ensure that client requests are totally ordered, so that the other replicas execute them in this order and maintain identical state.

Transient faults can corrupt the replica state and views and hence lead correct (non-Byzantine) replicas in *stale*, conflicting replica states and perceived views. This is a *serious challenge* not faced by [6] and the other works on BFT (with the exception of [4]), since they assume a consistent initial state from which the replication progresses. Furthermore, a faulty (Byzantine) primary may hinder the replication progress, and in this case it must be changed by the correct processors (by changing the view). Identifying a faulty primary becomes even more challenging in the presence of stale information (due to transient faults), as faulty replicas can take advantage of correct processors' corrupted information and mislead them in believing that the primary is behaving correctly, or accuse a correct primary of being faulty.

Our Contribution. We provide a novel asynchronous self-stabilizing BFT SMR service that addresses the above challenges (Sect. 3). Diverging from [4], we do not use clock synchronization and timeouts, but rather, we base our solution on a self-stabilizing failure detector (discussed below) and an automaton-based coordination technique (Sect. 3.1). This is the first work to combine self-stabilization and BFT replication that is based on failure detectors, thus encapsulating weaker synchronization guarantees than [4]. In view of [1], we consider our work as an important step towards realizing self-stabilizing BFT-based infrastructure for blockchain systems.

Overview of Our Solution. Our solution is composed of three *modules*:

View Establishment Module: This is the most critical and challenging module. It establishes a consistent view (and state) among $n - f$ replicas, where n is the number of replicas and f an upper bound on the number of faulty replicas. Managing convergence to a consistent view in the presence of Byzantine processors injecting arbitrary messages, and in the existence of other stale information in local states and communication channels is very demanding and it is impossible without a series of assumptions [3]. To this respect, we present an automaton-based solution where convergence requires a fragment of the computation to be free of failures (still, note that even under this constraint, view establishment is very challenging as one infers from Sect. 3.1). In Sect. 4, we relax this constraint by introducing, for the first time, an *event-driven* (unreliable) failure detector that can be tuned to ensure (in all reasonable executions) that enough responses from non-byzantine processors are received.

Replication Module: The replication module follows the replication scheme by Castro and Liskov [6], but adjusted to also cope with stale information. When there appears to be a common view and hence a primary, the replicas progress . the replication. In case an inconsistent replica state is detected (due to stale information), then this module requests a view establishment and falls back to a default state (whereas the scheme in [6] would require human intervention). In [6], clients sent requests to the primary using signed messages, so that the primary cannot tamper with the content of the message and hence affect the replication progress. Then the primary essentially coordinated a three-phase-commit protocol for establishing the request ordering. As we prefer to use information theoretically secure schemes, rather than computationally cryptographic secure schemes based on message signing, we require that clients contact all replicas. The primary is still the one to decide the order, but the replicas, through a self stabilizing all-to-all exchange procedure, validate the requests suggested to be processed by the primary; a request is *valid* if it has been seen by a strong majority of correct $((n - f)/2 + f)$ replicas (see Sect. 3.2).

Primary Monitoring Module: The primary is monitored by a view change mechanism, employing a failure detector (FD) to decide when a primary is suspected and, thus, a view change is required (Sect. 3.3). A faulty primary may hinder the replication progress by (i) not responding to messages from the replicas, (ii) not processing requests, or (iii) processing invalid requests. In case the primary is found to impede the replication progress, it is considered faulty and the module proceeds to change the primary, by installing the next view. Following the approach of Baldoni et al. [2], we propose an implementation of a self-stabilizing FD that checks both the responsiveness of the replicas (including the primary), and whether the primary is progressing the state machine. Our responsiveness FD can be seen as a self-stabilizing version of the muteness FD given in [11] but adapted to an asynchronous environment following the technique discussed in [15]; our self-stabilizing implementation follows [5].

To cover case (iii) above, our FD monitors the requests processed by the primary to check whether they are valid or not. Stale information gives rise to certain subtle issues that our implementation needed to address (see Sect. 3.3.1). Note that the work in [6] did not use a FD, but instead, for (i) and (ii) timeouts where used (under a liveness assumption that communication delay between the primary and the replicas is bounded); for (iii), as mentioned above, signed messages were used. The liveness assumption imposed on our FD (to overcome the FLP result [12]) is that a correct primary is never suspect by a majority of correct $((n - f)/2)$ replicas.

Following the approach of [1], to facilitate the presentation of our solution, in Sect. 3 we consider $n = 5f + 1$ replicas. In Sect. 4 we explain how our solution can be adjusted to obtain the optimal resilience of $n = 3f + 1$.

2   System Settings

Consider an asynchronous message-passing system with a fixed set of processors P, where $|P| = n$, and each processor $p_i \in P$ has a unique integer identifier, $i \in [0, n-1]$. At most $f = (n-1)/5$ processors may exhibit malicious (Byzantine) behavior, i.e., fail to follow the protocol. This includes fail-stop failures (crashes). (In Sect. 4 we increase f to $(n - 1)/3$.) A *correct* processor is one not exhibiting malicious behavior. Processors have hard-coded (incorruptible) knowledge of n and f. Transient faults (i.e., short-term violations of system assumptions) may also take place. These may corrupt local variables and program counters of any number of processors, as well as data links, thus introducing *stale information*.

Communication. The network topology is a fully connected graph, with links of bounded capacity *cap* in message packets. Processors exchange low-level messages called *packets* to enable a reliable delivery of high level *messages*. Packets may be lost, reordered, or duplicated but not arbitrarily created, although channels may initially (after transient faults) contain stale packets. Due to the boundedness of the channels, the number of stale packets in the system's communication links is also bounded $O(n^2 cap)$. We assume the availability of self-stabilizing protocols for reliable FIFO end-to-end message delivery over bounded-capacity unreliable channels, such as the ones of [9]. We also assume that messages reaching p_j from p_i are guaranteed to have originated from p_i, unless they are the result of a transient fault. I.e., a malicious processor p_k (where $i \neq j \neq k$) cannot impersonate p_i sending a message to p_j [8].

When processor p_i sends a packet, pkt_1, to processor p_j, the operation *send()* inserts a copy of pkt_1 into the FIFO queue representing the communication channel from p_i to p_j. To respect the channel's capacity bound, packets may be lost. Nevertheless, a packet sent infinitely often is received infinitely often. Packet pkt_1 is retransmitted until more than *cap* acknowledgments arrive, and then pkt_2 starts being transmitted. This forms an abstraction of token carrying messages between the two processors, implementing a token exchange. This token exchange technique facilitates a *heartbeat* to detect whether a processor is responsive; if its is not, the token will not be returned. Applications may piggyback information on tokens.

The Interleaving Model and Self-stabilization. A program is a sequence of *(atomic) steps*. Each step starts with local computations and ends with a communication operation, i.e., packet *send* or *receive*. We assume the standard interleaving model where at most one step is executed in every given moment. An input event can either be the arrival of a packet or a periodic timer triggering p_i to resend. The system is asynchronous and the rate of the timer is totally unknown. The *system state*, s_i, consists of p_i's variable values and the content of p_i's incoming communication channels. A step executed by p_i can change the state of p_i. We name the tuple of states $(s_1, s_2, \cdots, s_n)$ as the *system state*. An *execution (or run)* $R = c_0, a_0, c_1, a_1, \ldots$ is an alternating sequence of states c_x and steps a_x, such that each c_{x+1} is obtained from c_x by the execution of a_x. An execution is *fair* when every correct processor that has an applicable step a_i infinitely often, executes a_i infinitely often. The system's task is a set of executions called *legal executions* (*LE*) in which the task's requirements hold. An algorithm is *self-stabilizing* with respect to *LE* when every (unbounded) execution of the algorithm has a suffix that is in *LE*. An *iteration* of an algorithm formed as a do–forever loop is a complete run of the algorithm starting at the loop's first line and ending at the last line, regardless of entering branches.

Complexity Metric. We define an *asynchronous round*, a.r., of an execution R as the shortest prefix of R in which every correct processor p_i completes an iteration I_i, and all messages p_i sent during I_i were received.

3 Self-stabilizing BFT State Machine Replication

Our solution comprises of three modules: *View Establishment*, *BFT Replication*, and *Primary Monitoring* (composed of the *Failure Detector* (FD) and the *View Change* mechanism). Along with their interfaces they appear in Fig. 1.

A *view* is a bounded integer counter in $[0, n-1]$. A processor in view i considers processor p_i as the view's primary. If stale information or conflicting views are detected, the system *establishes* a view by moving to a known hard-coded view (e.g., view 0). A view common to $3f+1$ processors is called *serviceable* and allows processors to proceed with replication. Upon detecting a replica state conflict, the replication module requires a view establishment while it moves to a default state. If $3f + 1$ processors suspect the primary, the view change mechanism drives the system to a view change. to view $i + 1$. There are at most f faulty processors, so there can be at most f faulty primary changes before reaching to a correct one, if such processors have consecutive identities.

Thresholds. Following [1], we define important thresholds required to take decisions, based on $n = 5f + 1$ processors. Section 4 suggests how these thresholds can be adjusted for the optimal $n = 3f + 1$. The benefit of $n = 5f + 1$ is that it gives a simpler solution requires less processors to progress the replication. A processor p_i considers a view v as: (i) *Establishable*, if $n - f$ processors require its establishment, the maximal possible support to be demanded. A view that was indeed established (and was not the result of an arbitrary initial state), must be held by at least $n - 2f$ correct processors, (ii) *Adoptable*, (implying that correct processors can adopt it) if it is supported by $\max(n - 2f, n/2)$ processors, (iii)

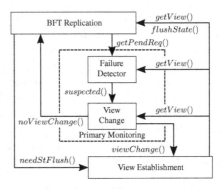

Fig. 1. (Information flow $\boxed{A} \xrightarrow{to} \boxed{B}$.) Calling $getView(i)$ returns p_i's current view pair or $\top$ if no such exists, and $getView(j)$ $(j \neq i)$, returns p_j's last reported view. If state conflict is detected, $needStflush()$ triggers a view establishment. $flushState()$ resets the state of the replication algorithm. $getPendReqs()$ returns a set of pending client requests that need to be executed by the current primary. $suspected()$ returns True if the primary is suspected. $noViewChange()$ returns False when a view change is taking place.

Serviceable, if a majority of correct processors appear to have this view, i.e., $\lceil (n - f)/2 \rceil + f$. This allows for the replication service to continue. Notably, for $n = 5f + 1$, the two properties coincide on $3f + 1$.

3.1 View Establishment

This module provides a unique view to the correct processors, and conducts view changes upon the instruction of the primary monitoring module (Sect. 3.3.2). We start with a description of the algorithm and continue the correctness outline.

3.1.1   Algorithm Description

Overview. The algorithm is implemented as a two state (or *phase*) automaton (see Fig. 2). Phase 0 is a monitoring phase that checks for view change requests or view and replica state conflicts, the latter of which are detected by the Replication module. Conflicts indicate the existence of stale information. Upon finding a conflict or seeing a view change instruction, the automaton moves to Phase 1. Upon seeing evidence that the new view or replica state is established, the processor moves back to Phase 0. This allows replication to continue.

Local Variables and Information Exchange. We use the subscript notation var_i to indicate a variable var of a processor p_i. If the owner of the variable is deduced from the description we omit it. The algorithm uses the variable type v*Pair* which is a pair of *views* $\langle cur, next \rangle$, where each view is an integer in $\{0, 1, \ldots, n-1\} \cup \{\bot\}$. Processors, have a common hard-coded fallback view DF_VIEW (say 0) and a reset vPair RST_PAIR = $\langle \bot, \text{DF\_VIEW} \rangle$ used in case of corruption. Processor p_i maintains an array of *vPairs* called $views_i[n][n]$.

The field $views_i[i][i]$ is p_i's current *vPair*. Specifically, $views_i[i][i].cur$ is p_i's current view and *next* only differs to *cur* if a view transition is taking place (i.e., the automaton is in Phase 1). Using the token passing mechanism described in Sect. 2, p_i sends $views_i[i][i]$ to the other processors. When p_i receives the token from p_j it stores p_j's piggybacked *vPair* in $views_i[i][j]$. In every iteration p_i sends $views_i[i]$ to the other processors and stores a received copy of $views_j[j]$ from p_j in $views_i[j]$. The phase array $phs[n]$ contains fields with 0 or 1 indicating automaton Phases 0 or 1. Field $phs_i[i]$ is p_i's phase and $phs_i[j]$ is the last reported phase by p_j sent. Boolean $vChange$ records a request by the primary monitoring mechanism to change the view when $vChange = $ True.

Witnessing. Algorithm 1 uses the alias $echo_i[j]$ for the triple $\langle views_i[i][j],\ phs_i[j],\ witnesses_i[j]\rangle$. The boolean array $witnesses_i[n]$ stores $witnesses_i[j] = $ True if p_j has sent a copy of $echo_j[i]$ identical to $echo_i[i]$ with regards to view pair and phase. A processor p_j that returns p_i's current values is called a *witness*. If the copy of $echo_j[i]$ is also the identical to $echo_i[j]$, i.e., towards the $witnesses_i[j]$ value p_j is also added to the *witnesSeen* set. The above

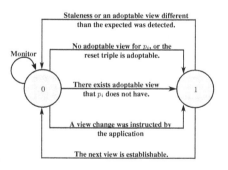

Fig. 2.   View establishment automaton for $p_i \in P$.

facilitates an echo mechanism, that p_i uses to learn if others are aware of its current values. The echoing mechanism of p_i, logs all those that have confirmed that they have witnessed p_i in its most recent view transition.

Detailed Description. The algorithm follows the states and transitions of the automaton of Fig. 2. When monitoring the existence of a unique view, the

---

Algorithm 1. Self-stabilizing View Establishment; code for processor p_i

---

1 **Macros:** $flushVE() = \{views_i[i][i] \leftarrow$ RST_PAIR; $flushState(); \}$
2 $updated(k) = $ **return** $(views_i[i][k] = views_i[k][k])$
3 $typeCheck(vPair\ vp) = $ **return** $((\forall x \in vp.\langle cur, next \rangle : x \in [0, n-1] \cup \{\bot\}) \wedge$
   $(vp.next \neq \bot))$
4 $legitPhsZero(vPair\ vp) = $ **return** $(((vp.cur = vp.next) \vee (vp = $ DF_VIEW$)) \wedge$
   $typeCheck(vp))$
5 $legitPhsOne(vPair\ vp) = $ **return** $((vp.cur \neq vp.next) \wedge typeCheck(vp))$
6 $stale(k) = $ **return** $((phs[k] = 0 \wedge \neg legitPhsZero(views[i][k])) \vee (phs[k] = 1 \wedge$
   $\neg legitPhsOne(views[i][k])))$
7 $staleM(m, k) = $ **return** $((m.phs = 0 \wedge \neg legitPhsZero(m.views[k])) \vee (m.phs = 1$
   $\wedge \neg legitPhsOne(m.views[k])))$
8 $goodVHold(view\ x, j) = $ **return** $(updated(j) \wedge \neg stale(j) \wedge (views_i[i][j].cur = x));$
9 $adoptSet(vPair\ vp) = \{p_j \in P : phs[j] = 0 \wedge goodVHold(vp.cur, j)\};$
10 $adoptable(vPair\ vp) = $ **return** $(|adoptSet(vp)| \geq 3f + 1))$
11 $transitAdopble(vPair\ vp) = $ **return** $(|adoptSet(vp) \cup \{p_j \in P : phs[j] = 1 \wedge$
   $goodVHold(vp.next, j)\}| \geq 3f + 1)$
12 $adopt(vPair\ v) = \{vp_i.\langle cur, next \rangle \leftarrow v.\langle cur, next \rangle \};$
13 $estable(vPair\ vp) = $ **return** $(|\{p_j \in P : goodVHold(vp.cur, j) \wedge (views_i[i][j].next =$
   $vp.next)\}| \geq 4f + 1)$
14 $existsCatchUpSet() = $ **return** $((|adoptSet(vp_i)| \geq 3f + 1) \wedge (|adoptSet(vp_i) \cup \{p_k \in P :$
   $goodVHold(vp_i.cur, k) \wedge phs[k] = 1 \wedge (views[i][j].next = vp_i.next)\}| \geq 4f + 1))$
15 $establish() = \{vp_i.cur \leftarrow vp_i.next\};$
16 $echo(j) = $ **return** $(echo[i] = echo[j])$
17 $echoNoWitn(k) = $ **return** $(\langle vp_i, phs[i] \rangle = \langle echo[k].views, echo[k].phs \rangle)$
18 $witnesSeen() = $ **return** $(|witnesSeen \cup \{p_i : witnesses[i]\}| \geq 4f + 1)$
19 $nextView() = \{views[i][i].next \leftarrow views[i][i].cur + 1\};$
20 $nextPhs() = \{\langle phs[i], witnesSeen, vChange \rangle \leftarrow \langle (phs[i] + 1 \mod 2), \emptyset, $ False$\rangle\};$
21 **Interface functions:** $viewChange() = \{vChange \leftarrow $ True$\};$
22 $getView(j) = \{$**if** $(j = i) \wedge (phs[i] = 0 \wedge witnesSeen())$ **then** $\{$**if** $(adoptable(view))$ **then**
   return $(view.cur)$ **else return** $(\top)\}$ **else return** $(views[i][j].cur)\};$

23 **do forever begin**
24 | **if** $stale(i)$ **then** $flushVE; phs[i] \leftarrow 0; witnesSeen \leftarrow \emptyset;$
25 | $witnesses[i] \leftarrow (\exists S \subseteq P : (p_k \in S \iff echoNoWitn(k)) \wedge |S| \geq 4f + 1);$
26 | $witnesSeen \leftarrow witnesSeen \cup \{p_j \in P : (echo(j) \wedge witnesses[j])\};$
27 | **if** $phs[i] = 0 \wedge witnesSeen()$ **then**
28 | | **if** $(needStFlush() \vee (adoptable(RST_PAIR) \vee ((\nexists vp' \in views[i] : adoptable(vp'))$
   $\wedge \neg transitAdoble(vp_i)))$ **then** $flushVE();$
29 | | **else if** $((\exists vp' \in views[i] : adoptable(vp')) \wedge (vp_i.cur \neq vp'.cur))$ **then**
   $adopt(vp'); nextPhs();$
30 | | **else if** $(vChange \wedge adoptable(vp_i))$ **then** $nextView(); nextPhs();$
31 | **else if** $phs[i] = 1 \wedge witnesSeen()$ **then**
32 | | **if** $((estable(vp_i) \vee existsCatchUpSet()) \wedge \neg stale(i))$ **then** $establish(); nextPhs();$
33 | | **else if** $stale(i) \vee (\exists vp' \in views[i] : adoptable(vp') \wedge vp'.cur \neq vp_i.next)$ **then**
   $flushVE(); nextPhs();$
34 | **foreach** $p_j \in P$ **do send** $\langle views[i], phs[i], witnesses[i], (views[i][j], phs[j],$
   $witnesses[j]\rangle);$

35 **upon receive** m **from** p_j **do if** $(\neg staleM(m, j))$ **then** $\langle views[j], phs[j], witnesses[j],$
   $echo[j]\rangle \leftarrow m;$

---

algorithm first checks whether p_i has a view that is not stale with respect to the phase and the view pair's integrity (line 24). It then proceeds to renew the set of witnesses, the processors that have echoed its most recent values of vp_i and $phs[i]$ (line 25), and then updates its $witnesSeen$ set with processors that have observed the whole $echo[i]$ (line 26). Before describing the phases, we note that every iteration completes with a broadcast of local variables (line 34).

Phase 0 – Monitoring. Lines 27–30 monitors the current view and initiates either a view establishment, or view change, given at least $4f + 1$ processors belong to its $witnesSeen$ set, i.e., they are aware of p_i's recent values and p_i is aware of this. If p_i considers that the RST_PAIR is adoptable or that no view has enough support, it moves to RST_PAIR, and Phase 1 (line 28). If it considers that there exists an adoptable view that it does not have, it adopts this and moves to Phase 1 (line 29). If a view change is requested by the view change module, and p_i has an adoptable view, then it increments its view and moves to Phase 1 (line 30).

Phase 1 – View Transition. Lines 31–33 control the completion of transiting to a new view and to Phase 0. Again *witnesSeen*() needs to hold (line 31). If the view being installed ($vp_i.next$) is establishable, or enough processors appear to be transiting to this, then p_i proceeds to $vp_i.next$ and Phase 0 (line 32). Otherwise, if it appears as the result of stale information, then p_i moves to RST_PAIR.

3.1.2 Correctness

Definitions. A fair execution R is *mal-admissible* if, throughout R, the set of faulty processors obeys the bound f. A fair execution is *mal-free* if throughout R every processor acts as a correct one. A message is a *threat* to $p_i \in P$ if its receipt can cause the overthrow of p_i's view. A *stable view* cannot be overthrown by malicious processors, and thus characterizes a legal system state.

Task Description. The view establishment task $\mathcal{VE}$ includes all the system states of mal-admissible executions in which there is a stable view, or where a stable view is followed by a view change leading to a new stable view.

Proof Outline. The proof starts by establishing a stale-free local state with a legit (possibly default) view. Within $O(n)$ a.r., and in the absence of a stable view (due to an initial arbitrary state), a mal-free execution reaches a safe system state $c \in \mathcal{VE}$, which has a stable view, either the default one or one that existed in the system. This concludes the *convergence* proof. We then prove *closure*. In the absence of view changes, no correct processor holding a stable view ever switches to a different view, but if there is a view change, then a new stable view is installed within $O(n)$ a.r.

Lemma 3.1. *Consider a mal-admissible execution R of Algorithm 1 starting in an arbitrary system state. Within $O(1)$ a.r., there are no messages from correct processors that encode a threat.*

Lemma 3.2. *Consider a mal-free execution starting without a stable view. Within $O(n)$ a.r., either a stable view $v \neq$ DF_VIEW is reached, or a view reset is initiated.*

Lemma 3.3. *In a mal-admissible execution starting with a stable view v, for every $p_k \in \mathcal{C} : views_k[k][k] \neq v$, within $O(1)$ a.r. $views_k[k][k] = v$ holds.*

Theorem 3.1 (Convergence). *Consider a mal-free execution of Algorithm 1 starting in an arbitrary state. Within $O(n)$ a.r., the system reaches a state with a stable view v, and every correct processor adopts view v.*

Theorem 3.2 (Closure). *Consider a mal-admissible execution R, starting with a state encoding a stable view v. Either, v remains stable throughout R, or* $\exists c \in R$, *which encodes a view change that results in a new stable view v'.*

3.2 State Replication Algorithm

The replication module (Algorithm 2) conducts SMR if there is a serviceable view. Our protocol follows Castro and Liskov [6], deviating only when catering for self-stabilization. In particular, (*i*) we introduce specific bounds for all our structures, (*ii*) we require that clients communicate their requests to all replicas. We proceed with a description of our solution and its correctness.

3.2.1 Algorithm Description

Clients and Requests. Processors receive requests from a known fixed set of clients C, where $|C| = K$. Following the typical well-formedness condition, clients do not send a new request before a previous one is complete, i.e., until it receives $f + 1$ identical responses. It is beyond the scope of this work to establish whether the content of a given request is malicious, as we concentrate on the server side [6].

The BFT Replication Task. Consider the set of correct processors $\mathcal{C} \subset P$, and a set of client requests $\mathcal{K} = \{\kappa_1, \kappa_2, \cdots, \kappa_K\}$. The BFT replication task requires all processors in $\mathcal{C}$ agree on a total order of execution of the requests of $\mathcal{K}$. Moreover, the client that issued the request eventually receives $f + 1$ identical request responses. After a transient fault takes place, safety (i.e., identical replica state) may be violated, until the system converges back to a legal state.

Ordering Requests. To impose a total order in the execution of requests, the primary assigns a unique sequence number $sq \in [0, \mathsf{MAXINT}]$ to each received request. This is an integer incremented from a practically inexhaustible counter e.g., a 64-bit one[1]. A transient fault may corrupt the counter to attain its maximum value abruptly. In this case we reset the view, the state and sq to 0. During view changes we do not reset the sequence number. While we cope with transient faults corrupting the request counter, we may still have a malicious primary that tries to propose arbitrarily high sequence numbers to the requests in order to exhaust the counter. We follow [6] in restricting a faulty primary from exhausting the counter by imposing an upper and lower bound on the sequence numbers that other processors will accept from the primary. We bound the sequence numbers sq that the primary can use for a request in any given instance of the execution to σK, where σ is a system defined integer constant. Under this bound, the primary can only assign a sequence number to a pending request if *(i)* this is the locally lowest unassigned one, and *(ii)* if this sequence number is not σK away from the sequence number of the last executed request.

---

[1] A 64-bit counter incremented per nanosecond, can last for 500 years (virtually an infinity).

Algorithm 2. Self-stabilizing Byzantine Replication; code for processor p_i

---

1 **Definitions:** lsn is the sequence number of the last executed request. $nextSeqn$ is the next
   unassigned sequence number. A request is *known* to p_i if another $3f$ processors report an
   identical copy of this request.

2 **Interface functions:**

3 $getPendReqs()$ = returns the set of pending requests reported to p_i by another $3f$
   processors;

4 $needStFlush()$ = **return** $(needFlush)$;

5 $flushState()$ = $\{flush \leftarrow \mathsf{True}\}$

6 **do forever begin**

7   **if** *Conflicting replica state or local stale information* **then** Reset the replica;
   $needFlush \leftarrow \mathsf{True}$;

8   **if** $flush = \mathsf{True}$ **then** Reset the replica;

9   **if** *No view establishment taking place* **then**

10   **if** *No view change* **then**

11   **if** p_i *is primary* **then**

12   **while** $\exists$ *known pending request* $r \wedge nextSeqn < lsn + \sigma K$ **do**

13   Assign $nextSeqn$ to r and move to status pre-prepared requests

14   **else**

15   **foreach** *known pending request* $r \wedge$ *primary has assigned seq. num.*
   $sq : sq < lsn + \sigma K$ **do** Place r in pre-prepared requests;

16   **foreach** *request* r *known as* pre-prepared **do** Move to status prepared;

17   **foreach** *request* r *known as* prepared **do** Move to status commit;

18   **foreach** *request* r *known as* commit $\wedge$ $q.sq = lsn + 1$ **do** Apply r and move to
   executed;

19   **else** Reset replica state;

20   **foreach** *Client* c **do** Send c's last processed request;

21   **foreach** *Processor* p **do** Send p_i's replica state and requests' queue;

22 **Upon receipt of message** m **from** $p_j \wedge$ No view establishment **do if** No view change
   then Store p_j's replica state and request queue **else** Only store replica state.

23 **Upon receipt of request** q **from client** c **store as pending;**

Description. Algorithm 2 implements the replication procedure by first checking and handling replica state conflicts (line 7). It then processes messages that have not been assigned a sequence number (line 12 for the primary and line 15 for non primary), and then proceeds to maintain the queues of prepared and committed requests (lines 16–17), and applying effects to the replica for committed requests (lines 18). Line 21 propagates its replica state to other processors and line 20 the last executed request of each client. It handles received information accordingly (lines 22–23).

3.2.2   Correctness

Outline. We first establish that locally detectable stale information is removed by resetting the local state to a default one. We then prove that correct processors sustain a consistent common state prefix, or if no such exists, they proceed to a view establishment (Lemma 3.4). Processors $p_i, p_j \in \mathcal{C}$ with replica states S_i, S_j have a *common state prefix* (CSP), if the processor state that applied the least number of transitions (say, p_i's) is a prefix of the other processor, i.e., $S_i = S'_j : S_j = S'_j \circ S''_j$. They also have a *consistent common state prefix* (CCSP) if they have a CSP, their transitions history is identical, and their state does not encode stale information that can make the two states to divert. A *safe system state* is one in which $3f + 1$ correct processors have a mutually known CCSP.

If a processor does not locally see such a state prefix, it declares a conflict via $needStFlush()$. The convergence theorem proof shows that Algorithm 2 manages to install a CCSP within $O(n)$ a.r. Closure ensures that the BFT replication task is satisfied once the system has converged.

Lemma 3.4. *Within $O(1)$ a.r. of Algorithm 2's execution, a processor $p_i \in C$ has no local stale information, and upon a conflict, it calls needStFlush().*

Theorem 3.3 (Convergence). *Consider an execution R of Algorithm 2. Within $O(n)$ a.r., the system reaches a safe system state.*

Theorem 3.4 (Closure). *Consider an execution R starting in a safe system state. The system remains in a safe state where it either conducts replication or it proceeds with a view change.*

3.3   Primary Monitoring

The primary is monitored by a view change mechanism, which employs a failure detector to decide when a primary is suspected and, thus, a view change is required. View change facilitates the liveness of the system, since if a malicious processor does not correctly progress the replication it is changed. We proceed to present the two parts of the module.

3.3.1   Failure Detection

We base our FD (Algorithm 3) on the token-passing mechanism described in Sect. 2, and follow the approach of [2] to check both: (i) the responsiveness of processors, (ii) that the primary is progressing the state machine.

Responsiveness Check (lines 8–10). Every processor p_i maintains a heartbeat integer counter for every other processor p_j. Whenever processor p_i receives the token from processor p_j over their data link, processor p_i resets p_j's counter to zero and increments all the counters associated with the other processors by one, up to a predefined threshold value T. Once the heartbeat counter value of a processor p_j reaches T, the FD of processor p_i considers p_j as unresponsive. In other words, the FD at processor p_i considers processor p_j to be live and connected if and only if the heartbeat associated with p_j is strictly less than T. Note that malicious processors can intentionally remain unresponsive and then become responsive again. A correct processor cannot distinguish this behavior from inaccuracies of the FD (due to packet delays) that make a correct processor appear briefly unresponsive but eventually appear as responsive when its delayed packets are received. Nevertheless, the use of the FD is to suggest which processors are responsive at a given time.

Primary Progress Check (lines 11–23). To achieve liveness, processors need to be able to check whether the primary is progressing the state machine by imposing order on the requests received. The responsiveness FD only suspects a non-responsive primary. A faulty primary can be very responsive at the level of

---

Algorithm 3. Self-stabilizing Failure Detector; code for processor p_i

---

1 **Constants:** T an integer threshold.
2 **Variables:** $beat[n]$ is an integer heartbeat array where $beat[j]$ corresponds to p_j's heartbeat
   and $beat[i]$ is unmodified and remains 0. $FDset$ is the set of processors that are responsive
   according to their heartbeat, cnt is a counter related to the primary or a proposed primary
   of p_i's current view. $primSusp[n]$ is a boolean array of {True/False} where $primSusp[j]$
   indicates whether processor p_j suspects the primary of its current view or not.
   $curCheckReq$ the requests' set (of size at most σK) that is currently being checked for
   progress. $prim$ holds the most recently read primary from the view establishment module.
3 The token passing mechanism (Sect. 2) piggybacks $FDset[i]$ and $primSusp[i]$ when sent to
   other processors, and updates fields $FDset[j]$ and $primSusp[j]$ upon receipt of the token
   from p_j.
4 $reset()$ sets all fields of $primSusp[\bullet]$ to False, set $curPendReqs$ to $\emptyset$ and $beat[\bullet]$ and cnt to 0.
5 **Interface function:**
6 $suspected() = (|\{p_j \in P : (getView(j) = getView(i)) \wedge (primSusp[j])\}| \geq 3f + 1))$

7 **Upon receipt of $token^j$ from p_j begin**
8 | $beat[j] \leftarrow 0; beat[i] \leftarrow 0$
9 | **foreach** $p_k \in P \setminus \{p_j, p_i\}$ **do** $beat[k] \leftarrow beat[k] + 1$;
10 | $FDset \leftarrow \{p_\ell \in P : beat[\ell] < T\}$;
11 | **if** $prim \neq getView(i)$ **then foreach** $p_j \in P$ **do** $reset()$;
12 | $prim \leftarrow getView(i)$;
13 | **if** $(noViewChange() \wedge (prim \in \{j : p_j \in P\}))$ **then**
14 | | **if** $(j = prim)$ **then**
15 | | | **if** $(\exists x \in curCheckReqs : x \notin getPendReqs()) \vee (curCheckReq = \emptyset)$ **then**
16 | | | | $cnt \leftarrow 0; curCheckReq \leftarrow getPendReqs()$
17 | | | **else** $cnt \leftarrow cnt + 1$;
18 | | | // if p_i, p_j in same view then store p_j's verdict on the primary
19 | | **else if** $(prim = getView(j))$ **then** $primSusp[j] \leftarrow token^j.primSusp$;
20 | | **foreach** $\{p_k \in P \setminus \{prim\}\}$ **do** $\{cnt \leftarrow 0\}$; // reset all counters except primary's;
21 | | **if** $prim = i$ **then** $cnt[i] \leftarrow 0$;
22 | | **if** $\neg(primSusp[i])$ **then**
   | | | $primSusp[i] \leftarrow ((prim \in \{k : p_k \in P\}) \wedge (prim \notin FDset) \wedge (cnt > T))$;
23 | **else if** $prim \notin \{j : p_j \in P\}$ **then** $curCheckReq \leftarrow \emptyset; cnt \leftarrow 0$;

---

packets, and thus evade suspicion by sending messages that are unrelated to the
requests of the clients. The primary can be detected to hinder progress when not
progressing requests that are locally reported to be known by $3f + 1$ processors.

To this end, the primary's progress check enhances the heartbeat FD to provide liveness but not safety as follows. The failure detector holds a set of requests
$curCheckReq$ that it drew from the replication module the last time the processor removed a request from $curCheckReq$ using the interface $getPendReqs()$
(line 16). If at least one request of $curCheckReq$ was removed from the currently pending requests, then the counter is reset to 0, and $curCheckReqs$ is
updated by $getPendReqs$ (line 16). Note that the approach is self-stabilizing,
since if $curCheckReqs$ is the result of an arbitrary initial state this is cleaned
within one iteration of the algorithm. If there was no progress then the primary's
counter cnt is incremented (line 17). Line 22 determines whether a primary p_j
is locally suspected if its counter is beyond the threshold for responsiveness or
request progress and sets $primSusp[j] = $ True. The interface $suspected()$ (line 6)
considers the primary as suspected if $3f + 1$ processors consider it as suspected.

Liveness Assumption. We assume that a correct primary is never suspected
by more than $2f$ correct processors.

Suspicion of Primary. If processor p_i suspects its primary, then $primSusp[i] = $
True permanently. So a malicious primary will either be suspected by correct

---

Algorithm 4. Self-stabilizing View Change; code for processor p_i

---

1 **Variables:** Tuple $vcm[n] = \langle vStatus, prim, needChange, needChgSet \rangle$ where $vStatus \in$ $\{$OK, $noService$, $vChange\}$, $needChange$ is a boolean and $needChgSet$ a set of processors that appear to require a view change. $vcm[i]$ holds p_i's values and $vcm[j]$ holds p_j's last gossiped value to p_i. DEF_STATE is a default value for $vcm = \langle$OK, $getView()$, False, $\emptyset \rangle$.

2 **Macros:** $cleanState() = \{$foreach $p_j \in P$ do $vcm[j] \leftarrow$ DEF_STATE; $\}$

3 $existsView(prim) =$ **return** $(prim \in \{j : p_j \in P\})$;

4 $supChange(x) =$ **return** $(\exists X \subseteq P : (\forall p_j, p_{j'} \in X : vcm[j].prim = vcm[j'].prim) \land$ $(|\bigcap_{p_k \in X} vcm[k].needChgSet| \geq 3f + 1) \land (|X| \geq x))$

5 **Interface function:** $noViewChange() = (vStatus =$ OK$)$;

6 **do forever begin**

7 | let $needChange \leftarrow suspected()$;

8 | if $existsView(getView(i))$ **then**

9 | **if** $(prim = getView(i)) \land (vStatus \neq$ vChange$)$ **then**

10 | $needChgSet \leftarrow needChgSet \cup \{p_j \in P : getView(i) = getView(j) \land$ $vcm[j].needChange =$ True$\}$;

11 | **if** $((|\{vcm[j].vStatus =$ noService$\}_{p_j \in P}| < 2f + 1))$ **then** $vStatus \leftarrow OK$;

12 | **if** $(vStatus =$ OK$) \land (supChange(3f + 1))$ **then** $vStatus \leftarrow noService$;

13 | **else if** $(supChange(4f + 1))$ **then** $vStatus \leftarrow$ vChange; $viewChange()$;

14 | **else if** $(prim = getView(i)) \land (vStatus =$ vChange$)$ **then** $viewChange()$;

15 | **else** $cleanState()$;

16 | **else** $cleanState()$;

17 | foreach $p_j \in P$ do send vcm;

18 **Upon receive** m **from** p_j **do** $vcm[j] \leftarrow m$;

---

processors, or be forced to progress the replication, and changing behavior is not tolerated by correct processors. Notice that the liveness assumption discussed above ensures that no more than $2f$ correct processors will set their suspicion to True for a correct primary.

Correctness Outline. The proof shows that Algorithm 3 is stabilizing. The responsiveness check stabilizes within $O(1)$ a.r., requiring an exchange with every one of the correct processors to reset the $beat[j]$ field for every correct processor p_j. If the primary progress check has stale information, then either (i) field $primSusp[i] =$ True, or (ii) set $curCheckRes$ contains requests that never existed. These do not force a stop of service, but both are reset after a call for a view change (possibly triggered by stale information). Thus the algorithm stabilizes within $O(1)$ a.r.

3.3.2 View Change upon Suspected Primary

Algorithm Outline. In a nutshell, a processor propagates messages about which processors have reported to require a view change. If $3f + 1$ processors appear to have suspected the primary, then the processor stops providing service even if itself has not suspected the primary itself. The above guarantees that since f of $3f + 1$ processors may be malicious, at least $2f + 1$ correct processors have firmly suspected the primary. The replication mechanism is left with $3f$ processors, which is not enough to make progress, so the view change is forced upon the system by the $2f + 1$ correct processors who are the majority of correct processors. The view change is initiated upon seeing that the intersection of those that require view change becomes $4f + 1$.

Detailed Description. Processor p_i executing Algorithm 1 first reads the FD (line 7), and then checks whether the view establishment module returns a current view (line 8). If not, it sets the algorithm's local variables to their default values (line 16). If the primary has changed (line 9), p_i resets the status and variables only if the status is not yet $vChange$. If any of these two conditions fail, it adds processors that have their $needChange$ flag to True to the $needChgSet$ (line 10). Line 11 resets the status to OK if there is no support to change the view, and it copes with arbitrary changes to the status. The algorithm then moves from status OK to noService if there are more than $3f + 1$ processors in $needChgSet$ (line 12). If the processor sees $4f+1$ processors in $needChgSet$ it moves to status $vChange$ and calls the $veiwChange()$ interface function of the view establishment module to initiate the view change procedure to the next view (line 13). While a view does not change, it holds a set of processors $needChgSet$, and it adds to this any processors that report having seen $suspected()$. While in status $vChange$ and the view not having changed, the algorithm renews its request to the view establishment module (line 14). Line 15 captures the case where the view change has finished and the local variables are set to their defaults. Lines 17 and 18 implement the communication between the processors.

Correctness. The convergence proof follows from careful observation of the algorithm suggesting that stale information in variables and data links is removed after $O(1)$ a.r. After this, there is no arbitrary initiation of a view change. We prove closure by suggesting that if the primary's activity (i.e., messages) lead $2f + 1$ processors to mutually reach to the noService status, then the system will move to a view change, and within the $O(n)$ a.r. required by the view establishment module, a new primary is installed.

Theorem 3.5 (Closure). *Consider an execution R where the primary p_j's activity leads to the encoding of a view change. Within $O(n)$ rounds, a new primary p_{j+1} is installed, and the view change mechanism returns to monitoring.*

4   Extensions

4.1   Relaxing the Assumption of Mal-Free Execution in View Establishment

Recall that the convergence proof for view establishment (Sect. 3.1) assumed mal-free executions (that is, a view is guaranteed to be established in the absence of Byzantine behavior). We now discuss how we can relax this assumption.

Tolerating Malicious Behavior During View Establishment. As expected, to be able to establish a view in the presence of both malicious behavior *and* stale information, stronger assumptions are required (cf. [3]).

Such a liveness assumption would be *k-admissibility:* Assume a ratio k between the fastest token round trips in a data link to the fastest non-faulty processors and the slowest non-faulty processors. In other words, under

k-admissibility, a ratio of k is assumed between the fastest and the slowest non-faulty processor (when a fast correct processor exchanges k tokens, then at least one token is exchanged by the slowest correct processor). We can consider an *event-driven FD* implemented as follows: When p_i broadcasts a message over the data-link token to all its neighbors, p_i resets a counter for each attached link and starts counting the number of tokens arriving back to it, until at least $n - 2f$ distinct counters reach a value that is at least k. Then, under k-admissibility, p_i can safely assume that values from all non-faulty processors (i.e., $n - f$) arrived.

In other words, given k, the above simple FD can be tuned to ensure correct processors get replies from all correct processors. (In a synchronous system k would be small, and as asynchrony increases, k would need to increase.) Also, if a solution takes decisions based on a threshold (fraction) of processors (as in our solution), then k can be reduced, hence making the liveness assumption requiring "less synchrony". In some sense, this FD can be considered as an *on-demand* failure detector than can be tuned based on the "level" of synchrony of the system. The introduction of this FD, allows, for the first time, to avoid the constant overhead of background bookkeeping. One can see that our view establishment convergence proofs hold (i.e., establishing a view under both Byzantine behavior and stale information is still guaranteed) in k-admissible executions, if our view establishment algorithm of Sect. 3.1 is equipped with the aforementioned FD (parameterized with k).

4.2 Optimality

Our solution presented in Sect. 3 assumes that at most $f = (n - 1)/5$ of the processors are faulty. As discussed there, to establish a view, we require the agreement of $n - f$ processors (i.e., $4f + 1$), to adopt an established view and correct processors to sustain it, $\max\{n - 2f, n/2\}$ processors must support it (i.e., $3f + 1$), whereas in order to make decisions regarding the progress of the replication (servicable view) we need $\lceil (n - f)/2 \rceil + f$ processors (i.e., $3f + 1$) to agree (strong majority of correct processors). This is also the threshold we need for a primary to be suspected, and $n - f$ to proceed to change the view.

We have parameterized our solution so that these thresholds can be adjusted for different ratios between faulty and correct processors (as explained, we used $n = 5f + 1$ as it makes the presentation easier to follow). In particular, for the *optimal resilience* (cf. [1]) ratio of $f' = (n - 1)/3$, and going over the correctness proofs of our solution, it follows that establishing a view will require agreement of $2f' + 1$ processors, adopting a view will require the support of $\frac{3}{2}f' + 1$ processors, whereas serviceabilty requires $2f' + 1$ processors to agree (the same threshold required in [6]).

5   Conclusion

We presented the first self-stabilizing BFT algorithm based on failure detection to provide liveness. The approach is modular and allows for suggested extensions and to achieving optimal resilience. The result paves the way towards self-stabilizing distributed blockchain system infrastructure.

References

1. Abraham, I., Malkhi, D.: The blockchain consensus layer and BFT. Bull. EATCS **3**(123), 74–95 (2017)
2. Baldoni, R., Hélary, J., Raynal, M., Tanguy, L.: Consensus in Byzantine asynchronous systems. J. Discrete Algorithms **1**(2), 185–210 (2003)
3. Beauquier, J., Kekkonen-Moneta, S.: Fault-tolerance and self-stabilization: impossibility results and solutions using self-stabilizing failure detectors. Int. J. Syst. Sci. **28**(11), 1177–1187 (1997)
4. Binun, A., Coupaye, T., Dolev, S., Kassi-Lahlou, M., Lacoste, M., Palesandro, A., Yagel, R., Yankulin, L.: Self-stabilizing Byzantine-tolerant distributed replicated state machine. In: Bonakdarpour, B., Petit, F. (eds.) SSS 2016. LNCS, vol. 10083, pp. 36–53. Springer, Cham (2016). https://doi.org/10.1007/978-3-319-49259-9_4
5. Blanchard, P., Dolev, S., Beauquier, J., Delaët, S.: Practically self-stabilizing Paxos replicated state-machine. In: Noubir, G., Raynal, M. (eds.) NETYS 2014. LNCS, vol. 8593, pp. 99–121. Springer, Cham (2014). https://doi.org/10.1007/978-3-319-09581-3_8
6. Castro, M., Liskov, B.: Practical Byzantine fault tolerance. In: Proceedings of the OSDI 1999, pp. 173–186 (1999)
7. Dolev, S.: Self-stabilization. The MIT Press, Cambridge (2000)
8. Dolev, S., Eldefrawy, K., Garay, J., Kumaramangalam, M.V., Ostrovsky, R., Yung, M.: Brief announcement: secure self-stabilizing computation. In: Proceedings of the PODC 2017, pp. 415–417 (2017)
9. Dolev, S., Hanemann, A., Schiller, E.M., Sharma, S.: Self-stabilizing end-to-end communication in (bounded capacity, omitting, duplicating and non-FIFO) dynamic networks. In: Richa, A.W., Scheideler, C. (eds.) SSS 2012. LNCS, vol. 7596, pp. 133–147. Springer, Heidelberg (2012). https://doi.org/10.1007/978-3-642-33536-5_14
10. Dolev, S., Welch, J.L.: Self-stabilizing clock synchronization in the presence of Byzantine faults. J. ACM **51**(5), 780–799 (2004)
11. Doudou, A., Garbinato, B., Guerraoui, R., Schiper, A.: Muteness failure detectors: specification and implementation. In: Hlavička, J., Maehle, E., Pataricza, A. (eds.) EDCC 1999. LNCS, vol. 1667, pp. 71–87. Springer, Heidelberg (1999). https://doi.org/10.1007/3-540-48254-7_7
12. Fischer, M.J., Lynch, N.A., Paterson, M.S.: Impossibility of distributed consensus with one faulty process. J. ACM **32**(2), 374–382 (1985)
13. Lamport, L.: Time, clocks, and the ordering of events in a distributed system. Commun. ACM **21**(7), 558–565 (1978)
14. Lamport, L., Shostak, R., Pease, M.: The Byzantine generals problem. ACM Trans. Program. Lang. Syst. **4**(3), 382–401 (1982)
15. Mostéfaoui, A., Mourgaya, E., Raynal, M.: Asynchronous implementation of failure detectors. In: Proceedings of DSN 2003, pp. 351–360 (2003)

Brief Announcement: Providing End-to-End Secure Communication in Low-Power Wide Area Networks

Ioannis Chatzigiannakis[1(✉)], Vasiliki Liagkou[2], and Paul G. Spirakis[3,4]

[1] Sapienza University of Rome, Rome, Italy
ichatz@dis.uniroma1.it
[2] Computer Technology Institute and Press "Diophantus", Patras, Greece
liagkou@cti.gr
[3] Computer Science Department, University of Liverpool, Liverpool, UK
p.spirakis@liverpool.ac.uk
[4] Computer Engineering and Informatics Department,
Patras University, Patras, Greece

Abstract. Recent technologies for low-rate, long-range transmission in unlicensed sub-GHz frequency bands enables the realization of Long-range Wide Area Network. Despite the rapid uptake of LPWANs, security concerns arising from the open architecture and usage of the unlicensed band are also growing. While the current LPWAN deployments include basic techniques to deal with end-to-end encryption there are specific security issues that arise due to the overall architecture and protocol layer design. In this paper, a new scheme to establish end-to-end secure communication in long-range IoT deployments is introduced. The advantages over the existing approaches and architectural design are presented in the context of typical smart cities application scenarios.

1 Introduction to Security Issues and Vulnerabilities in LPWAN

In the past few years, the approach of exploiting sub-GHz was proposed in order to increase the transmission range of nodes by trading-off data transmission rate while keeping power consumption at low levels [1]. This so-called Low-Power Wide Area Networks (LPWANs) allow IoT devices to connect to Concentrators (also called a *collector*) over distances in the range of several kilometres. Concentrators forward data received from the IoT devices to a Network Server (over for example Ethernet or 3G/4G/5G) that manages all the decoding of the packets and handles redundant transmissions. Overall, LPWANs are considered promising candidates for IoT applications, since they allow *high energy autonomy* of the connected devices, *low device and deployment costs, high coverage capabilities* and support *large number of devices* [2].

Recently some technical papers concentrated on the security vulnerabilities in LPWANs providing alternative solutions for the used cryptographic

© Springer International Publishing AG, part of Springer Nature 2018
I. Dinur et al. (Eds.): CSCML 2018, LNCS 10879, pp. 101–104, 2018.
https://doi.org/10.1007/978-3-319-94147-9_8

primitives [3], focus on application server vulnerabilities [4] or introduce alternative key management [5].

In LPWANs the encryption of the payload is by default enabled in every transmission. The data frame of an end-node has a 32-bit identifier, a 7-bit network identifier and a 25-bit network address and the maximum payload is 250 Bytes. Since IoT devices are not assigned to a specific concentrator, the data frames do not include any concentrator identifier. In this way, it is possible for anyone to receive the encrypted data packets. In order to prevent from replaying packets, a frame counter is used both for upstream and downstream messages which will block a transmission from being sent more than once.

Two different 128-bit AES keys are used for a two-step message chain for both upstream and downstream message exchanges. In the first step, an Application Session Key (AppSKey) is used to encrypt the data frame between the IoT device and the application server. In the second step, a Network Session Key (NwkSKey) is used to verify the authenticity of the nodes. The data frame exchanged between the IoT device and the Network server is encrypted with the NwkSKey. Therefore, each message is encrypted by using the XOR operation with the corresponding key.

Currently in LPWAN there are specific security issues that arise due to the overall architecture and protocol layer design:

Keys Storage. Keys need to be safely stored by the IoT devices, the Network Server and the Application Server. Moreover in LPWAN network the IoT device is placed to an unprotected external or internal environment for very long time thus its impractical and costly to increase the physical security level of the IoT devices.

Symmetric Encryption Factors. AES is operating in counter mode (CTR) and not in electronic codebook (ECB) mode. In this mode of operation, IoT devices generate cyphertexts which are output of the XOR procedure on the string that contains a counter, the AppSkey and the plaintext. As a result, encryptions are vulnerable to chosen cyphertext attack since if an attacker changes the payload data she can figure out which bit position in the encrypted payload corresponds to the same bit position in the plaintext. Major security flaw.

Authentication. The Network Server and the intermediate concentrator (or an attacker on the intermediate network) are in a position to modify the encrypted payload without the Application server being able to notice the change. If an adversary could posses the session key, then he can generate a LoRaWAN message that will pass the signature checking procedure at the network server.

Compromised IoT Device. LPWANs are suitable for large deployments of battery operated static IoT devices that remain for long periods of times (in many cases spanning several years) in semi-controlled environments or even uncontrolled areas.

Untrusted Concentrators. Traffic passing through this point can be easily recorded and even manipulated.

2   An End-to-End Secure Communication Scheme

The LoRa LPWAN architecture is extended by introducing the so-called *Median Server* that complements the functionality of the Network Server and Application Server by taking over the role of the *Registration Authority* of the system both for IoT devices and concentrators. A PKI Credential Authority (CA) is introduced to ensure that only authenticated IoT devices interact with the system and connect only to an authenticated concentrator that issued their certificates.

The overall security is further reinforced by establishing a VPN network for the communication between the concentrators, the median server and the network server. The VPN connections use SSL sessions with bidirectional authentication (i.e., each side must present its own certificate). A block cypher and fingerprint (hash value) for encrypting/decrypting packets are activated along with the HMAC construction to authenticate them. In this way, passive attacks (packet sniffing, eavesdropping) are eliminated. However, even if packet encryption is unbreakable, it does not prevent active attackers to insert into a communication channel and add, modify or delete packets. Active attacks are thwarted by embedding Device Identifier (DevEUI) (timestamps) on packets and make IoT devices able to keep track of timestamps in order to make sure that they never accept a packet with the same timestamp twice.

Furthermore, the critical data like symmetric keys, private-public keys and IoT device credentials are protected using a *HMAC* before they are stored into the Network server and Application server to further assure their integrity. In particular, the HMAC-MD5 is used within the Application server on IoT device credentials (username, password). In this way, critical data disclosure is prevented in situations like database server thefts or unscrupulous administrators.

In terms of *preventing modifications on payload data*, a MAC is used to authenticate transmitted payload data against any modification. The Application server verifies that the message was received from an authenticated IoT device and subsequently decrypts it and locks it in order to detect possible post-modifications and illicit manipulations.

A fundamental requirement for the proposed model is to strictly link IoT device tasks with system's application data. The proposed architecture is associated with a workflow mechanism that guarantees data transmission thought heterogenous parties whereas supervising user's device interaction. In LPWAN the data rate transitions between the IoT device and LPWAN infrastructure is low and makes the security synchronization interaction mechanisms impractical thus the flow control determines a certain lifecycle for payload application data, from its insertion into the LPWAN till the time that is ready to be stored and utilized by the application server. IoT device payload passes through certain phases introduced by the mechanism. Each phase has its own predefined tasks committed by the user. The mechanism introduces associations between phases (i.e., each phase depends on the successful completion of its previous one) and executes them in a linear fashion (1st, 2nd, ...), making a discrete workflow for each payload.

References

1. Centenaro, M., Vangelista, L., Zanella, A., Zorzi, M.: Long-range communications in unlicensed bands: the rising stars in the IoT and smart city scenarios. IEEE Wirel. Commun. **23**, October 2016
2. Chatzigiannakis, I., Vitaletti, A., Pyrgelis, A.: A privacy-preserving smart parking system using an IoT elliptic curve based security platform. Comput. Commun. **89–90**, 165–177 (2016)
3. Kim, J., Song, J.: A simple and efficient replay attack prevention scheme for LoRaWAN. In: ICCNS (2017)
4. Michorius, J.: What's mine is not yours: Lora network and privacy of data on publishing devices (2016)
5. Naoui, S., Elhdhili, M.E., Saidane, L.A.: Enhancing the security of the IoT LoRaWAN architecture. In: 2016 International Conference on Performance Evaluation and Modeling in Wired and Wireless Networks (PEMWN), pp. 1–7, November 2016

Privacy via Maintaining Small Similitude Data for Big Data Statistical Representation

Philip Derbeko$^{(\boxtimes)}$, Shlomi Dolev, and Ehud Gudes

Computer Science Department, Ben-Gurion University of the Negev,
Beer-Sheva, Israel
philip.derbeko@gmail.com, {dolev,ehud}@cs.bgu.ac.il

Abstract. Despite its attractiveness, Big Data oftentimes is hard, slow and expensive to handle due to its size. Moreover, as the amount of collected data grows, individual privacy raises more and more concerns: "what do they know about me?" Different algorithms were suggested to enable privacy-preserving data release with the current de-facto standard differential privacy. However, the processing time of keeping the data private is inhibiting and currently not practical for every day use. Combined with the continuously growing data collection, the solution is not seen on a horizon.

In this research, we suggest replacing the Big Data with a much smaller similitude model. The model "resembles" the data with respect to a class of query. The user defines the maximum acceptable error and privacy requirements ahead of the query execution. Those requirements define the minimal size of the similitude model. The suggested method is demonstrated by using a wavelet transform and then by pruning the tree according to both the data reduction and the privacy requirements. We propose methods of combining the noise required for privacy preservation with noise of similitude model, that allow us to decrease the amount of added noise thus, improving the utilization of the method.

Keywords: Big Data · Privacy · Wavelets · Differential privacy
Similitude model

1 Introduction

Data privacy is an important requirement and continues to grow in importance and public awareness as more data is gathered. A current leading method of privacy is differential privacy [9]. The idea of differential privacy is to hide the existence of a single record in the dataset, such that the adversary who queries the data cannot tell with high probability whether any given record is present in the database or not. A common method of providing differential privacy is adding a random noise to the result of the query or to the data itself. The level of the added noise should be large enough to hide the absence or presence of any

© Springer International Publishing AG, part of Springer Nature 2018
I. Dinur et al. (Eds.): CSCML 2018, LNCS 10879, pp. 105–119, 2018.
https://doi.org/10.1007/978-3-319-94147-9_9

record and it is related to the sensitivity of the function [10], i.e. how much does the function change if the database differs by a single record.

Data reduction is a technique that is used to decrease the amount of data required for calculations. Obviously, the goal of data reduction is to minimize the size while at the same time maximizing the utilization of the saved data. For instance, in streaming data, the goal of is to keep a small synopsis of the data to answer a previously defined query. The synopsis uses a limited-sized memory to approximate potentially unlimited streaming data [11].

There is a substantial difference between data compression and reduction. While both techniques reduce the size of the saved data, the difference is in the amount of work required to answer a query. Compressed methods first require to inflate all or part of the data to its original size. This process has a computation complexity proportional to the dataset size. On the contrary, the data reduction technique builds a similitude model that represents the dataset and uses it to answer given queries. Thus, queries are answered much faster with processing time proportional to the similitude model size. On the other hand, data reduction introduces errors into the query results that are not present in lossless compression methods.

The wavelet transformation decomposes the data into a set of decreasing coefficients. If only the large coefficients are retained, the data is represented using a limited and smaller transformation model. Different number of techniques were developed to choose which coefficients to retain depending on the error measure. For instance, [22] shows that retaining the largest coefficients minimizes Mean Square Error (MSE), while using L_∞ error measure requires different techniques.

Similarly to differential privacy, data reduction introduces an error in query results. While differential privacy adds a random noise to hide the presence of specific records, the data reduction techniques introduce noise as a result of dropping part of the records. Thus, in a naïve privacy-preserving data release of similitude model, the noise is added twice which reduces utilization. The idea of the paper it to use one source of error complimentary for the second one, and thus, to improve the utilization of the similitude model.

The similitude model concept, a demonstration of a practical technique to implement the data reduction concept and a method of reducing the amount of required noise for privacy-preserving data release of similitude model are the main themes of the paper. The rest of the paper is structured as follows. Section 2 defines a problem and presents the basic terminology. Section 3 overviews the relevant existing research. Our basic approach is presented in Sect. 4 and discusses how to handle various types of queries. Lastly, the paper is concluded in Sect. 5.

2   Problem Definition

The input is a set of data points $X = x_1, x_2, \ldots x_n$ where the data points are numerical and belong to a specific range: $x_i \in [a, b]$. Our goal is to replace the full dataset with a reduced model of the data. The reduced model should represent the entire dataset and it is used to answer statistical queries. The model is called the similitude model of the data.

Since every data reduction loses information, the goal is to build a model such that the error of data representation is minimized. In addition, the privacy of the data should be protected, thus the similitude model should ensure privacy-preserving data release.

Privacy requirements are defined using ϵ-differential privacy, i.e. for any two data sets X and Y that differ in a single record

$$Pr(Q(x) = O) \leq e^{\epsilon} \cdot Pr(Q(y) = O), \tag{1}$$

where O is an output of queries Q.

Our contributions in this paper are:

- The observation that a similitude model is sufficient to answer statistical queries, which saves space and processing time. At the same time, keeping only similitude model also reduces the risk of (non statistical) information leakage.
- The idea of combination of data reduction (using wavelet transform) with differential privacy to reduce overall added noise.
- A specific and provable method to combine the above mentioned techniques.

3   Related Work

As our work relates to a number of fields, this section briefly covers relevant previous work.

Wavelets are a good fit for data reduction [13, 22]. While the total number of wavelet coefficients equals to the size of the data, the larger coefficients are making the largest impact on the accuracy of the transform while all other coefficients are approximated with zero. Thus, it is possible to keep only a limited number of coefficient and effectively reduce the data size while keeping the error small.

Another data reduction (compression) technique is the Fourier transform. The usage of Fourier transform for compression of time-series data is initiated in [21]. This approach was extended to histograms by [3] where histograms were presented as time-series data. The data was first lossy compressed and then sanitized. The method takes advantage of the reduced sensitivity of compressed data. The compression is done by using Fourier transform, suited for real-valued data, and it shows utility improvement by a factor of 10 compared to the standard method where noise is added to each entry in the frequency matrix [10]. [21] presented a similar idea for time-series data distributed across multiple locations. The noise is added to the first k coefficients of Fourier Perturbation Algorithm and in many practical cases $k << n$, where n is the size of the dataset.

An approach similar in spirit to our work, i.e. building a smaller representation of the data, was presented in [28]. The work suggests to deal with a high-dimensionality of the data by constructing a low-dimensional

Bayesian network, which models correlations between data attributes. The method is shown to be effective for private data release and achieves good utilization/accuracy as compared to previous methods, such as addition of Laplace noise to each dataset row [10] and adding noise to the result of Fourier transform [5].

A significant research considered non-interactive, private data release for range queries [3,17,20,27]. Privacy-preserving data release research focused mainly on the minimization of the added noise, regardless of the amount of the released data. In [27] it is shown that the amount of added noise can be reduced, which increases the utility of the method, by performing wavelet transformation, adding random noise to the wavelet coefficients, and then mapping back to the data. Since the wavelet coefficients are a linear combination of the data, the noise added to them is shared, which allows the reduction of the amount of added noise by polylogarithmic factor. A similar method is privacy-preserving release of histograms [17]. Hierarchical methods for differentially private histograms were considered in [20]. The paper considered one-dimensional and multi-dimensional histograms with different privacy budget allocations per level.

The requirements of differential privacy can be met by releasing a *synthetic data* that has similar statistical properties as the real dataset [6,7]. Unfortunately, in the worst case scenario, those algorithms require a very high preprocessing time, linear in the number of possible kinds of records which is exponential in a number of data attributes [23,24]. Despite the theoretical results, there have been a number of practical attempts of differential data release based on synthetic data, see below.

An algorithm that generates synthetic data was described in [15]. The algorithm starts with a crude approximation of the data, which is improved iteratively using Multiplicative Weights approach [16]. In each iteration, the algorithm chooses the worst query, i.e. a query with the biggest mistake with respect to the original data. The algorithm (MWEM) is based on the combination of Multiplicative Weights update rule and Exponential Mechanism that was first considered in [15]. Improvements are made by finding a query whose answer on a real data differs considerably from its answer on an approximate data. The query is then incorporated into data approximation with adaptive weights mechanism. Even though the algorithm performs quite well from computational point of view, it still depends on the size of the query domain, which grows exponentially with the number of data attributes.

To deal with high-dimensional data, a variant of the algorithm was proposed in [12]. The method, DualQuery, is *dual* to MWEM in a sense that it maintains distribution over queries instead of distribution over data. DualQuery also uses MW approach to update the maintained distribution. While experimentally, MWEM outperforms the DualQuery algorithm for low-dimension data, DualQuery complements MWEM algorithm by allowing a private analysis of high-dimensional data.

MWEM and DualQuery both output synthetic dataset from the same domain as an input data. This is a major difference from our proposed method of building a similitude model, which is not confined to be from the same domain.

Opposite to other methods, our suggested method works both on single-node queries, adjacent range queries and general multi-range queries. Another distinct requirement is to keep reduced data, thus, an error is added prior to the addition of a privacy-driven random noise.

Despite an extensive ongoing research of privacy-preserving data release, we believe that our method uniquely answers some specific needs. While previous research compressed or replaced an original data with a model, most of those works did not consider privacy-preserving release. On the other hand, privacy-preserving data release does not consider a reduction of the amount of data. Even methods that generate synthetic data, do not usually generate a smaller and data representative model. Our proposal of using similitude model while ensuring privacy of the data combines the above approaches in a practical and efficient way.

4   Privacy Preserving Data Reduction

This section discusses the privacy preserving data reduction that uses wavelet transformation. We start with a short explanation of Haar wavelet transform that is used in the section. As the presented techniques are not specific to the specific type of transform, we use Haar transform both because it is intuitive to understand and fast to compute.

4.1   Haar Wavelet Transform

Wavelet transform is a method of decomposing data into a hierarchical set of functions. Each function approximately describes the data. The approximation changes from coarse to fine-grained as the hierarchy advances. We use Haar wavelet where at each step the data is replaced with an "average" value and the deltas from the average using convolution with high-pass and low-pass filters. The delta values are the coefficients of the wavelet. The process then is repeated for deltas of high-pass and for deltas of low-pass until the entire data is replaced with wavelet coefficients.

In the following examples, we use the Adult dataset from UCI [18] as an example dataset, data processing packages Pandas [1], PyWavelet [2] for the wavelet processing and Graphviz [4] for drawing. The dataset contains 32,562 records with partial missing values from the census bureau, where each record has data about a different person. As the data itself is only used as an example, the missing values are cleaned by simply dropping the rows with any such values resulting in 30,162 records. For the simplicity, the person's age values (continuous) were used as a data dimension for the wavelet transform.

Figure 1 presents an example of wavelet decomposition of the first 16 values from the dataset. Only 16 values were chosen to show the wavelet, as larger trees

Fig. 1. Wavelet Coefficients tree of the 16 first entries of the Adult dataset. Age is used as a wavelet transform dimension. The root value is the approximation coefficient and the values on the edges are detail coefficients.

are harder to understand. The root value is the approximation coefficient and the values on the edges are detail coefficients. Only those values are saved in the wavelet transform. The values in other nodes are reconstructed from the root and detail coefficients.

The nodes are numbered from the top to bottom and from the left to right. Thus, the root node is v_1, its left child is v_2 and its right child is v_3, etc. The tree is constructed bottom-up, where each couple of values is convolved with high and low filters. The result of convolution is the value of the node at the next level (c_i) and the difference of the data from the convolution results is kept as a coefficient. Notice that the difference is the same for the nodes with a different sign, thus, it is enough to keep only the absolute value: ϕ_i for node v_i. The process continues until a single value is left, which is the root coefficient: c_1. In order to reconstruct the values, the process is reversed and starts from the top. Each pair of values is reconstructed with its difference applied, with a relative sign, to the root node. In the tree-like hierarchical representation of the data, each level provides an approximation of the data with improving error, i.e. an approximation error at the top level is higher and it gets refined with the node depth.

The amount of wavelet coefficients equals to the amount of the data points, n in our case. Thus, the wavelet transform in itself does not result in data reduction. To reduce the amount of saved data, only a part of the wavelet coefficients have to be kept. By definition of the transform, each subsequent level of the coefficients have smaller and smaller contribution to the accuracy of the estimation. This property of wavelet coefficients was used to approximate the answering of point and range queries in [8,14,19]. We use a similar technique while connecting the decision on which coefficients to keep to the differential privacy requirements.

As the differential privacy mechanism is built for statistical queries, the rest of the paper deals with counting queries. Range (count) queries are an extremely common statistical and data analysis tool. Given a range of values, those queries count the number of items with values in this range. The common way of handling range queries is to build a frequency matrix (F) or a histogram, where each value, or a range of values in case of a histogram, has a separate cell that contains the number of appearance of this value.

For the sake of clarity, we have limited the number of bins of frequency matrix to 16. Table 1 shows a frequency matrix for the ages column of the entire Adult dataset. As the queries are performed on F, the wavelet transform is also performed on the frequency matrix.

Table 1. Frequency matrix of the Adult dataset

Age range	Count	Age range	Count	Age range	Count
17–21	1998	21–27	4415	27–31	3184
31–37	4993	37–41	3170	41–47	4374
47–51	2336	51–57	2597	57–61	1289
61–67	1077	67–71	345	71–77	244
77–81	65	81–87	37	87–90	38

4.2 Single-Node Queries: Pruning with Differential Privacy in Mind

The first type of queries that we consider is a single-node query, i.e. the counting queries that are answered by a single node of a wavelet transform.

At first, a frequency matrix F (see table above) is build from the input data X. Then it is transformed by a wavelet to get a set of coefficients C. The data reduction is then performed by pruning of the tree at a tree level according to the required error. The value of the coefficients is highest at the root and then it decreases towards the leaves. For node v the pruning is replacing its direct children by an average value and keeping the range of the node. Thus, the error of pruning is:

$$e(v) = \sum_{i=1}^{l} E(v_i), \tag{2}$$

where $v_1, v_2, \ldots v_l$ are children nodes of v.

Every node v is pruned, i.e. replaced with its average value if the $e(v)$ introduced by the pruning is less than λ, which is an error threshold required by differential privacy. As opposite to [25] for sparse data and [26] for dense data, our threshold is derived from privacy and not from storage requirements.

Differential privacy requires to add a random noise to "hide" the presence or absence of any individual record. According to [10], adding noise according to Laplace distribution, with standard deviation of $S(f)/\epsilon$ is enough to satisfy differential privacy depends. The density function of Laplace distribution with a 0 mean is $f(x) = \frac{1}{2\lambda}e^{-|x|/\lambda}$ and λ being a standard deviation or a scale parameter. $S(f)$ is defined to be the sensitivity of a function f - as a smallest number that is larger than the maximum changes in the function value as a result of a deletion of a single input record:

$$\|f(x) - f(x')\| \leq S(f), \tag{3}$$

where x and x' differ in a single entry and $S(f)$ is the sensitivity of f. In the case of a single-range query over F, the sensitivity of the function is 1, and thus, adding noise according to $Lap(1/\epsilon)$ will ensure differential privacy.

As the pruning of the wavelet transform tree introduces noise to the data, our idea is to use the error introduced by the data reduction to reduce the amount of added noise in order to preserve differential privacy. The smaller the sensitivity of the function, the less noise is added and the higher is the utilization of the released data. Data reduction in wavelet trees prunes a sub-tree, or in other words, replaces it with the average values. The impact of a single value on an average of a pruned sub-tree is smaller than the impact without pruning. Moreover, the larger is the pruned tree, the smaller is the impact (sensitivity) and the smaller is the added noise.

Let $T(v_i)$ be a sub-tree of a given node v_i and $t = |T(v_i)|$ be the size of the sub-tree. If $T(v_i)$ is pruned, the coefficient of v_i is replaced by an average value: $\widetilde{\phi}_i$. The sensitivity of the nodes in $T(v_i)$ is $S(T(v_i)) = \frac{max_{v_j \in T(v_i)}|V(v_j)-V(\widetilde{v_j})|}{t}$, where $V()$ is the re-constructed value of a node in a wavelet transform tree. In case of counting queries the sensitivity reduces to $S(T(v_i)) = \frac{1}{t}$. The total added error for a pruned node v_i then will be:

$$E(v_i) = Lap(S(T(v_i))/\epsilon) + \sum_{v_j \in T(v_i)} |V(v_j) - V(\widetilde{\phi_j})|. \tag{4}$$

where the data reduction savings are $DR(v_i) = t \cdot |v_i|$.

Now, we can define the pruning algorithm.

1. Iterate over all leaf nodes in the transform tree. For each node v_i:
2. Calculate $T(v_i)$, $S(T(v_i))$, $DR(v_i)$ and $E(v_i)$.
3. Draw X_i according to Laplace distribution: $X_i \sim Lap(S(T(v_i))/\epsilon)$.
4. If the sum of the pruning error and X_i is larger then required error, then the node v_i is pruned [27].

Notice that the noise is added to all leaf nodes. The added noise is either pruning noise together with a random noise or only a random noise.

Lemma 1. *For a data set X, the described above data reduction preserves ϵ-differential privacy.*

Proof: The proof is trivial, as the random noise is added according to the sensitivity function as in [10], the resulting tree is differentially private.

4.3   Simulation Results

Figure 2 shows the tree of a wavelet transformation of a 16-bins frequency matrix of the Adult dataset with $\epsilon = 0.1$. We chose to show 16-bins matrix in the paper for the sake of clarity of the graph. In reality the amount of bins will be much larger. This is the same frequency matrix as shown in Table 1 with added empty range of ages 0 to 17. Notice that the frequency matrix in itself provides a compression of the data, as items in the same bin are replaced by an average value that represents the bin. Each node shows both the pruning error of its sub-tree (PErr) and the amount of required ransom noise (RNoise). In the paper we use L_1 error measure. The combined error in this case is clearly dominated by the pruning error, however, keep in mind that this 16-bins frequency matrix represents a dataset of 30,162 records. As each bin contains many records, pruning of any node results in a big error.

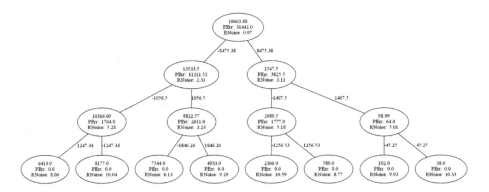

Fig. 2. Wavelet transform tree of 16-bins frequency matrix of the Adult dataset. Each node shows the wavelet coefficient, L_1 pruning error (PErr) and privacy-preserving random noise (RNoise with $\epsilon = 0.1$).

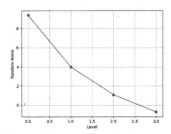

Fig. 3. Average privacy-preserving noise per level of the wavelet transform tree. The root level is marked as 3 and the leaves level is 0 with $\epsilon = 0.1$

Figure 3 shows the amount of the added noise per wavelet transform level. The shown amount of the noise is the average over the nodes of that level. The graph shows the success of our method of noise reduction as the amount of noise reduces considerably towards the root of the tree.

Figure 4 compares the amount of added noise by MWEM algorithm with the amount of added noise by our similitude model. It can be seen that both algorithms has significant error, again due to the size of the dataset and the

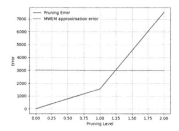

Fig. 4. Comparison between Pruning error of wavelet similitude model and MWEM algorithm.

fact that frequency matrix is already compressed. Similitude model presents a better error than MWEM even when the first level is pruned. After that the error increases dramatically. This example shows an addition weak point of MWEM algorithm. As the algorithm iterates over the domain of possible queries, the size of the domain is significant for convergence of the algorithm. However, in a case of frequency matrix, as can be seen in the Adult dataset, the size of the matrix is small and thus, also the number of single-node queries. Thus, MWEM algorithm does not have enough data to converge and has a relatively big error.

The following section generalizes the idea of combining data reduction noise with random noise for differential privacy for multi-node range queries.

4.4 Multi-range Counting Queries

In this section we first consider a single range query that covers a number of adjacent nodes, and then we extend the results to multi-node range queries that count the number of records in a multiple non-adjacent ranges. The major distinction of queries, which are answered by multiple nodes from single-node range queries, is that the error of the query is the sum of errors of all participating nodes. This decreases the utility of such queries and also makes the error dependent on the number of participating nodes, which is not a desired characteristic.

For example, consider a range that covers k leaf node in the wavelet coefficient tree. Performing a sum over their values will add noise of $\mathrm{Lap}(1/\epsilon) \cdot k$ to the answer, considerably reducing the utility of the data release. For the sake of simplicity, in the following we assume that queries relate to entire nodes or pruned nodes sub-trees. Cases where the query relates to a part of the pruned sub-trees are handled easily as the added error is the same for all nodes in the sub-tree. However, they also make the equation less clear and thus, were not included in this paper.

Range Queries. Consider a range query q and let $V(q) = v_k, v_{k+1} \ldots v_l$ be a set of nodes that are summed to answer q. The error added to the answer of the query q is

$$E(q) = \sum_{i=k}^{l} E(v_i) = \sum_{i=k}^{l} Lap(S(T(v_i))/\epsilon) + \sum_{v_j \in T(v_i)} |V(v_j) - V(\widetilde{phi_j})|.$$

Denote the noise added to a node v_i to preserve privacy as $n(v_i)$. Clearly, the noise increases with the range of the query and more records participate in the query. Ultimately, the amount of added noise should not be directly related to the number of nodes that participate in a given query. To achieve that, the noise added to individual records is increased. The increase is then traded off to probabilistically decreased noise in multi-range queries.

Noise Added to a Range Query. Numbering leaf nodes from the beginning of the data to the end (in Fig. 1 from left to right), define E_v and O_v to be sets of nodes with even and odd indices accordingly, i.e. $V(q) = E_q \cup O_q$. Node $v_i \in E_v$ will be added noise according to the following simple rule:

$$n(v_i) = Lap(1/\epsilon) + \begin{cases} 1/\epsilon & \text{for } v_i \in E_v \\ -1/\epsilon & \text{for } v_i \in O_v \end{cases}$$

In other words, the mean of the noise added to even nodes is $1/\epsilon$, and to odd nodes is $-1/\epsilon$. The basis of the idea is as multi-range counting query sums the errors, the alternating noise signs will cancel one another with high enough probability. First, we show that the summation of added random noise for differential privacy is small, and later we add the noise of data reduction.

The amount of added random noise for a query q then is

$$\sum_{i \in V(q)} n(v_i) = \sum_{i \in E_q} n(v_i) + \sum_{j \in O_q} n(v_j). \tag{5}$$

The right side of the equation is a sum of Laplace distributed, independent random variables.

Since the added noise value is not bounded and independent for every node, we utilize Chernoff's inequality to bound the sum of the added random noise to the nodes participating in a query. Lets assume that $k = |V(q)|$ participate in the query.

$$P(|S_k - E(S_k)| \geq t) \leq \frac{Var(S_k)}{t^2}. \tag{6}$$

Substituting the value of the variance of added noise, which is the same for all nodes: $Var(S_i) = 2(1/\epsilon)^2$ into the equation.

$$P(|S_k - E(S_k)| \geq t) \leq \frac{Var(S_k)}{t^2} = \frac{2k(1/\epsilon)^2}{t^2}.$$

The total added noise in Eq. 5 then becomes:

$$P(|\sum_{i \in V(q)} n(v_i) - E(n(v_i))| \geq 2t) = \tag{7}$$

$$P(|\sum_{i \in E_q} n(v_i) + \sum_{j \in O_q} n(v_j) - (E_{v_i \in E_q}(n(v_i)) + E_{v_j \in O_q}(n(v_j)))| \geq 2t)$$

$$\leq P(|\sum_{i \in E_q} n(v_i) - 1/\epsilon| \geq t) + P(|\sum_{j \in O_q} n(v_j) + 1/\epsilon| \geq t)$$

$$\leq \frac{2|E_q|(1/\epsilon)^2}{t^2} + \frac{2|O_q|(1/\epsilon)^2}{t^2} \tag{8}$$

$$= \frac{2}{\epsilon^2 t^2}(|E_q| + |O_q|). \tag{9}$$

Notice that

$$E(n(v_i)) = E_{v_i \in E_q}(n(v_i)) + E_{v_j \in O_q}(n(v_j)) = 1/\epsilon - 1/\epsilon = 0.$$

Therefore, the above inequality becomes:

$$P(|\sum_{i \in V(q)} n(v_i)| \geq 2t) = \leq \frac{2}{\epsilon^2 t^2}(|E_q| + |O_q|). \tag{10}$$

Thus, as the number of nodes participating in a multi-range query increases, the amount of added random noise converges to 0. The above calculations hold for both cases where $|E_q| = |O_q|$ and $|E_q| \neq |O_q|$.

The added noise clearly preserves differential privacy, as the amount of noise added to each node is according to the sensitivity of the function. Moreover, twice the required amount of random noise is added to the pruned nodes. However, this extra noise is later traded to reduce the amount of aggregated random noise in the range queries.

To summarize, in pruned wavelet trees with privacy requirements, the error of a query that is answered by $k = |V(q)|$ nodes is:

$$E(V(q)) = \sum_{V_i \in V(q)} Lap(S(T(v_i))/\epsilon) + \sum_{v_j \in T(v_i)} |V(v_j) - V(\widetilde{phi_j})|$$

$$\leq \frac{2}{\epsilon^2 t^2}(|E_q| + |O_q|) + \sum_{V_i \in V(q)} \sum_{v_j \in T(v_i)} |V(v_j) - V(\widetilde{phi_j})|. \tag{11}$$

4.5   General Multi-range Queries

Now, we generalize the range query mechanism for multi-range counting queries that consider non-adjacent records. We do not distinguish between explicit multi-range queries: count a number of people who are 17–20 or 45–52 years old, or implicit queries, when the query uses a different dimension of the data than that used for wavelet transform. For instance: count a number of people with income over 50 K, which will result in multiple participating ranges when transform was done on the age of the persons. The equation derived for single-range query is extended to a general case by showing that the difference between the number of even and odd nodes in a general case will not be large with a high probability.

Using the same mechanism for noise addition, as for range queries, see Sect. 4.4, leads to the same result in Eq. 10. Assuming that the amount of odd and even nodes is roughly equal, the noise cancellation effect will work. However, while for range queries the following holds: $abs(|E_q| - |O_q|) \leq 1$, in a general case, this is not necessarily true, i.e. it is possible that only even nodes participate in the given equation.

To limit the probability of such event, each query is modeled by a random process of drawing a node without a replacement from the wavelet tree. Define draw of an even node to be a "success". Due to no replacement, the distribution of a number of successes is hyper-geometric with a mean value of $d = k\frac{|E|}{|O|}$, where $|E|$ and $|O|$ are number of even and odd nodes in the tree respectively. As the data is padded to contain even number of nodes before performing wavelet transform, those numbers are equal. Thus, $d = k\frac{|E|}{|O|} = \frac{k}{2}$. Using Chebyshev's inequality and substituting hyper-geometric distribution variance we get:

$$P(|D - E(d)| \geq t) \leq \frac{Var(D)}{t^2} = \frac{k(N-k)}{4(N-1)t^2}, \tag{12}$$

where D is a random variable of d and N is the total number of nodes. The inequality shows that the larger the number of nodes participating in a query, the closer the behavior of a general multi-range query to the behavior of a single-range query.

5   Conclusion

An exploding amount of collected data has led to a variety of data handling techniques. In parallel, a concern for individual privacy and security continues to grow, especially considering frequent data breaches.

In this paper, we present a data reduction technique by building a similitude model of the data and using it to answer queries. The technique strives to capture the essence of a BigData, which is easier to manage and faster to work with than with an entire dataset. We connect data reduction and privacy-preserving data release by adapting a known method of data reduction, using wavelet transform to satisfy differential privacy requirements. In addition, we show a way to reduce the amount of added ransom noise for both single-range and for multi-range queries. The techniques take into consideration the added error of the data reduction in order to reduce the added noise.

A more comprehensive comparison of the technique with other methods, and adaptation of the method for streaming data are the goals of our future research.

Acknowledgement. The research was partially supported by the Rita Altura Trust Chair in Computer Sciences; the Lynne and William Frankel Center for Computer

Science; the Ministry of Foreign Affairs, Italy; the grant from the Ministry of Science, Technology and Space, Israel, and the National Science Council (NSC) of Taiwan; the Ministry of Science, Technology and Space, Infrastructure Research in the Field of Advanced Computing and Cyber Security; and the Israel National Cyber Bureau.

Authors are grateful to John Ullman for the fruitful discussions of the paper ideas and differential privacy.

References

1. Pandas - python data analysis library. http://pandas.pydata.org
2. Pywavelets - wavelet transforms in python. https://github.com/PyWavelets/pywt
3. Ács, G., Castelluccia, C., Chen, R.: Differentially private histogram publishing through lossy compression. In: 2012 IEEE 12th International Conference on Data Mining, pp. 1–10 (2012)
4. AT&T and Contributers. Graphviz - graph visualization software. http://graphviz.org
5. Barak, B., Chaudhuri, K., Dwork, C., Kale, S., McSherry, F., Talwar, K.: Privacy, accuracy, and consistency too: a holistic solution to contingency table release. In: Proceedings of the Twenty-Sixth ACM SIGMOD-SIGACT-SIGART Symposium on Principles of Database Systems, PODS 2007, pp. 273–282. ACM, New York (2007)
6. Blum, A., Dwork, C., Mcsherry, F., Nissim, K.: Practical privacy: the SulQ framework. In: PODS, pp. 128–138. ACM (2005)
7. Blum, A., Ligett, K., Roth, A.: A learning theory approach to non-interactive database privacy. In: Proceedings of the Fortieth Annual ACM Symposium on Theory of Computing, STOC 2008, pp. 609–618. ACM, New York (2008)
8. Chakrabarti, K., Garofalakis, M., Rastogi, R., Shim, K.: Approximate query processing using wavelets. VLDB J. **10**(2–3), 199–223 (2001)
9. Dwork, C.: Differential privacy. In: Bugliesi, M., Preneel, B., Sassone, V., Wegener, I. (eds.) ICALP 2006. LNCS, vol. 4052, pp. 1–12. Springer, Heidelberg (2006). https://doi.org/10.1007/11787006_1
10. Dwork, C., McSherry, F., Nissim, K., Smith, A.: Calibrating noise to sensitivity in private data analysis. In: Halevi, S., Rabin, T. (eds.) TCC 2006. LNCS, vol. 3876, pp. 265–284. Springer, Heidelberg (2006). https://doi.org/10.1007/11681878_14
11. Aggarwal, C.C. (ed.): Data Streams: Models and Algorithms. Springer, New York (2007). https://doi.org/10.1007/978-0-387-47534-9
12. Gaboardi, M., Arias, E.J.G., Hsu, J., Roth, A., Wu, Z.S.: Dual query: practical private query release for high dimensional data. In: Xing, E.P., Jebara, T. (eds.) Proceedings of the 31st International Conference on Machine Learning. Proceedings of Machine Learning Research, vol. 32, pp. 1170–1178. PMLR, Bejing, 22–24 June 2014
13. Garofalakis, M., Kumar, A.: Deterministic wavelet thresholding for maximum-error metrics. In: Proceedings of the Twenty-Third ACM SIGMOD-SIGACT-SIGART Symposium on Principles of Database Systems, PODS 2004, pp. 166–176. ACM, New York (2004)
14. Gilbert, A.C., Kotidis, Y., Muthukrishnan, S., Strauss, M.J.: Optimal and approximate computation of summary statistics for range aggregates. In: Proceedings of the Twentieth ACM SIGMOD-SIGACT-SIGART Symposium on Principles of Database Systems, PODS 2001, pp. 227–236. ACM, New York (2001)

15. Hardt, M., Ligett, K., Mcsherry, F.: A simple and practical algorithm for differentially private data release. In: Pereira, F., Burges, C.J.C., Bottou, L., Weinberger, K.Q. (eds.) Advances in Neural Information Processing Systems 25, pp. 2339–2347. Curran Associates Inc. (2012)

16. Hardt, M., Rothblum, G.: A multiplicative weights mechanism for privacy-preserving data analysis, pp. 61–70, May 2010

17. Hay, M., Rastogi, V., Miklau, G., Suciu, D.: Boosting the accuracy of differentially private histograms through consistency. Proc. VLDB Endow. **3**(1–2), 1021–1032 (2010)

18. Lichman, M.: UCI Machine Learning Repository (2013)

19. Matias, Y., Vitter, J.S., Wang, M.: Wavelet-based histograms for selectivity estimation. SIGMOD Rec. **27**(2), 448–459 (1998)

20. Qardaji, W.H., Yang, W., Li, N.: Understanding hierarchical methods for differentially private histograms. PVLDB **6**, 1954–1965 (2013)

21. Rastogi, V., Nath, S.: Differentially private aggregation of distributed time-series with transformation and encryption. In: Proceedings of the 2010 ACM SIGMOD International Conference on Management of Data, SIGMOD 2010, pp. 735–746. ACM, New York (2010)

22. Stollnitz, E.J., Derose, T.D., Salesin, D.H.: Wavelets for Computer Graphics: Theory and Applications. Morgan Kaufmann Publishers Inc., San Francisco (1996)

23. Ullman, J.: Answering n2+O(1) counting queries with differential privacy is hard. In: Proceedings of the Forty-Fifth Annual ACM Symposium on Theory of Computing, STOC 2013, pp. 361–370. ACM, New York (2013)

24. Ullman, J., Vadhan, S.: PCPs and the hardness of generating private synthetic data. In: Ishai, Y. (ed.) TCC 2011. LNCS, vol. 6597, pp. 400–416. Springer, Heidelberg (2011). https://doi.org/10.1007/978-3-642-19571-6_24

25. Vitter, J.S., Wang, M.: Approximate computation of multidimensional aggregates of sparse data using wavelets. SIGMOD Rec. **28**(2), 193–204 (1999)

26. Vitter, J.S., Wang, M., Iyer, B.: Data cube approximation and histograms via wavelets. In: Proceedings of the Seventh International Conference on Information and Knowledge Management, CIKM 1998, pp. 96–104. ACM, New York (1998)

27. Xiao, X., Wang, G., Gehrke, J.: Differential privacy via wavelet transforms. In: 2010 IEEE 26th International Conference on Data Engineering (ICDE 2010), pp. 225–236 (2010)

28. Zhang, J., Cormode, G., Procopiuc, C.M., Srivastava, D., Xiao, X.: Privbayes: private data release via Bayesian networks. ACM Trans. Database Syst. **42**(4), 25:1–25:41 (2017)

Highway State Gating for Recurrent Highway Networks: Improving Information Flow Through Time

Ron Shoham$^{(\boxtimes)}$ and Haim Permuter$^{(\boxtimes)}$

Ben-Gurion University, 8410501 Beer-Sheva, Israel
`ronshoh@post.bgu.ac.il`, `haimp@bgu.ac.il`

Abstract. Recurrent Neural Networks (RNNs) play a major role in the field of sequential learning, and have outperformed traditional algorithms on many benchmarks. Training deep RNNs still remains a challenge, and most of the state-of-the-art models are structured with a transition depth of 2–4 layers. Recurrent Highway Networks (RHNs) were introduced in order to tackle this issue. These have achieved state-of-the-art performance on a few benchmarks using a depth of 10 layers. However, the performance of this architecture suffers from a bottleneck, and ceases to improve when an attempt is made to add more layers. In this work, we analyze the causes for this, and postulate that the main source is the way that the information flows through time. We introduce a novel and simple variation for the RHN cell, called Highway State Gating (HSG), which allows adding more layers, while continuing to improve performance. By using a gating mechanism for the state, we allow the net to "choose" whether to pass information directly through time, or to gate it. This mechanism also allows the gradient to back-propagate directly through time and, therefore, results in a slightly faster convergence. We use the Penn Treebank (PTB) dataset as a platform for empirical proof of concept. Empirical results show that the improvement due to Highway State Gating is for all depths, and as the depth increases, the improvement also increases.

Keywords: Deep learning · Machine learning
Recurrent Highway Network · Recurrent Neural Networks
Sequential learning

1 Introduction

Training very deep neural networks has become very common in the last few years. Both theoretical and empirical evidence points to the fact that deeper networks can represent more efficiently specific functions (Bengio et al. [1], Bianchini and Scarselli [2]). Some commonly used architectures for deep feed-forward networks are Resnet [6], Highway Networks [17] and Dense-Net [9]. These architectures can be structured with tens, and sometimes even hundreds of layers. Unfortunately, training a very deep Recurrent Neural Network (RNN) still remains a challenge.

© Springer International Publishing AG, part of Springer Nature 2018
I. Dinur et al. (Eds.): CSCML 2018, LNCS 10879, pp. 120–128, 2018.
https://doi.org/10.1007/978-3-319-94147-9_10

Zilly et al. [19] introduced the Recurrent Highway Network (RHN) in order to address this issue. Its main difference from previous deep RNN architectures, was incorporating Highway layers inside the recurrent transition. By using a transition depth of 10 Highway layers, RHN managed to achieve state-of-the-art results on several benchmarks of word and character prediction. However, increasing the transition depth of a similar RHN, does not improve the results significantly.

In this paper, we first analyze the reasons for this phenomena. Based on the results of our analysis, we suggest a simple solution which adds a non-significant number of parameters. This variant is called a *Highway State Gating* cell or a *HSG*. By using the HSG mechanism, the new state is generated by a weighted combination of the previous state and the output of the RHN cell. The main idea behind the HSG cell is to provide a fast route for the information to flow through time. That way, we also provide a shorter path for the back-propagation through time (BPTT). This enables the use of a deeper transition depth, together with significant performance improvement on a widely used benchmark.

2   Related Work

Gated-Recurrent-Units (GRUs) [3] were suggested in order to reduce the number of parameters of the traditional and commonly used Long-Short-Term-Memory (LSTM) cell (Hochreiter and Schmidhuber [8]). Similarly to HSG, in GRUs the new state is a weighted sum of the previous state and a non-linear transition of the current input and the previous state. The main difference is that the transition is of a depth of a single layer and, therefore, less robust.

Kim et al. [11] introduced a different variant of the LSTM cell which is inspired by Resnet [6]. They proposed adding to the LSTM cell a residual connection from its input to the reset gate projection output. By that they allowed another route for the information to flow directly through. They managed to train a net of 10 residual LSTM layers which outperformed other architectures. In their work, they focused on the way that the information passes through layers in the feed-forward manner, and not on the way it passes through time.

Wang and Tian [18] used residual connections in time. In their work they talked about the way information passes through time. They managed to improve performance on some benchmarks, while reducing the number of parameters. The difference is that they needed to work with a fixed residual length that is a hyper-parameter. Also, their work focused on cells with a one layer transition depth.

Another article, relating to Zoneout regularization (Krueger et al. [12]) also relates to information flow through time. The authors introduced a new regularization method for RNNs, where the idea is very similar to dropout [16]. The difference is that the dropped neurons in the state vectors get their values in the former time-step, instead of being zeroed. They mentioned that one of the benefits of this method is that the BPTT skips a time-step on its path back through time. In our work, there is a direct (weighted) connection between the current state and the former one, which is used similarly both for training and inference.

Another relevant issue is the *slowness* regularizers (Hinton [7], Földiák [4], Luciw and Schmidhuber [13], Jonschkowski and Brock [10], Merity et al. [15]) which add a penalty for large changes in state through time. In our work we do not add such a penalty, but we allow a direct route for the state to pass through time-steps, and therefore we 'encourage' the state not to change when it is not needed.

3   Revisiting Vanilla Recurrent Highway Networks

Let L be the transition depth of the RHN cell, and $x^{[t]} \in \mathbb{R}^m$ be the cell's input at time t. Let $W_{H,T,C} \in \mathbb{R}^{n \times m}$ and $R_{H_l,T_l,C_l} \in \mathbb{R}^{n \times n}$ represent the weight matrices of H nonlinear transforms and the T and C gates at layer $l \in \{1, \ldots, L\}$. The biases are denoted by $b_{H_l,T_l,C_l} \in \mathbb{R}^n$, and let $s_l^{[t]}$ denote the intermediate output at layer l at time t, with $s_0^{[t]} = s_L^{[t-1]}$. The gates T and C utilize a sigmoid (σ) non-linearity and "$\cdot$" denotes element-wise multiplication. An RHN layer is described by

$$s_l^{[t]} = h_l^{[t]} \cdot t_l^{[t]} + s_{l-1}^{[t]} \cdot c_l^{[t]}, \tag{1}$$

where

$$h_l^{[t]} = \tanh(W_H x^{[t]} \mathbb{I}_{\{l=1\}} + R_{H_l} s_{l-1}^{[t]} + b_{H_l}), \tag{2}$$

$$t_l^{[t]} = \sigma(W_T x^{[t]} \mathbb{I}_{\{l=1\}} + R_{T_l} s_{l-1}^{[t]} + b_{T_l}), \tag{3}$$

$$c_l^{[t]} = \sigma(W_C x^{[t]} \mathbb{I}_{\{l=1\}} + R_{C_l} s_{l-1}^{[t]} + b_{C_l}), \tag{4}$$

and $\mathbb{I}$ is the indicator function. A very common variant for this is coupling gate C to gate T, i.e. $C = 1 - T$. Figure 1 illustrates the RHN cell.

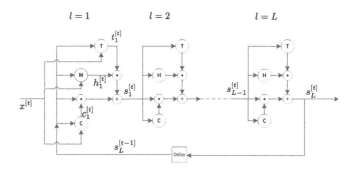

Fig. 1. Schematic showing RHN cell computation. The Feed-Forward route goes from bottom to top through L stacked Highway layers. On the right side there is the memory unit, followed by the recurrent connection.

According to Zilly et al. [19], one of the main advantages of using deep RHN instead of stacked RNNs, is the path length. While the path length of L stacked RNNs from time t to time $t + T$ is $L + T - 1$ (Fig. 2), the path length of a RHN of depth L is $L \times T$ (Fig. 3). The high recurrence depth can add significantly higher modeling power.

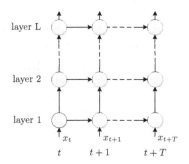

Fig. 2. The figure illustrates an unfolded RNN with L stacked layers. Here the path length from time t to time $t + T$ is $L + T - 1$.

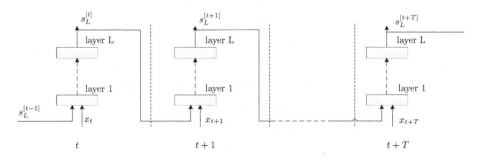

Fig. 3. The figure illustrates an unfolded RHN with L layers. Here the path length from time t to time $t + T$ is $L \times T$.

We believe that its power might, sometimes, also be its weakness. Let us examine a case where information that is relevant for a large number of time steps is given at time t; for example in stocks forecasting, where we expect a sharp movement to occur in the next few time steps. We would like the state to remain the same until the event happens (unless any dramatic event changes the forecast). In this case, we probably prefer the net state to remain stable without dramatic changes. However, when using a deep RHN, the information must pass through many layers, and that might cause an unwanted change of the state. For example, with a RHN of depth 30, the input state at time t has to pass 300 layers in order to propagate 10 time steps. To the best of our knowledge, there is no use of a feed-forward Highway Network of this depth in any field. This fact also affects the vanishing gradient issue using BPTT. The fact that the gradient needs to back-propagate through hundreds of layers causes it to vanish and not be effective. The empirical results support our assumption, and it seems like a performance bottleneck occurs when we use deeper nets.

4   Highway State Gate in Time

We suggest a simple, yet efficient, solution for the depth-performance bottleneck issue. Let $W_{R,F} \in \mathbb{R}^{n \times n}$ represent the weight matrices, and let $b_G \in \mathbb{R}^n$ be a bias vector. Let $s_L^{[t]}$ represent the output of the RHN cell at time t. $\hat{s}^{[t]}$ is the output of the HSG cell at time t. The HSG cell is described by

$$\hat{s}^{[t]} = g \cdot \hat{s}^{[t-1]} + (1 - g) \cdot s_L^{[t]}, \tag{5}$$

where

$$g^{[t]} = \sigma(W_R \hat{s}^{[t-1]} + W_F s_L^{[t]} + b_G). \tag{6}$$

A scheme of the HSG cell and an unfolded RHN with HSG is depicted in Figs. 4 and 5, respectively. The direct outcome of adding an HSG cell is giving the information an alternative and fast route to flow through time.

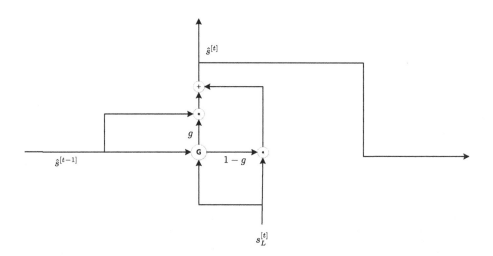

Fig. 4. The figure illustrates a zoom into the HSG cell.

Since gate g utilizes a Sigmoid, its values are in the range $[0, 1]$. When $g = 0$, i.e. HSG is closed, $\hat{s}^{[t]} = s_L^{[t]}$. When $G = 1$, i.e. the gate is opened, $\hat{s}^{[t]} = \hat{s}^{[t-1]}$. In the first case, the net functions as a vanilla RHN. In this case the information from the former state passes only through the functionality of the RHN. This

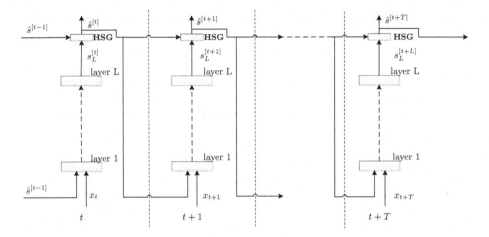

Fig. 5. A macro scheme of an unfolded RHN with HSG cell. The state feeds both the RHN and the next time-step HSG cell.

means that the functionality of a regular RHN can be achieved easily even after stacking the HSG layer.

One of the strengths of this architecture is that each state neuron has its own stand-alone gate. This means that some of the neurons can pass information easily through many time-steps, whereas other neurons learn short time dependencies.

Now let us examine the example we mentioned above, when using RHN with the HSG cell. The net depth is 30, and a state needs to propagate 10 time-steps. In this case, the state has multiple routes to propagate through. The propagation lengths are now $10 + 30j$, with $j \in \{0, 1 \ldots 10\}$. This means that the information has multiple routes, and even if we use a really deep net, it still has a short path to flow through. For this reason, we expect our variant to enable training deeper RHNs more efficiently. The results below support our claim.

5   Results

Our experiments study the benefit of adding depth to a RHN with and without stacking HSG cells at its output. We conducted our experiments on the Penn Treebank (PTB) benchmark.

PTB: The Penn Treebank[1], presented by Marcus et al. [14], is a well known data set for experiments in the field of language modeling. The goal is predicting the next word at each time step, based on the past. Its vocabulary size is 10k unique words. All words that are not in the vocabulary are labeled to a single token. The database is structured of 929k training words, 73k validation words, and 82k test words.

---

[1] http://www.fit.vutbr.cz/imikolov/rnnlm/simple-examples.tgz.

We used a hidden size of 830, similarly to that used by Zilly et al. [19]. For regularization, we use variational dropout [5], and L2 weight decay. The learning rate exponentially decreased at each epoch. An initial bias of -2.5 was used for both the RHN and the HSG gates. That way, the gates are closed at the beginning of training. We tried RHN depths from $\{10, 20, 30, 40\}$. Results are shown in Table 1. It can be well seen from the results that a performance bottleneck occurs when adding more layers to the vanilla RHN. However, adding more layers to the RHN network with the HSG cell results in a steady improvement. Figure 6 also illustrates the difference between both architectures during training. It can be seen that not only does the net with HSG achieve better results, it also converges a bit faster than the vanilla one. Another interesting aspect is the histogram of the gate values of the HSG cell in Fig. 7. It can be seen that most of the gates are usually closed (small valued). However, in a significant number of cases the gates open, which means that the model passes a very similar state to the next time step.

Table 1. Single RHN model test and validation perplexity of the PTB dataset

RHN	Validation set		Test set	
	With HSG	W/o HSG	With HSG	W/o HSG
Depth = 10	67.5	67.9	65.0	65.4
Depth = 20	65.6	66.4	62.9	63.2
Depth = 30	64.8	66.4	62.0	63.4
Depth = 40	64.7	66.7	61.7	63.6

[a]Note that the HSG is more significant as the depth of the RHN increases.

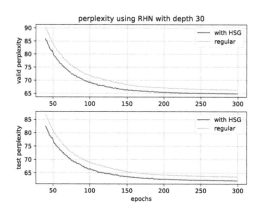

Fig. 6. Comparison of the learning curve between RHN with (green) and without (red) HSG cell. The upper and the lower graphs show the perplexity on the validation and test sets respectively. (Color figure online)

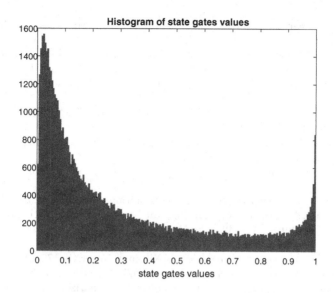

Fig. 7. Histogram of HSG cell gates values. The values were drawn from a trained RHN of depth 30, with a hidden size of 830. There are 66400 values from a 80 random time steps. The gates utilize a Sigmoid function and, therefore, the values are in the range of $[0, 1]$. We see that in most of the cases the gate values are relatively low, which means that the state gates are closed, and the new state is generated in a feed-forward manner. However, for a substantial number of times, the values are high, which means that the information flows directly through time.

6   Conclusion

In this work, we revisit a widely used RNN model. We analyze its limits and issues, and propose a variant for it called Highway State Gate (HSG). The main idea behind HSG is to generate an alternative fast route for the information to flow through time. The HSG uses a gating mechanism to assemble a new state out of a weighted sum of the former state and the RHN output. We show that when using our method, training deeper nets results in better performance. To the best of our knowledge, this is the first time in the field of Recurrent Nets that adding layers to this scale resulted in a steady improvement.

References

1. Bengio, Y., LeCun, Y., et al.: Scaling learning algorithms towards AI (2007)
2. Bianchini, M., Scarselli, F.: On the complexity of neural network classifiers: a comparison between shallow and deep architectures. IEEE Trans. Neural Netw. Learn. Syst. **25**(8), 1553–1565 (2014)
3. Cho, K., Van Merriënboer, B., Bahdanau, D., Bengio, Y.: On the properties of neural machine translation: Encoder-decoder approaches. arXiv preprint arXiv:1409.1259 (2014)

4. Földiák, P.: Learning invariance from transformation sequences. Neural Comput. **3**(2), 194–200 (1991)
5. Gal, Y., Ghahramani, Z.: A theoretically grounded application of dropout in recurrent neural networks. In: D. Lee, D., Sugiyama, M., Luxburg, U.V., Guyon, I., Garnett, R. (eds.) Advances in Neural Information Processing Systems, vol. 29, pp. 1019–1027. Curran Associates Inc. (2016)
6. He, K., Zhang, X., Ren, S., Sun, J.: Deep residual learning for image recognition. arXiv preprint arXiv:1512.03385 (2015)
7. Hinton, G.E.: Connectionist learning procedures. In: Machine Learning, vol. 3, pp. 555–610. Elsevier (1990)
8. Hochreiter, S., Schmidhuber, J.: Long short-term memory. Neural Comput. **9**, 1735–1780 (1997)
9. Huang, G., Liu, Z.: Densely connected convolutional networks (2016)
10. Jonschkowski, R., Brock, O.: Learning state representations with robotic priors. Auton. Robots **39**(3), 407–428 (2015)
11. Kim, J., El-Khamy, M., Lee, J.: Residual LSTM: design of a deep recurrent architecture for distant speech recognition. arXiv preprint arXiv:1701.03360 (2017)
12. Krueger, D., Maharaj, T., Kramár, J., Pezeshki, M., Ballas, N., Ke, N.R., Goyal, A., Bengio, Y., Courville, A., Pal, C.: Zoneout: regularizing RNNs by randomly preserving hidden activations. arXiv preprint arXiv:1606.01305 (2016)
13. Luciw, M., Schmidhuber, J.: Low complexity proto-value function learning from sensory observations with incremental slow feature analysis. In: Villa, A.E.P., Duch, W., Érdi, P., Masulli, F., Palm, G. (eds.) ICANN 2012. LNCS, vol. 7553, pp. 279–287. Springer, Heidelberg (2012). https://doi.org/10.1007/978-3-642-33266-1_35
14. Marcus, M.P., Marcinkiewicz, M.A., Santorini, B.: Building a large annotated corpus of English: the Penn Treebank. Comput. Linguist. **19**(2), 313–330 (1993). ISSN 0891–2017
15. Merity, S., McCann, B., Socher, R.: Revisiting activation regularization for language RNNs. arXiv preprint arXiv:1708.01009 (2017)
16. Srivastava, N., Hinton, G., Krizhevsky, A., Sutskever, I., Salakhutdinov, R.: Dropout: a simple way to prevent neural networks from overfitting. J. Mach. Learn. Res. **15**, 1929–1958 (2014). http://jmlr.org/papers/v15/srivastava14a.html
17. Srivastava, R.K., Greff, K., Schmidhuber, J.: Highway networks. arXiv preprint arXiv:1505.00387 (2015)
18. Wang, Y., Tian, F.: Recurrent residual learning for sequence classification. In: Proceedings of the 2016 Conference on Empirical Methods in Natural Language Processing, pp. 938–943 (2016)
19. Zilly, J.G., Srivastava, R.K., Koutník, J., Schmidhuber, J.: Recurrent highway networks. arXiv preprint arXiv:1607.03474 (2016)

Secured Data Gathering Protocol for IoT Networks

Alejandro Cohen$^{(\boxtimes)}$, Asaf Cohen$^{(\boxtimes)}$, and Omer Gurewitz$^{(\boxtimes)}$

Department of Communication Systems Engineering,
Ben-Gurion University of the Negev, 84105 Beer-Sheva, Israel
{alejandr,coasaf,gurewitz}@bgu.ac.il

Abstract. Data collection in Wireless Sensor Networks (WSN) and specifically in the Internet of Things (IoT) networks draws significant attention both by the industrial and academic communities. Numerous Medium Access Control (MAC) protocols for WSN have been suggested over the years, designed to cope with a variety of setups and objectives. However, most IoT devices are only required to exchange very little information (typically one out of several predetermined messages), and do so only sporadically. Furthermore, only a small subset (which is not necessarily known *a priori*) intends to transmit at any given time. Accordingly, a tailored protocol is much more suited than the existing general purpose WSN protocols. In many IoT applications securing the data transmitted and the identity of the transmitting devices is critical. However, security in such IoT networks is highly challenging since the devices are typically very simple, with highly constrained capabilities, e.g., limited memory and computational power or no sophisticated algorithmic capabilities, which make the utilization of complex cryptographic primitives unfeasible. Furthermore, note that in many such applications, securing the information transmitted is not sufficient, since knowing the transmitters identity conveys a lot of information (e.g., the identity of a hazard detector conveys the information that a threat was detected).

In this paper, we design and analyze an efficient secure data collection protocol based on information theoretic principles, in which an eavesdropper observing only partial information sent on the channel cannot gain significant information on the transmitted messages or even on the identity of the devices that sent these messages. In the suggested protocol, the sink collects messages from up to K sensors simultaneously, out of a large population of sensors, without knowing in advance which sensors will transmit, and without requiring any synchronization, coordination or management overhead. In other words, neither the sink nor the other sensors need to know who are the actively transmitting sensors, and this data is *decoded* directly from the channel output. We provide a simple secure codebook construction with very efficient and simple encoding and decoding procedures.

© Springer International Publishing AG, part of Springer Nature 2018
I. Dinur et al. (Eds.): CSCML 2018, LNCS 10879, pp. 129–143, 2018.
https://doi.org/10.1007/978-3-319-94147-9_11

1   Introduction

The Internet of Things (IoT) is an innovative communication paradigm that aims at interconnecting a large number of devices (things) in order to enable the realization of systems that improve every day life, ranging from smart cities, smart homes, pervasive health care, assisted living, environmental monitoring, surveillance, etc. A great percentage of the heterogeneous devices comprising the IoT are expected to be small, with very constrained processing and storage resources and with highly limited energy resources and network capabilities. Accordingly, ensuring security and privacy while maintaining reliability and satisfactory performance (e.g., latency) in such dense and highly constrained environment, is very challenging [1–4].

Numerous Medium Access Control (MAC) protocols, and in particular ones for Wireless Sensor Networks (WSN) have been suggested over the years (a short representative overview is provided in Sect. 3). The challenge is the need to support any kind of data needed to be conveyed in a transmission in a highly diverse environment which includes a large variety of topologies, a wide range of traffic patterns and high heterogeneity of the devices. To cope, most of these MAC protocols have followed the traditional layering approach, in which there is complete separation between the various mechanisms involved in each layer of the stack, such that eventually the information sent is encapsulated within the MAC frame, paying an overhead and trading performance (e.g., throughput and delay) for generality. For example, a frame which conveys only few bits of information needs to include control information such as source and destination address, message type, etc., which involves a large overhead even when overlooking the higher layers overhead (e.g., an 802.11 ACK frame which contains only one bit of relevant information DATA was successfully received, is 14 bytes plus physical layer encapsulation). However, in many IoT applications, the majority of the traffic is one directional (upstream) to an arbitrary sink node (its particular identity is not relevant) which collects the information and processes it or forwards it (e.g., to the cloud), and even more importantly each such device needs to report only a limited amount of information, and only infrequently. For example, a device may be required to send a keep-alive message periodically, informing the sink node that its battery has not drained out. In addition, each device can occasionally send one of several possible reports depending on the device mission (e.g., a motion detector reports. Note that for all these applications, sending the device ID and an index, pointing to one of the limited number of predetermined messages is sufficient to convey all the information it needs to send. Accordingly, in order to cope with the challenge of a huge number of highly constrained devices competing for very limited wireless resources, one can take into account that they only need to transmit sporadically and only a limited amount of information.

As previously mentioned securing the data transmitted and the identity of the transmitting devices is a great challenge which is critical to many applications. All the more so, since as previously mentioned, most of these IoT devices have highly constrained capabilities, hence utilizing complex cryptographic primitives or hiding the transmission is unfeasible. Furthermore, note that the fact that a device has very limited information to send, which can be an advantage for communication, is a great burden on security, as knowing the transmitters identity conveys a lot (and sometimes all) of the information sent.

In this paper, we design, analyze and evaluate a secure and highly efficient MAC protocol specially designed to privately collect information from a large number of devices such that an eavesdropper (or an unintended receiver) which observes only some of the transmissions on the channel is not only unable to decode any of the information sent but cannot even identify the identity of the devices sending the information. The suggested protocol utilizes information theoretic concepts and novel signaling and decoding techniques which allow us to jointly optimize all layers together. We assume the devices are very simple, with highly constrained capabilities. Thus, the key idea that our protocol relies on is that instead of the typical frame mechanism using data encapsulation, each sensor is assigned a unique *transmission pattern* for each of its messages, which conveys both the information and the sensor's ID. Whenever a sensor wishes to transmit a report, it waits to receive a predefined periodic preamble sent by a sink, and then transmits a sequence of *impulses* according to the transmission pattern which corresponds to the report it wishes to send. The sink node *receives several simultaneous transmissions in a way that it can recognize both the senders and the information sent* from the aggregated channel output received. This is done using a carefully designed codebook, and a matching decoding algorithm that identifies both the sensors which transmitted and their codewords. It is important to note that unlike Code-Division Multiple-Access Channels (CDMA), the sink node can rely on a simple energy detection in order to decode, and not on the exact received power or any power adaptation mechanism, thus dramatically improving robustness. Besides the channel utilization efficiency of the suggested protocol, the suggested protocol also ensures secrecy and privacy, in particular an eavesdropper receiving only parts of the channel output, is kept ignorant both of the content of the message and of the identity of the senders.

We illustrate the basic protocol and outline its basic concepts via the toy example depicted in Fig. 1. In this simple example, we assume a network of N devices transmitting to a single sink. Each device has a set of messages it can transmit. We denote by C_i the number of messages of device i. Each one of the messages is assigned a unique pattern, comprising high and low level elements which are known to all nodes in the network, including potential eavesdroppers. After receiving a predefined beacon from the sink, which initiates a conceptual set of mini-timeslots, all the devices that are awake and waiting start transmission according to the pattern assigned to the message to be transmitted. Specifically, each transmitting device emits energy in the minislots which correspond to a high level in the message pattern and stays idle in the other minislots. In the above example sensors 2, k and $N - 1$ are awake and ready to transmit messages 5, 1 and 7, respectively. The received signal at the sink is a combination of all the transmitted sequences. The sink performs an energy detection procedure on the received signal according to a predefined threshold, identifying

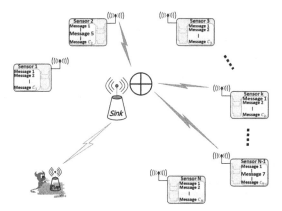

Fig. 1. A schematic illustration of an IoT network. Data is aggregated by a sink, when K sensors, out of a large population of N sensors in the network have a message to transmit over the wireless channel.

which minislots were busy (at least one of the devices has emitted energy) and which were idle (below the energy threshold). Note that the sink considers only the Boolean sum of the "bit" patterns used by the transmitting sensors i.e., all the sink needs to identify is on which minislots the energy is above the noise level (above the threshold). Accordingly, each minislot should be at a length sufficient to identify whether there was transmission on that minislot or not (decode one bit). Based on the filtered sequence (the exact sequence of busy and idle minislots) the sink deduces which set of transmitted sequences (which corresponds to a set of messages with their corresponding devices) have generated the sequence received. Note that in order for the code to succeed, only a single set of sequences should match any possible received pattern, and any other combination of sequences but the correct one, would result in a mismatch between the expected and the actual minislot energy levels. An eavesdropper which observes only parts of the channel output (can identify the channel output of only a subset of the minislots), is kept ignorant both of the content of the messages and of the identity of the senders. I.e., its received pattern corresponds to many possible transmitted-sequence sets, and practically taking into account all the possible transmitting-sequence sets, covers all the devices as potential transmitters and all sequences as possible messages being transmitted. Note that it is not only that the eavesdropper cannot uniquely identify the transmitting devices and their transmitted messages, but it cannot even eliminate any of the devices as a potential transmitter nor any of the messages as potential transmitted message. Note that a simple Time Division Multiple Access (TDMA), which assigns each device several minislots (based on the number of its possible messages), is highly inefficient, i.e., *requires many minislots (linear in the number of sensors) regardless of the actual number of sensors waiting to transmit in each round.* Our suggested solution, relying on the moderate number of sensors expected to transmit at each transmission opportunity, requires *only a logarithmic number of minislots.* More importantly such TDMA is not secured whatsoever, i.e., at least the identity of the transmitting device is revealed based on the time allocated to it.

In particular, the contribution of this work is threefold: (i) We present a new secured MAC protocol, for data collection from dense wireless networks, in which the devices are expected to transmit only sporadically, and only a predefined amount of information (one out of a bank of possible messages per device). In the suggested protocol, the sink can collect up to K reports simultaneously without any management or scheduling. (ii) To support the protocols, we provide a codebook construction with a very simple encoding and decoding procedure, such that not only the code is efficient but also the transmitted codewords are self-contained and do not require headers, trailers or sender identity. We further suggest a corresponding effective decoding algorithm which is based on Column Matching [5]. The suggested protocol is secured, such that an eavesdropper observing only part of the channel output cannot decode the messages or even identify the senders.

2   System Model

We model the IoT network as a dense wireless sensor network that collects information from the area covered by the network. Throughout the paper, we will interchange "device" with "sensor". The network can contain one or multiple sinks (cluster heads) each collecting reports from a large set of sensors independently. Throughout the paper, we will focus on a single such cluster consisting of a large set of wireless sensors (devices), denoted by $\mathcal{N}$, and a single sink. We denote by $N = |\mathcal{N}|$ the total number of sensors in the network (cluster). In addition, we assume that one or more non collaborating eavesdroppers are present, each observing a noisy version of what is received by the sinks. Since the eavesdroppers do not collaborate, throughout the paper we will concentrate on a single arbitrary eavesdropper.

We assume that all sensors are connected to the sink node, *but only in the following limited sense*: First, beacons by the sink should be heard by all sensors. Then, it is sufficient for the sink to be able to detect only whether there were transmissions (one or more) on the channel. For example, whether the received SNR is above a predefined threshold which is above the noise floor. The sink *does not necessarily need to decode any information from a single transmission from a single sensor*, and the received SNR is not necessarily above a decodable threshold. We assume the sink should have a sufficiently high SNR, yet only to be able to decide whether there was energy on the channel in each mini-slot or not. Decoding will be done only based on these binary values, and the exact amount of energy detected is not important.

We assume that each sensor i can transmit out of a bank of C_i different messages. There are no restrictions on message length or content. For simplicity we will assume throughout the paper that $C_i = C, \forall i$. Nonetheless, an extension to different C_i's is straightforward. We further assume that each sensor needs to transmit a message sporadically. In particular, we will assume that the sensors employ a duty cycle mechanism in which they randomly wake up and transmit a message unless they have an urgent message they need to transmit, in which case they wake up instantly waiting for transmission opportunity. We assume that the wake-up times including the urgent report instances are arranged such that the probability that more than K sensors are awake at the same time waiting for transmission opportunity is very low. Note that while we set no restriction on K (i.e., $K \leq N$), the suggested protocol is more efficient the smaller K is, compared to N.

The eavesdropper cannot receive all transmissions on the channel, and manages to detect the correct amount of energy only on a fraction δ of the mini-slots. That is, we assume that out of the T mini-slots, the eavesdropper has information only on δT mini-slots on average. Note that this model is similar to the common erasure channel.

3   Related Works

We divide the discussion on related works into two parts. In the first, we provide a brief overview of several multipurpose WSN MAC protocols. Then, since our protocol is inspired by the classical Group Testing (GT) approach, in the second part we give a brief overview of related GT results.

WSN MAC Protocols. Since on the one hand, one of the foremost objectives of WSN is energy conservation, and on the other sensor nodes are expected to report only sporadically (and many of the reports can tolerate a short delay), most of the MAC protocols which were designed for WSN over the past decade and a half rely on a duty cycling technique in which each sensor node turns its radio on only periodically, alternating between active and sleeping modes [6–8]. Such protocols took different approaches to address the rendezvous challenge in which a sender and a receiver should be awake at the same time in order to exchange information. In the synchronous approach, nodes' active and sleeping periods are aligned, i.e., all sensor nodes are active at the same time intervals and are required to contend for transmission opportunities during these intervals, e.g., [9–11]. The asynchronous approach allows sensor nodes to choose individual wakeup times, maintaining unsynchronized duty-cycles, and employing various strategies to detect transmissions in the network and enable rendezvous between senders and receivers, e.g., [12–14]. However, all these protocols are designed to support various types of traffic patterns, diverse topologies (e.g., single and multi-hop topologies) and most importantly, to support any kind of information exchange between the sensors. Accordingly, they have adopted the traditional approach in which the proposed channel access mechanism is independent from the message payload exchange between the sensor nodes, *at the price of data encapsulation and signaling overhead*. In this work, since the topology is limited to a single hop topology and, more importantly, since the information each sensor needs to convey is limited to one out of a number of known messages, we take a cross-layer design in which the coding and the channel access algorithm are intertwined.

Group Testing. Classical group testing was used during World War II in order to identify syphilis infected draftees while dramatically reducing the number of required tests, by examining pooled tests of mixed blood samples [15].

The concept was adopted later for multi access protocols. Specifically, MAC protocols adopted the GT philosophy for Collision Resolution Protocols (CRP). The basic idea behind these protocols is to resolve collisions whenever multiple users are trying to access the channel simultaneously, e.g., the binary-tree CR, the epoch mechanism or the Clipped Binary-Tree Protocol. An extensive survey of such protocols is given in [16, Chap. 5]. However, all these protocols utilized the GT concept *only as a collision resolution mechanism*, i.e., in case a collision occurred, they used the concept in order to decide who should contend for the channel next and when. Data was decoded successfully only when a single node transmitted without collision.

The main contribution of this paper is in suggesting a protocol which decodes all data, from all simultaneously transmitting sensors, using a novel extension to this concept, namely, analysing the location of the colliding and non-colliding minislots in order to identify both the senders and the data sent.

4   Secure WSN Data Collecting Protocol Design

In the suggested protocol, the sink periodically transmits a predefined beacon, which starts a report transmission interval. We term this beacon RFR (Request For Reports). The RFR is then followed by a sequence of T minislots. Figure 2 provides an illustration of the protocol operation. Note that there is no need for

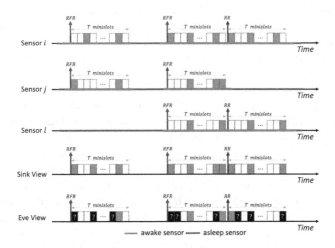

Fig. 2. Basic protocol operation.

synchronization or for each sensor to keep track of the minislot boundaries at all times, as the awake sensors waiting for transmission can synchronize based on the received RFR. Denote by τ the maximum propagation delay between any sensor and the sink, and by η the time required by the sink to identify that there is a transmission going on. Note that η only corresponds to the duration required by the sink to sample the channel and identify that there is a transmission going on (there are no headers, preambles or data involved), hence can be very short. We assume that the minislot duration is longer than $2\tau + \eta$, which is sufficient for all sensors to receive the RFR, start a transmission of duration η and for the sink to receive all the transmissions starting in this minislot. I.e., as far as the sink is concerned, the time duration of a minislot is such that there is no transmission that can start at the current slot and leak to the next slot. It is reasonable to assume that $\eta \geq \tau$, hence the minislot duration is greater than 3τ. After transmitting the RFR, the sink node switches to receive mode and identifies whether there was a transmission at each of the following T minislots. Recall that the sink detection is binary, i.e., it can only recognize whether there was a transmission in a minislot or not. It does not try to detect how many sensors transmitted during an occupied minislot.

Each sensor is assigned a unique sequence of ones and zeroes of length T, for each of its messages. The construction of the sequences is given in Sect. 6. A sensor intending to send a report wakes up and waits for the RFR. After receiving the RFR, the sensor follows the sequence associated with the message it intends to transmit, transmitting "energy" of duration η (in the form of a pre-defined signal) at each minislot in which the corresponding bit in the sequence is one.

After the T minislots interval, the sink has a sequence of ones and zeroes of length T, indicating at which of the minislots it identified transmission. The eavesdropper (Eve) has only a δ fraction of the sequence. We denote by $Y(t)$ and by $Z(t)$ the sequence observed by the sink and the eavesdropper, at the t-th interval. Based on the observed $Y(t)$ and $Z(t)$ the sink and the eavesdropper try to decode the messages transmitted by the sensors, respectively (we provide two decoding algorithms in Sect. 6 and [17]). Note that since the sequences are unique, each sequence indicates the identity of the sender and the message sent. If the sink is not able to decode the received sequence $Y(t)$, which, as we prove in Sect. 6, can only happen if the number of transmitting sensors in the interval was greater than the expected number K for which the sequences were designed, it transmits another beacon, termed Retransmission Request (RR). This starts the exact same procedure as the RFR only this time sensors waiting for transmission participate in the following interval only with some probability. The probability for participating in the following interval is predefined and can prioritize different messages (e.g., messages with high urgency will receive high probability and messages that can tolerate delay will be assigned low probabilities and sensors with non-urgent messages can go back to sleep waiting for their next wakeup time). The discussion on the collision resolution probabilities is beyond the scope of this paper, as our novelty lies in the transmission protocol and its ability to allow multiple transmission of messages without the various MAC and upper layers overheads. The collision resolution procedure is rather standard and can be repeated multiple times.

It is important to note that the suggested protocol can be *interleaved within traditional wireless sensor protocols*, in which some or all sensors are required to send occasionally a regular report. For example, the suggested protocol can be incorporated within the operation of RI-MAC [13], such that occasionally the sink transmits an ordinary RI-MAC beacon, which is different from the RFR beacon, to initiate a RI-MAC operation interval, i.e., the RI-MAC beacon will be followed by ordinary DATA transmissions according to the ordinary RI-MAC protocol. In the same manner, the suggested protocol can be combined with transmitter initiating protocol such as X-MAC [12], such that a sensor wishing to report a typical DATA packet transmits a sequence of short preambles prior to DATA transmission according to the X-MAC protocol.

5 Model Formulation and Transmission Process

In this section, we formalize the *SWSN* model that is used throughout the paper. We denote the set of wireless sensors by $\mathcal{N}$. We will concentrate on a time instance right after an *RFR* has been sent by the sink and a subset of unknown sensors comply with the *RFR*, and transmit their reports. We denote this unknown subsets by $\mathcal{K}$. We denote by $N = |\mathcal{N}|$ and $K = |\mathcal{K}|$ the total number of sensors, and the number of active sensors at the same time slot, respectively. In the analytical part of the paper we will assume that the number K of active sensors is known a-priori. The algorithms and results presented hereby can be easily adopted to the case where only an upper bound on K is known, and the actual number is smaller. The case where more than K transmit was briefly described at the end of Sect. 4. The sink objective is to determine which subset of the sensors were active and what is the information (messages) they transmitted. Throughout the paper, logarithms are in base 2. Figure 3a gives a graphical representation of the secure model.

Each of the sensors has its own set of C independent messages. We denote each such message by $M_{n,c}$, $n \in \mathcal{N}, c \in \{1, \ldots, C\}$, and sensor $n \in \mathcal{N}$ message list by:

$$\mathbf{M}_n = [M_{n,1}; M_{n,2}; \ldots; M_{n,C}]$$

We consider all the possible sets of K active sensors, and denote by $W \in \{1, \ldots, \binom{N}{K}\}$ the index of the *subset* $S \subset \{1, \ldots, N\}$ *of sensors active and transmitting at the same time*. Thus, S_w denotes the w-th subset of size K out of the N sensors. We assume that W is uniformly distributed, that is, there is no *a-priori* bias to any specific subset of active sensors. We further denote by

$$\mathbf{M}_W = [M_{w_1,c_1}; M_{w_2,c_2}; \ldots; M_{w_K,c_K}]$$

the K messages transmitted by the active sensors (members of S_w) to the sink. Note that each row in $\mathbf{M}_W$ corresponds to a separate message, that is, M_{w_i,c_i} is the c_ith message of the w_ith sensor in the active set.

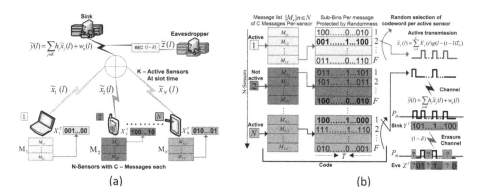

Fig. 3. (a) Secure wireless sensors network model. (b) Encoding, transmission and detection in the suggested protocol.

As previously described, for each message a sensor has a unique sequence, which in the sequel we refer to as codeword. That codeword is the one transmitted whenever the sensor intends to send this message. In other words, if sensor n is set to send $M_{n,c}$, it uses the codeword X_n^T associated with this message. Each message has a different codeword associated with it (we drop the message index for clarity). Given the particular messages sensors intend to transmit, we define a *transmission* matrix $\mathbf{X} = [X_1^T; X_2^T; \ldots; X_N^T] \in \{0,1\}^{N \times T}$, where each row corresponds to a codeword, describing the message a sensor may transmit *if it is active*.

Assuming that sensors use Power Amplitude Modulation (PAM), we denote by

$$\tilde{x}_j(l) = \sum_{t=1}^{T} X_j(t)g(l - (t-1)T_0), \quad 0 \le l \le T \cdot T_0,$$

the signal transmitted by the j-th sensor, where minislot t is defined by $(t-1)T_0 \le l \le tT_0$ and $g(l), 0 \le l \le T_0$ denotes the PAM pulse. The channel output signal $\tilde{y}(l)$ is given by

$$\tilde{y}(l) = \sum_{j \in \mathcal{K}} h_j \tilde{x}_j(l) + w_n(l),$$

where h_j is a channel fade for the jth active sensor and w_n is an additive noise at the sink. Note that the fade herein is fixed for the entire transmission only for simplicity of exposition, and it may depend on t as well. Figure 3b depicts an example.

We denote by P_{th} the power threshold of the sink's hard decision mechanism. Hence, the outcome vector $Y_T = (Y(1), \ldots, Y(T))$ at the sink is binary, with 1 in a minislot t if

$$\int_{(t-1)T_0}^{tT_0} \tilde{y}(l)g(l - (t-1)T_0)dl \ge P_{th},$$

and 0 otherwise. In this paper, we assume that the noise in the channel cannot produce errors at the sink. In the full version of this work [17], we extend the model, and consider the case in which noise can produce positive and negative errors.

We assume eavesdropper can observe a noisy vector $\tilde{z}(l)$, generated from the output signal $\tilde{y}(l)$. In this paper we consider an erasure channel at the eavesdropper, with erasure probability of $1 - \delta$, i.i.d. That is, on average, $T\delta$ mini-slots of the channel output are not erased and are available to the eavesdropper. We assume the eavesdropper uses the same hard decision mechanism, hence observes $Z^T \in \{0, 1, ?\}^T$. While those are an un-necessary restrictions on Eve, the hard decision detection mechanism, and the erasure channel. Such that, Eve may use any decision mechanism, and the noisy vector $\tilde{z}(l)$, may be generated from a channel with false positive or false negative possible errors using the mutual information analysis given in [18, Sect. 6], or for both possible errors as given in [19, Sect. 2]. It simplifies the *technical* aspects and allows us to focus on the key methods.

We denote by $\hat{W}$ and $\hat{\mathbf{M}}_{\hat{W}}(Y^T)$ the set of sensors estimated by the sink to be the transmitting set, and the estimated set of messages sent by them, respectively, according to the received signal Y^T. We refer to the possible transmission matrix, together with the decoder as a *SWSN algorithm*. The following definition and secrecy constrain lays out the goals of the SWSN algorithm.

Definition 1. *A sequence of SWSN algorithms with parameters N, K, C and T is asymptotically* reliable *and secure if,*

(1) Reliable: *At the sink, observing Y^T, we have*

$$\lim_{N \to \infty} P(\hat{\mathbf{M}}_{\hat{W}}(Y^T) \neq \mathbf{M}_W) = 0,$$

i.e., the error probability both in the active set and the message associated with this set, goes to zero as the number of sensors goes to infinity.

(2) Secure: *at the eavesdropper, observing Z^T, we have*

$$\lim_{T \to \infty} \frac{1}{T} I(\mathbf{M}_W; Z^T) = 0.$$

That is, asymptotically guarantee zero mutual information, namely, Eve cannot decode anything from the set of messages.

In the next section, we construct a code (for parameters N, K and C) which associates a codeword of length T to each of the $N \cdot C$ messages (C per sensor) and a decoding algorithm $\hat{\mathbf{M}}_{\hat{W}}(Y^T)$, such that observing Y^T, the sink will identify the subset of active sensors and the messages transmitted by them with desired high probability, yet, observing Z^T in either of the setups, the eavesdropper will not be able to identify the subset of active sensors and the messages transmitted.

6   Code Construction and Decoding at the Sink

In order that the sink will identify the subset of active sensors and the messages transmitted and keep the eavesdropper ignorant, for each sensor $n \in N$, we create a sub-bin with several codewords for each message. We then *randomly* map each message that the sensor wants to transmit, $c \in \{1, \ldots, C\}$, to a codeword in the corresponding sub-bin, that is, the $\{n, c\}$-th sub-bin, which contains F codewords of size T. Figure 3b depicts an example.

More precisely, we assume a source of randomness $(\mathcal{R}, p_R)$, with known alphabet $\mathcal{R}$ and known statistics p_R is available to the encoder. It is important to note that this source of randomness does not have to be shared with any other party. A *stochastic encoder* [20], at each active sensor $j \in \mathcal{K}$, selects uniformly at random one codeword $x^T(c_j, f)$, $1 \leq c \leq C$ and $1 \leq f \leq F$ from his sub-bin, that is, it maps a selected message and the source of randomness to a transmission codeword $\mathbf{X}_n^T$. This mapping, using the randomness, is intended to confuse the eavesdropper regarding the sensor transmitting and the message sent. Hence, over the MAC channel, we still have a transmission matrix $\mathbf{X}_{S_w}^T$, each of its rows corresponding to a different active sensor in the index set S_w, and that transmission matrix contains K codewords of size T, yet now there is no $1:1$ mapping between messages and codewords, and each message corresponds to F codewords.

6.1   Codebook Generation

For each sensor we generate a bin, containing several sub-bins. The number of such sub-bins *corresponds* to the number C of messages that each sensor has. The number F of codewords in each sub-bin corresponds to $T\delta$, the number of un-erased mini-slots that the eavesdropper may obtain, yet normalized by the number of active sensors. The codebook is depicted in the central side of Fig. 3b. Let $P(x) \sim Bernoulli(\ln(2)/K)$. Using a distribution $P(X^T) = \prod_{i=1}^T P(x_i)$, generate a bin of $C \cdot F$ independent and identically distributed codewords. Then, we split each bin to sub-bins of F codewords $x^T(c, f)$, $1 \leq c \leq C$ and $1 \leq f \leq F$. Hence, for each message, $c \in \{1, \ldots, C\}$, there are F possible codewords correspond in the $\{n, c\}$-th sub-bin. During the encoding process, only one codeword from the $\{n, c\}$-th sub-bin will be randomly selected for transmission. Reveal the codebooks to the sensors and the sink, *we assume Eve may have this codebook as well*.

6.2   Decoding at the Sink

In this paper, the decoder we suggest is the optimal decoder, Maximum Likelihood (ML). This decoder will declare the right set of K messages transmitted (and the active sensors) with a high probability if T satisfies the bound Lemma 1 below. However, such a decoding algorithm is complex. Hence, in the full version of this work [17], we consider a computationally efficient algorithm.

As described in Sect. 5, in the first decoding step, the sink uses a hard decision mechanism to achieve the binary channel output vector Y^T. After Y^T is obtained, the ML decoder looks for a collection of K codewords $\hat{\mathbf{X}}_{S_{\hat{w}}}^T$, *each one taken from a separate sub-bin under different bins*, for which Y^T is most likely. Namely, $P(Y^T | \hat{\mathbf{X}}_{S_{\hat{w}}}^T) > P(Y^T | \mathbf{X}_{S_{\hat{w}}}^T), \forall \hat{w} \neq \hat{w}$.

Due the randomness in the encoding procedure of the SWSN algorithm, there is no $1:1$ mapping between messages transmitted and codewords, and each message corresponds to a few codewords. Hence, the sink looks for both the set $\hat{w}$, and the codewords $\hat{\mathbf{X}}_j^T, \forall j \in S_{\hat{w}}$, which are most likely. Then, the sink declares $\hat{W}(Y^T)$ as the set of active sensor, where $\hat{w}$ is the set of bins in which the codewords reside and maps the selected codewords $\hat{\mathbf{X}}_{S_{\hat{w}}}^T$ back to the messages $\hat{c}_j$ according to the corresponding sub-bins.

6.3 Reliability

The following lemma is a key step in proving the reliability of the decoding algorithm.

Lemma 1. *If the size of the codewords satisfies*

$$T \geq \max_{1 \leq i \leq K} \frac{1}{1-(1+\varepsilon)\delta} \frac{1+\varepsilon}{i/K} \log \binom{N-K}{i} C^i,$$

then, under the codebook above, as $N \to \infty$ the average error probability approaches zero.

Note that using the upper bound $\log \binom{N-K}{i} \leq i \log \frac{(N-K)e}{i}$, the maximum in Lemma 1 is easily solved, and we have

$$T \geq \tfrac{1+\varepsilon}{1-\delta} K \log(N-K)Ce.$$

That is, assuming a SWSN algorithm with the parameters N, K and C, for any $0 \leq \delta < 1$, if size of the codewords $T = \Theta\left(\frac{K \log NC}{1-\delta}\right)$, for some $\varepsilon \geq 0$, then there exists a sequence of SWSN algorithms which are reliable and secure (under the conditions given in Definition 1).

The proof of Lemma 1 extends the results given in [18, Theorem 3.1] to the codebook required for SWSN. Specifically, we may interpret the analysis in [18, Sect. 2] as analogous to the non secure case where each sensor has *only one message in its bin*. However, in the SWSN protocol suggested herein, each sensor has C messages in its bin, and the decoder has $\binom{N}{K} C^K F^K$ possible subsets of codewords to choose from, $\binom{N}{K}$ for the number of possible sensors, C^K for the number of possible messages in each bin and F^K for the number of possible codewords to take in each sub-bin. Thus, when fixing the error event, there are $\binom{N-K}{i} C^K F^K$ subsets to confuse the decoder. Specifically, to obtain the bound on the size of the codewords given in Lemma 1, the error probability analysis in the ML decoder extends the bound given in [17, Lemma 3], by considering $C \cdot F$ codewords per sensor, and by considering multiple error events, as given in [21], for the analysis of the secure GT error probability bound. E.g., events where the decoder chooses the wrong codeword for some sub-bin, yet identified the sensors and the messages transmitted correctly (since the sub-bins were correctly identified), and events where the codeword selected was from a wrong sensor sub-bin (hence resulted in an error).

To analyze the information leakage at the eavesdropper, we note that the Eve's channel can be viewed as (binary) Boolean multiple access channel, followed by a $BEC(1 - \delta)$. The sum capacity to Eve cannot be larger than δ. In fact, since the codebook is randomly i.i.d. distributed, Eve can obtain from each active sensor a rate of at most δ/K. Consequently, from each codeword Eve sees at most a capacity of δ/K. However, for each possible message in the codebook suggested each sensor have sub-bin with $F = 2^{T\delta/K}$ codewords, from which during the encoding process, only one codeword will be randomly selected for transmission to confuse the eavesdropper. On average the eavesdropper with a capacity of at most δ/K will have the same number of candidacies (possible codewords) in each sub-bun, such that, will keep ignorant. The formal analysis of the information leakage, to prove the security constraint is met, is a direct consequence of the proof given in [21, Sect. 5.B], where we protect not only on the information of which of the sensors was transmitting a message but also protect the messages as well. Hence, in the same way as given in [21], yet, instead of showing that $I(W; Z^T)/T \to 0$, we can show that $I(\mathbf{M}_W; Z^T)/T \to 0$.

7   Conclusions

In this paper, we design a secured highly efficient WSN protocol for IoT networks, which collects information from a large number of devices, such that an eavesdropper which has access to a noisy version of the sink output, will be kept completely ignorant both regarding the messages sent and the subset of devices that transmitted them. This secured protocol relies on the fact that only a small unknown subset of devices out of the dense device population will attempt transmission at each transmission opportunity, and on the fact that each device has only a limited amount of information it needs to convey. We provide a simple secure codebook construction with very efficient and simple encoding and decoding procedures.

Acknowledgment. This research was partially supported by the Israeli MOITAL NEPTUN consortium and in part by the European Union Horizon 2020 Research and Innovation Programme SUPERFLUIDITY under Grant 671566.

References

1. Shipley, A.: Security in the internet of things, lessons from the past for the connected future. Security Solutions, Wind River, White Paper (2013)
2. Díaz, M., Martín, C., Rubio, B.: State-of-the-art, challenges, and open issues in the integration of internet of things and cloud computing. J. Netw. Comput. Appl. **67**, 99–117 (2016)
3. Zhou, J., Cao, Z., Dong, X., Vasilakos, A.V.: Security and privacy for cloud-based iot: Challenges. IEEE Commun. Mag. **55**(1), 26–33 (2017)
4. Mehmood, Y., Ahmad, F., Yaqoob, I., Adnane, A., Imran, M., Guizani, S.: Internet-of-things-based smart cities: recent advances and challenges. IEEE Commun. Mag. **55**(9), 16–24 (2017)

5. Chan, C.L., Jaggi, S., Saligrama, V., Agnihotri, S.: Non-adaptive group testing: explicit bounds and novel algorithms. IEEE Trans. Inf. Theory **60**(5), 3019–3035 (2014)

6. Polastre, J., Hill, J., Culler, D.: Versatile low power media access for wireless sensor networks. In: Proceedings of the 2nd International Conference on Embedded Networked Sensor Systems (SenSys), pp. 95–107. ACM (2004)

7. Ye, W., Heidemann, J., Estrin, D.: An energy-efficient MAC protocol for wireless sensor networks. In: Proceedings of IEEE Twenty-First Annual Joint Conference of the IEEE computer and communications societies, INFOCOM 2002, vol. 3, pp. 1567–1576. IEEE (2002)

8. Huang, P., Xiao, L., Soltani, S., Mutka, M.W., Xi, N.: The evolution of MAC protocols in wireless sensor networks: a survey. IEEE Commun. Surv. Tutor. **15**(1), 101–120 (2013)

9. Lin, J., Ingram, M.A.: SCT-MAC: a scheduling duty cycle MAC protocol for cooperative wireless sensor network. In: 2012 IEEE International Conference on Communications (ICC), pp. 345–349. IEEE (2012)

10. Liu, C.-J., Huang, P., Xiao, L.: TAS-MAC: a traffic-adaptive synchronous MAC protocol for wireless sensor networks. ACM Trans. Sens. Netw. (TOSN) **12**(1), 1 (2016)

11. Kakria, A., Aseri, T.C.: Survey of synchronous MAC protocols for Wireless Sensor Networks. In: 2014 Recent Advances in Engineering and Computational Sciences (RAECS), pp. 1–4. IEEE (2014)

12. Buettner, M., Yee, G.V., Anderson, E., Han, R.: X-MAC: a short preamble MAC protocol for duty-cycled wireless sensor networks. In: Proceedings of the 4th International Conference on Embedded Networked Sensor Systems (SenSys), pp. 307–320. ACM Press, New York (2006)

13. Sun, Y., Gurewitz, O., Johnson, D.B.: RI-MAC: a receiver-initiated asynchronous duty cycle MAC protocol for dynamic traffic loads in wireless sensor networks. In: Proceedings of the 6th ACM Conference on Embedded Network Sensor Systems (SenSys), pp. 1–14. ACM (2008)

14. Tang, L., Sun, Y., Gurewitz, O., Johnson, D.B.: EM-MAC: a dynamic multichannel energy-efficient MAC protocol for wireless sensor networks. In: Proceedings of the Twelfth ACM International Symposium on Mobile Ad Hoc Networking and Computing (MobiHoc), p. 23. ACM (2011)

15. Dorfman, R.: The detection of defective members of large populations. Ann. Math. Stat. **14**(4), 436–440 (1943)

16. Rom, R., Sidi, M.: Multiple Access Protocols: Performance and Analysis. Springer, New York (1990). https://doi.org/10.1007/978-1-4612-3402-9

17. Cohen, A., Cohen, A., Gurewitz, O.: Data aggregation over multiple access wireless sensors networks. arXiv preprint (2017)

18. Atia, G.K., Saligrama, V.: Boolean compressed sensing and noisy group testing. IEEE Trans. Inf. Theory **58**(3), 1880–1901 (2012). A minor corection appered in, vol. 61, no. 3, p. 1507, 2015

19. Sejdinovic, D., Johnson, O.: Note on noisy group testing: asymptotic bounds and belief propagation reconstruction. In: 2010 48th Annual Allerton Conference on Communication, Control, and Computing (Allerton), pp. 998–1003. IEEE (2010)

20. Bloch, M., Barros, J.: Physical-Layer Security: From Information Theory to Security Engineering. Cambridge University Press, Cambridge (2011)

21. Cohen, A., Cohen, A., Jaggi, S., Gurewitz, O.: Secure group testing. arXiv:1607.04849 (2016)

Towards Building Active Defense Systems
for Software Applications

Zara Perumal and Kalyan Veeramachaneni[✉]

Data to AI Lab, MIT LIDS, Cambridge, MA 02139, USA
{zperumal,kalyanv}@mit.edu

Abstract. Over the last few years, cyber attacks have become increasingly sophisticated. PDF malware – a continuously effective method of attack due to the difficulty of classifying malicious files – is a popular target of study within the field of machine learning for cybersecurity. The obstacles to using machine learning are many: attack patterns change over time as attackers change their behavior (sometimes automatically), and application security systems are deployed in a highly resource-constrained environments, meaning that an accurate but time-consuming machine learning cannot be deployed.

Motivated by these challenges, we propose an *active defender* system to adapt to evasive PDF malware in a resource-constrained environment. We observe this system to improve the f_1 score from 0.17535 to 0.4562 over five stages of receiving unlabeled PDF files. Furthermore, average classification time per file is low across all 5 stages, and is reduced from an average of 1.16908 s per file to 1.09649 s per file. Beyond classifying malware, we provide a general *active defender* framework that can be used to deploy decision systems for a variety of applications operating under resource-constrained environments with adversaries.

1 Introduction

In recent years, cyber attacks have increased dramatically in both scale and sophistication. Last spring, the WannaCry ransomware attack crippled computers around the world [9]. Soon after, attacks on the Equifax credit reporting agency compromised the personal information of millions of users [20]. In addition, banks and Bitcoin exchanges have been subject to an increasing number of attacks. Despite the wide-ranging nature of these attacks, a few commonalities exist. First, most of these attacks enter an enterprise network through an application endpoint, generally when a user unknowingly lets a file with "malware" inside the network - for example, by downloading a malicious "*pdf*" (file) that was delivered *via* an email (application). Second, the most recent attacks are increasingly attributed to Nation-State actors, or Nation-State sponsored cyber-gangs [7,19]. These powerful attackers often target individuals or small-scale enterprises. Such large adversaries can devote many more resources to attacking a system than their targets can devote to preventing such attacks– an asymmetry that presents a challenging problem [17].

© Springer International Publishing AG, part of Springer Nature 2018
I. Dinur et al. (Eds.): CSCML 2018, LNCS 10879, pp. 144–161, 2018.
https://doi.org/10.1007/978-3-319-94147-9_12

The increasing complexity and scale of software applications makes it even more difficult to monitor their use, find their vulnerabilities, and defend them against attacks. Application developers have to constantly make trade-offs, balancing between the usability, effectiveness and security of the application. Simple rule-based or signature-based defense systems, while quick to respond to attacks, are not robust enough to provide true protection. Meanwhile, the fact that large amounts of data are constantly being collected has led developers to seek machine learning-based solutions [3].

However, many significant challenges stand in the way of using machine learning for cyber security. First of all, the evolving nature of cyber attacks breaks the assumption that the historic attacks used to train a predictive model would resemble what will actually arrive when the system is deployed. Instead of trying the same type of attack over and over, attackers design automated evasive algorithms specifically to evade these deployed models and create new variants [23,30].

The second challenge stems from the complex dynamics of the security ecosystem. The actors in a given security problem generally include sophisticated attackers, overburdened security analysts, enterprises who want to defend themselves but not forgo the efficacy of their function, and end-users with a limited knowledge of how to protect themselves (and, subsequently, the enterprise). Complications might include certain detection strategies being public knowledge, or the limited availability of real-time computational resources to run sophisticated detection approaches. Existing solutions fall short in a number of ways. For instance, a highly optimized and accurate attack detection solution could be useless if it also delays people's ability to access and use the application it is defending.

To mitigate these problems, we present an *active defender* system aimed at providing accurate detection in a resource-constrained, adversarial environment.

An Active Defender System: As shown in Fig. 1, our active defender utilizes a "Synthesize-Model-Decide-Adapt" (SMDA) framework to maintain high accuracy while reducing classification time and resource usage. In this paper, we focus on a use case involving PDF malware and test how our approach performs in presence of evasive adversaries.[1].

Our Contributions Through This Paper are as Follows:

1. We present a general-purpose *Synthesize-Model-Decide-Adapt* framework to enable the building, evaluating, deploying and subsequently adapting machine learning-based application security systems.
2. We propose a multi detector based hierarchical *decision making* system. We tune the system using Bayesian optimization methods to optimize the usage of the detectors.
3. We present a simple Max-Diff approach, which we show evades even the most sophisticated attack classification system.

---

[1] Several recent studies suggest that PDF malware is evading classification using various automated methods [8,11,23,30].

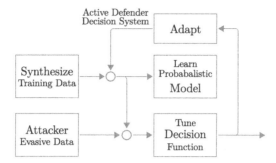

Fig. 1. Active Defender Sytem: The Active Defender system uses the "Synthesize-Model-Decide-Adapt" framework. First the system is initialized by *Synthesizing* training data, then learning several machine learning-based detection *models*, and tuning the *decision system*. After the system is deployed it is used to *decide* on new data, including evasive data generated by the attacker. After a decision is made on newly received data, the system *adapts* to update the *models* and *decision* system

The paper is organized as follows: Sect. 2 presents the use case we focus on in this paper. Section 3 presents our method for synthesizing training examples, Sect. 4 presents the classifiers we use, Sect. 5 presents a tunable decision system, and Sect. 6 presents how we adapt our system. Section 7 presents the experimental settings and our results. Section 8 presents the conclusions and discusses future work. Rather than dedicate a separate section to related work, we have included it in context across different sections.

2   Malware Through PDFs

Of all the different file types available, users trust portable document formats the most. People use portable document formats – PDFs for short – to upload everything from academic conference paper submissions to government tax forms. They are also often passed through emails as attachments. Despite their popularity, these unassuming documents contain a powerful format that enables attackers to embed and hide malicious code, spy on users, or encrypt an end user's computer in a ransomware attack [2]. In the next subsection we present multiple ways to detect malware embedded in PDFs.

2.1   PDF Malware Detection

Network Detection: Network detection aims to prevent the delivery of malicious content by intercepting it before a user has a chance to download it. Through email analysis such as spam detection or network frequency methods, enterprises can filter out anomalous behaviour and limit the phishing emails and attached malware that make it to the end user. This is usually done in combination with static or dynamic analysis.

Static Classifiers: These classifiers use static features of the PDF to quickly detect anomalies before passing the document on to the users. These methods are preferred for their low latency, but have higher error rates. Static classification methods include signature-based detection methods, which can search a received file for the unique bit strings of known malicious files. Other methods attempt to utilize higher-level features, such as n-gram analysis or JavaScript pattern recognition [16, 18]. The most successful static feature-based classifiers have been *PDFRate* and *Hidost*. The PDFRate classifier extracts 135 features based on the structure and metadata of a PDF document [25] and uses them to train a machine learning classifier (see [4, 24]). Once the model is trained, it works fast, taking less than a second to classify each new PDF. However, the classifier can be evaded by adversarial algorithms using genetic programming methods [30].

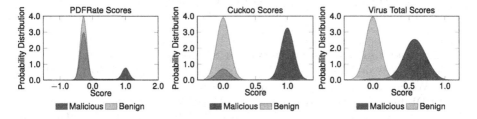

Fig. 2. Kernel density estimation approximation of Probability Density for scores generated using the PDFRate classifier (left), Cuckoo (center) and VirusTotal (right). This plot shows the classifiers' scores for malicious PDFs in pink and benign PDFs in blue. The KDE plot was generated with a Gaussian kernel of width 0.1. A higher overlap indicates an inability to accurately detect the *malicious* PDF. Predictably, out of the three methods, VirusTotal is the best (with the least overlap), and PDFRate is the worst (with the most overlap). (Color figure online)

Dynamic Behavioural Analysis: The Cuckoo Sandbox runs each PDF dynamic analysis sandbox on an isolated "sandboxed" environment. A Cuckoo server runs on the host computer, receives files, and sends them to a virtual machine for analysis. In the virtual machine, Cuckoo simulates opening PDFs in a vulnerable version of Adobe Acrobat, collects information, and compares this information to a set of known behavioural signatures. Cuckoo is fairly accurate for known malicious signatures, and usually requires 30 s of simulation time in a virtual machine. For our experiments we used virtual machines set up in VMWare, running Windows XP and a vulnerable version of Adobe Acrobat Reader 8.1.1. Using the results of the Cuckoo classification, we recorded the list of known malicious behavioural signatures observed on the virtual machine when the PDF was processed, and a Boolean indicating if any signatures were observed.

Publicly Available APIs: VirusTotal is an API-based detection system. After a user uploads a PDF, it is checked using up to 59 static, dynamic and anti-virus classifiers, and the results of this classification are returned. Some of the classifiers used by VirusTotal include Endgame, Kaspersky Antivirus, Symantec, and Sophos. Aggregating these results can result in a highly accurate classification (see Fig. 2), but can also be time- and resource-intensive: a single classification can take up to two minutes, or longer in times of high server load. Furthermore, corporations may be rate-limited by the API and may have to pay for uploads. We collect the aggregate percentage of antivirus engines that classified the uploaded PDF as *malicious*, and we also collect the following classification attributes for each of the scanners used by VirusTotal: version of scanner used, scan result, and malicious classification.

Human Expert Analysis: Using human analysts to inspect malware samples is the most accurate form of detection. However, the sheer rate of incoming PDFs requires faster methods. Analysts can compare samples through a variety of methods, including comparing network calls and memory access, running the sample on a hardware sandbox, or comparing activity on a device through a firewall.

Our Dataset and Method Evaluation Strategy: To demonstrate the efficacy of different detection techniques, we accumulated a repository of 207,119 PDFs. We created this dataset from a combination of existing, externally provided PDF files, and variations of these PDF files generated via a process called *mutation* (we will describe this briefly in Sect. 3). We gathered these PDFs from the following sources:

- Contagio: The Contagio dataset provided a corpus of 9,000 benign PDFs and 10,597 malicious files [1].
- EvadeML: EvadeML data provided by Weilin Xu contains 16440 malicious PDF files developed using the "EvadeML" algorithm. These files are based off of 500 malicious files in the Contagio data set and are designed to confuse the PDFRate classifier [30].
- Self-generated: PDFs can be generated from existing PDFs according to the "Random-Mutation","Evade-ML" [30] or "Max-Diff" algorithms. Both the "Evade-ML" and "Max-Diff" algorithms are based on genetic programming. These algorithms create pools of samples, score them, and mutate the best-scoring samples to create more malicious files. In this data set, we generated 8232 malicious files using the "Max-Diff" algorithm and 35,680 benign files using random mutation.

We thus have a total labeled set of 79,949 files: 35,269 malicious and 44,680 benign.

3   Synthesizing Training Data

To develop an adaptive machine learning solution, we need labeled training examples from the past and an ability to generate evasive versions over time.

In recent years, machine learning has been used to automatically create malware samples.

Machine Learning to Create Malicious Samples: Beyond the direct methods used to inject malicious code and create PDF files, attackers can *mutate* existing PDF malware in order to create new ones. Many recent studies have focused on methods for generating adversarial PDF files to evade machine learning classifiers.

The mimicus framework presents a method for manipulating PDF classification, through modifying mutable features and through gradient descent methods using attributes of the model [4,15]. The EvadeML framework presents a black box genetic programming approach to evade a classifier when the classification score is known [30]. The EvadeHC method evades machine learning classifiers without knowledge of the model or classification score [11]. The Seed-ExploreExploit framework presents another evasion method for deceiving black box classifiers by allowing adversaries to prioritize level diversity and accuracy to generate samples [23].

Other methods operate on the feature space and generate evasive features that could confuse classifiers; however, it is not always clear how to convert evasive features back into malicious files [13]. Many additional methods have been presented to deceive machine learning classifiers based on stationarity assumptions that do not hold in an adversarial environment [5,6,10–12,14,21,22,27]. These attacks often focus on complex classifiers, such as deep learning systems, which can be overfit to rely on features that are correlated with malware, rather than those that are necessary for malware. In [29], Wang et al. show that complex classifiers can be evaded if even one unnecessary feature is present.

EvadeML: EvadeML uses a genetic programming method to produce variants of the malicious files using a method they call *mutate*.

$f_{mutate}()$ The malicious files are mutated using components from the pool of benign files S_b. The mutation method is implemented using a modified version of the PDFRW software package[2] and works as follows:

Step 1: Load all PDF files into a tree structure.
Step 2: Mutate each malicious PDF by randomly doing one of the following:
  - Insert randomly selected sub-tree from a benign file.
  - Swap a randomly selected sub-element with a randomly selected sub-tree from the benign file.
  - Delete a randomly selected element in the malicious tree representation.
Step 3: Write the mutated tree back as a PDF file.

These variants are tested against the Cuckoo sandbox to ensure that their malicious nature is preserved, then scored using static classification scores [30]. EvadeML found variants that received classification scores of <0, with PDFRate classifier. For this classifier, scores for benign are supposed to be closer to −1. A negative score for malicious files indicate that the method was able to evade the classifier. In summary the algorithm works as follows:

---

[2] https://github.com/mzweilin/pdfrwructure.

Step 0: Start with an empty set $S_{evade} = \{\}$

Step 1: Create a set of mutant files using f_m using the set of S_m. Call this set $S_{mutants}$.

Step 2: Check which among the mutants are malicious using the oracle function $o()$. In our case this is the Cuckoo classifier. Call this set S^m_{mutant}

Step 3: Apply PDFRate classifier to the set S^m_{mutant} and generate classification scores.

Step 4: Select the mutants that have classification scores less than the *cutoff*. Add these to the set S_{evade}. These files are the ones that evade the PDFrate classifier.

Step 5: Repeat steps 1–4 until $|S_{evade}| \geq n_v$, where n_v is number of evasive variants desired.

Max-Diff Approach: We propose the *Max-Diff* approach as an alternative method for generating malicious files. It is similar to the EvadeML algorithm in that it uses a malicious and benign pool of variants, scores the malicious variants, mutates the best-scoring variants, adds them to the pool of malicious files, and then continues. However, unlike the Evade-ML algorithm, it does not seek to find files that receive a classification score less than the *cutoff* for a single classifier. Instead, it selects for files that receive different classification scores from different classifiers in the system. For the *active defender* system, Max-Diff targets files that evade PDFRate or VirusTotal. The algorithm works are follows:

Step 0: Start with an empty set $S_{evade} = \{\}$

Step 1: Create a set of mutant files using f_m and using the set of S_m. Call this set $S_{mutants}$.

Step 2: Check which among the mutants are malicious using the oracle function $o()$. In our case this is the Cuckoo classifier. Call this set S^m_{mutant}

Step 3: Apply PDFRate classifier to set S^m_{mutant} and generate classification scores as p_1

Step 4: Upload the set to the VirusTotal website and generate VirusTotal classifier scores for S^m_{mutant} as well as classification scores. Collect these scores in p_3

Step 5: Select the mutants that have abs($p_1 - p_3$) greater than a threshold specified by *cutoff*. Add these to the set S_{evade}. These files represent files that receive different scores from PDFRate vs. VirusTotal, and could confuse a classification system.

Step 6: Repeat steps 1–4 until $|S_{evade}| \geq n_v$.

Evasive Performance: We analyze these approaches based on their capability to evade PDFRate (the fastest classifier) and VirusTotal (the most accurate classifier). As we see in Fig. 3, malicious files generated using the *Evade-ML* are effective in evading PDFRate classification. Additionally, we observe that evasive files generated using the *Max-Diff* approach are especially effective at evading the VirusTotal classifier. In comparing these results we see that the more time-consuming classifier, VirusTotal, does achieve higher accuracy against evasive variants than the PDFRate classifier. However, even Virus Total is not foolproof, which motivates the need for use of human analysts.

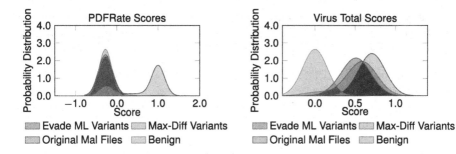

Fig. 3. KDE approximation of Probability Density for the PDFRate (top) and Virus-Total (bottom) scores. This plot shows the classification scores for different types of files. The benign files are shown as pink, the Contagio malware samples are shown in purple, the EvadeML variants are shown in blue, and the Max-Diff variants are shown in green. The KDE plot was generated with a Gaussian kernel of widths 0.15, 0.15, 0.17, and 0.17 for the Benign, Contagio, EvadeML, and Max-Diff files respectively. In this case, the Evade-ML, Max-Diff, and Benign files had very similar probability densities, and differing width kernels were used in order to distinguish them. (Color figur online)

4   Learning Models from Training Data

In our *active defender* system, we use the training data provided to us in the form of file sets S_m and S_b. We divide the classifiers into two types, *primary* and *secondary*. Primary classifiers/models take the files as input and produce a probabilistic score p. Secondary classifiers/models take the output of the primary classifiers and deliver a probabilistic score. Although all secondary models are machine learning models, not all primary models are.

Primary Classifiers: The *active defender* system uses the classifiers described in Sect. 2. These are: PDFRate (denoted as C_1), the Cuckoo classifier (denoted as C_2, which returns scores p_2 indicating if known behavioural signatures of malicious files were detected) and a VirusTotal (C_3) classifier that outputs the percentage of VirusTotal classifiers that classify the file as malicious (p_3)

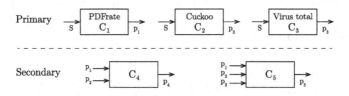

Fig. 4. Active Defender Classifiers: Primary classifiers (C_1,C_2,C_3) receive samples and produce probabilistic scores (p_1,p_2,p_3). Secondary classifiers use these probabilistic scores as inputs. Secondary classifier C_4 uses inputs (p_1,p_2) to produce the probabilistic score p_4, and C_5 uses inputs (p_1,p_2,p_3) to output the probabilistic score p_5.

Secondary Classifiers: Secondary classifiers are designed to take in output scores from the primary classifiers and learn a machine learning model. Two secondary classifiers are developed in our *active defender* system. They are:

- C_4 uses the outputs of PDFRate (C_1) and Cuckoo (C_2) as inputs and produces a probabilistic score (p_4).
- C_5 uses the outputs of PDFRate (C_1), Cuckoo (C_2), and Virus Total (C_3) as inputs and produces a probabilistic score (p_5) (Fig. 4).

5   A Tunable Decision System

In real time, in order to determine whether or not a new input file s is malicious, we apply a hierarchical decision system that *adaptively* makes use of multiple classifiers. In developing such a decision system, we considered the following goals:

- Increase throughput: We would like to make decisions about PDFs as fast as possible. Because PDFRate is the fastest in giving us the prediction (and VirusTotal is the slowest), we would like to use PDFRate as much as possible.
- Maintain accuracy: While it is easiest to increase our throughput by choosing to use the PDFRate classifier every time, this will lead to a lot of false positives if we have to maintain a high recall of 90%, and we would have to augment with Cuckoo or VirusTotal.

To achieve the goals above, we propose the following:

- a bi-level decision function for classifiers, described in Sect. 5.1,
- a hierarchical, tunable decision system, described in Sect. 5.2,
- A cost function that evaluates the efficacy of a given decision system, described in Sect. 5.3,
- a tuning algorithm that produces a viable decision system, described in Sect. 5.4.

5.1   Bi-level Decision Function

Given a classifier C_i and its output score p_i, a bi-level decision function allows us to make a decision D_i based on two decision thresholds, t_i^1 and t_i^2, as depicted in Fig. 5 and more formally given by:

$$D_i = \begin{cases} \text{Benign} & \text{if } p_i < t_i^1 \\ \text{Uncertain, output } p_i & \text{if } p_i \geq t_i^1 \text{ and } p_i < t_i^2 \\ \text{Malicious} & \text{if } p_i \geq t_i^2 \end{cases} \quad (1)$$

This allows us to make a decision when we are absolutely confident, and enables us to postpone the decision in regions where we are uncertain.

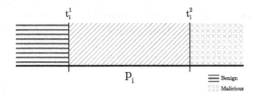

Fig. 5. Bi-level decision function. With input of p_i, the bi-level decision returns a result if it is certain of the classification. It classifies an input as benign if $p_i < t_i^1$ and malicious if $p_i \geq t_i^2$. If $t_i^1 < p_i < t_i^2$, the function returns p_i, as the result is uncertain.

5.2   Hierarchical Tunable Decision System

The hierarchical decision system is shown in Fig. 6. This system determines a final classification result (y) and a probabilistic score (P_{final}) for each input sample using layers of $bi-level$ classifiers. The P_{final} score is calculated using the output of the last classifier (p_{last}), the threshold used in the last decision

$$P_{final} = \frac{N_{pc\_used}}{N_{pc\_total}} * \text{abs}(p_{last} - t_{last})$$

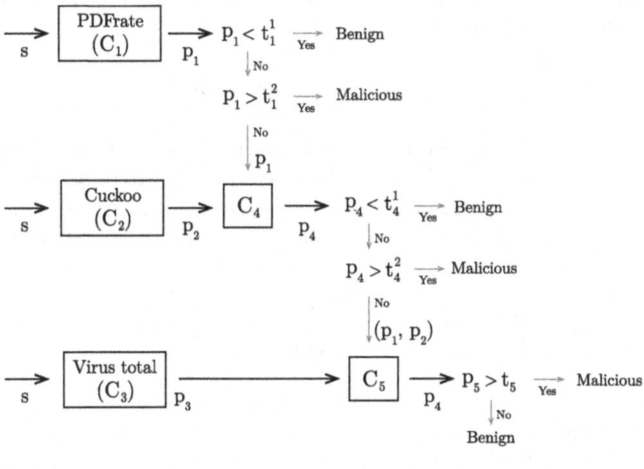

Fig. 6. Active Defender Hierarchical Decision System: A PDF is first sent to the PDFRate classifier (C_1). Based on the output of PDFRate, p_1, a decision is made whether to return a result or send the file to the Cuckoo classifier (C_2). If the file is sent to the Cuckoo classifier, the results from PDFRate (p_1), and Cuckoo (p_2) are sent to the secondary classifier C_4 and a decision is made as to whether to return a result or sent the file to VirusTotal (C_3). If the file is sent to the VirusTotal classifier, classification scores from the PDFRate (p_1), Cuckoo (p_2), and VirusTotal (p_3) classifiers are sent to the C_5 secondary classifier and a final decision is made.

(t_{last}), the number of primary classifiers used (N_{pc_used}), and the total available primary classifiers (N_{pc_total}) as shown below:

$$P_{final} = \frac{N_{pc\_used}}{N_{pc\_total}} * \text{abs}(p_{last} - t_{last}) \tag{2}$$

5.3   Cost Function

The cost function expresses the two objectives we specified above – the throughput, and whether the desired accuracy is achieved. Given a fully specified decision system, with classifiers $C_{1...5}$, decision thresholds $t_1^1, t_1^2, t_4^1, t_4^2, t_5$, and a set of files S, the cost c incurred by the system is evaluated as:

$$\mathbb{c} = -\gamma * g(\hat{Y}, Y) + (1 - \gamma) * \frac{1}{|S|} \sum_i r_i \tag{3}$$

where $\hat{Y}$ is the set of predicted labels, Y are the corresponding true labels, $g(.)$ measures the accuracy of the predicted labels, $\frac{1}{|S|} \sum_i r_i$ is the average classification time taken to make these decisions based on the subset of models used for each file in the set per sample, and γ is a weight associated with each of the factors.

$g(.)$ **function.** The $g(.)$ function describes the accuracy of a system. We provide two methods of characterizing system accuracy. In $g_1(.)$, the f1 score is optimized to improve precision and recall.

$$g_1(.) = f_1(predicted, true\_labels) \tag{4}$$

In $g_2(.)$, the function requires a minimal threshold of precision and then optimizes for recall. This function is especially applicable for malware detection, as allowing an additional malicious file to enter the system can be very costly, but is required to keep false rejection of benign files below a certain specified rate for user happiness.

$$g_2 = \begin{cases} \text{recall}(\hat{Y}, Y) & \text{if precision}(\hat{Y}, Y) \geq 0.9 \\ 0 & \text{otherwise} \end{cases} \tag{5}$$

5.4   Tuning Algorithm

Since the *active defender* system utilizes a set of thresholds to determine the decision for an input sample, *tuner* optimizes these thresholds based on their effect on a cost function. The tuning algorithm uses additional training data to optimize.

Tuner comprises two main steps. First, the tuner algorithm enumerates an initial set of classifier thresholds using an enumeration function $e()$ to generate a set of thresholds T, and scores them with the cost function g. This is done in two steps.

- First, we produce a 2 lists of possible "threshold pairs" for each pair of thresholds: $\ell_{t_1}, \ell_{t_4}, (t_1^1, t_1^2), (t_4^1, t_4^2)$ respectively.
- For the last threshold t_5 we produce a list of possible thresholds ℓ_{t_5}.
- Finally, we create ℓ_T using all possible combinations of threshold pairs across lists ℓ_{t_1}, ℓ_{t_4} and ℓ_{t_5}.

In our enumeration function $e()$ we produce threshold pairs using the 0%, 20%, 40%, 60%, 80%, and 100% percentile values of previous classification scores for that classifier. For example, if previous PDFRate classification scores p_1 were observed between 0.0 and 0.5, then:
$\ell_{t_1} \equiv \{(0.0, 0.1), (0.1, 0.2), (0.3, 0.4), (0.4, 0.5)\}$. The last threshold list is (ℓ_{t_5}) a list of the 0% , 20%, 40%, 60%, 80%, and 100% percentiles for respective score p_5. More complex enumeration functions can be developed to capture a more expressive range of thresholds.

Enumerating a large threshold set is important in systems with complex cost functions such as $g_2(.)$ which are not monotonic. If too few initial thresholds are enumerated, optimization can result in thresholds that find a local rather than a global minimum cost function value.

Second, *tuner* uses a maximum of $n_{iterations}$ of Bayesian hyperparameter tuning[3] to propose an additional candidate threshold sets, evaluate it using g, add it to the threshold set T, and find the thresholds that minimize the cost function g. In iterative tuning, ϵ specifies the minimum distance between successive minimum scores to stop optimization [26].

6   Adapting over Time

One of the most important aspects of the *active defender* system is the ability of the entire system to adapt over time, enabling it to overcome attackers who build evasive variants. Known as active learning, this adaptation can happen over time, simply by adding *verified* labeled training examples.

In [28], Veeramachaneni and Arnaldo study the use of active learning in cybersecurity. They use multiple outlier detection systems, send suspicious activity to analysts and seek their input in order to be able to provide more training examples to the machine learning model over time. We build on this idea, generating more labeled training data using a variety of possible methods:

- higher accuracy classifiers: we can incorporate predictions from VirusTotal as possible source of true labels.
- human analysts: we can send some examples to humans to get their analysis. This is expensive, but still doable.
- synthetically generated evasive variants: From time to time, we can create evasive confirmed malicious variants using the machine learning methods we described in Sect. 3.

---

[3] We make use of the open source library: https://github.com/HDI-Project/BTB.

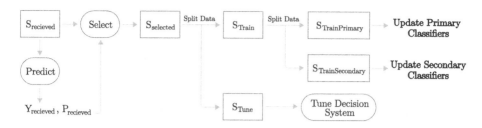

Fig. 7. Data flow diagram of how new training examples are used to adapt the system. (1) Input data $S_{received}$ sent through the decision system to produce predicted labels $Y_{received}$ and probabilities. (2) Samples are selected in using probabilities $P_{received}$. (3) The selected data $S_{selected}$ is split into S_{train} and S_{tune}. (4) The training data S_{train} is split into $S_{primary}$ used to train the primary classifiers and $S_{secondary}$ used to update the secondary classifiers. (5) The tuning data S_{tune} is used to tune the decision system

Adapt in Active Defender: In an active learning scenario, we use additional data to *update* the models and *tune* the decision system. For unlabeled data, the system generates labels and final probabilities using the predictions from the previous learned models and the decision system. The adapt algorithm uses the following steps also shown in Fig. 7.

- SELECT: chooses the files that are above a set minimum probability threshold (α) to be used as malicious training examples.
- UPDATE: uses a subset of the selected files specified by (λ) to learned model. This subset is further divided into two parts: one used to train the primary and the second used to train secondary classifiers, specified by parameter μ. In the *active defender* system, the *PDFRate* classifier (C_1) is the only primary classifier that can be retrained.
  The second part is used to update the secondary classifiers, C_4, C_5 using the new predictions of PDFRate for the labeled data.
- TUNE: uses the remaining files to tune the decision function given the enumeration function, $e()$, maximum number of tuning iterations ($n_{iterations}$), and difference between successive minimum scores ϵ.

7   Experimental Setup

In order to understand the performance of the *active defender* system, we analyze its accuracy and resource use as it adapts. In the experimental design, we first split the data into two data sets, as shown in Fig. 8.

Training Data: D_1 corresponds to data used to train the initial system. In our experimental setup, D_1 consists of the 10,597 Contagio malware files and 10,597 benign PDFs randomly selected from the 44,680 benign files discussed in Sect. 2.

Adaptation Data: D_2 is data received by the system after it is deployed. The adaptation data, D_2, consists of the evasively generated malware and remaining

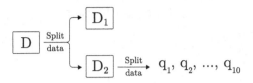

Fig. 8. Splitting Experimental Data. In the following experiment, the data is split into data sets D_1 and D_2. D_1 is used to initialize the decision system. D_2 represents data received by the system after it is deployed. D_2 is split into subsets q_i, representing the files received in each successive stage.

$$\boxed{D_1} \longrightarrow DS_1 \longrightarrow DS_2 \longrightarrow DS_3 \quad \cdots$$
$$\qquad\qquad \uparrow \qquad\qquad \uparrow$$
$$\qquad\qquad q_1 \qquad\quad q_2$$

Fig. 9. Updating the decision system. In the experiment, training data D_1 is used to initialize the decision system. After the system is deployed, it receives additional data. After each additional received dataset q_i, the decision system *adapts*.

benign PDF files. As shown in Fig. 8, this data is split into subsets $q1$ through $q5$ and is sent to the decision system across 5 stages or time periods.

Settings: We perform 25 random trials. In each trial, the order of the files is randomized giving D_1 and D_2 different files across trials. As a result the subsets $q_1 \ldots q_5$ are also different.

In setting up the decision system, we set the following parameters for the tuning algorithm described in Sects. 5 and 6. The cost function is $g_1()$; a γ value of 0.9 is used to prioritize accuracy over resource usage and an epsilon value of $\epsilon \equiv 0.1$ is set (Fig. 9).

7.1   Results

Overall, we see that the system is able to adapt to achieve high accuracy in the presence of evasive adversaries, and to reduce resource usage over time.

Accuracy: As shown in Fig. 10 and Table 1, we observe the performance of the decision system when classifying successive sets of received files. We characterize accuracy by observing the f_1 score, comparing truth versus labeled data. As evasive variants are introduced in stage 1, we observe a low f_1 score. However, as the stages progress, we observe that the system is able to adapt to improve accuracy over time.

Resource Usage: For the purposes of this experiment, we characterize resource usage by studying the average time used to classify each file. As shown in Fig. 11 and Table 1, when the system is initialized, classification time is relatively low, at around 1(s) per file. However, we observe that the classification time continues to decrease over time, indicating that the PDFRate static classifier is improving

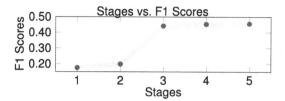

Fig. 10. Accuracy over Adaptation: In this figure, we observe the f_1 score vs. the experimental stage over time. We plot the mean f_1 score as points and show the standard deviation in the surrounding band. We observe poor results in Stage 1, when evasive samples are introduced. Over time, we observe that the f_1 score increases as the system adapts to evasive samples.

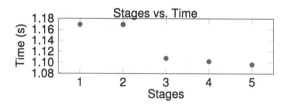

Fig. 11. Time taken to classify as the system runs for several stages: In this figure, we observe the estimated classification time per file at each stage. We plot the mean time as points and show the standard deviation in the surrounding band. Here we see that the average classification time is pretty low, around 1.16 s, and decreases throughout the course of the experiment. The deviation in time is small per stage, and is not observable due to the estimation function.

Table 1. Experimental data: 25 trials of the Active Defender System performance over 5 stages. Column μ_{F1} corresponds to the average f_1 score across all trials. Column σ_{f1} corresponds to the standard deviation in f_1 score across all trials. Column $\mu_{Time\,per\,File}$ corresponds to the average estimated classification time per file. Column $\sigma_{Time\,per\,File}$ corresponds to the standard deviation in approximated classification time per file.

Stage	μ_{f1}	σ_{f1}	$\mu_{Time\ per\ File}$	$\sigma_{Time\ per\ File}$
1	0.17535	0.01003	1.16908	<0.0001
2	0.19852	0.01459	1.16908	<0.0001
3	0.44201	0.01804	1.10766	<0.0001
4	0.45301	0.01829	1.10208	<0.0001
5	0.4562	0.02082	1.09649	<0.0001

and being utilized. In calculating the estimated classification time, we model the PDFRate as taking 1 s, Cuckoo as taking 25 s and VirusTotal as taking 90 s. These have come from our own experience running three classifiers for several thousands of PDFs. Notably, the standard deviation in classification time is too small to observe using four decimals of precision. This is likely due to the majority of files being classified by the static classifier and our estimation function limiting the variability in time.

8   Discussion and Future Work

We were able to make four contributions through this paper. First, we developed a method to use machine learning in application security in a resource-constrained environment. Second, we developed algorithms that use active learning to improve fast classifiers in the presence of adversaries. Third, we provided an extensible framework to facilitate building, evaluating and deploying decision systems in an adversarial and resource-constrained environment. Fourth, we provided a simple evasive algorithm that was shown to confuse automated classifiers.

While studying the adversarial and resource-constrained problem of detecting evasive PDF malware and building these solutions, we identified a few takeaways that motivate future work.

Evasive Approaches Motivate Adaptation Over Time: In studying the available classifiers, we were surprised to see that the *max-diff* approach was effective in evading the VirusTotal classifier. VirusTotal is a powerful classification system that has been acquired by Google and was considered to be one of the best products of 2007. If this genetic programming-based algorithm can cause confusion in malicious and benign files, this suggests that adversaries are more than capable of deploying their own evasive algorithms to evade automated classifiers. This motivates the need for human-in-the-loop systems and systems that adapt over time.

Active Defender: In studying the behaviour of the *active defender* decision system, we identified aspects of the decision and adaptation methods that could be improved upon in future work. Exploring different methods for tuning the decision system could reduce the tuning time necessary to achieve high accuracy. Studying different ways of using available primary classifiers could decrease classification time. Using randomization to select a small number of files to be sent to the most accurate classifiers can make the system more robust to files that can completely evade simple classifiers. Finally, studying more complex methods could improve adaptation – for example, automated synthesis to create new samples to improve confidence in predictions after adaptation.

8.1   Conclusion

As motivated attackers use more and more computational resources and state-of-the-art algorithms to persistently attack smaller corporations, it is necessary to figure out how to allow detection methods to *adapt* in a resource-constrained environment. As enterprises collect more and more data, machine learning can be an asset to application security; however, each institution looking to defend their system will have different limitations on the resources they can devote to analyzing this data. In this paper, we propose an active defense system that utilizes the SMDA framwork. This system can be tailored for different resource limitations and environments. Furthermore, we believe that this software framework and algorithms can generalize beyond PDF malware detection, enabling researchers and corporations to work together to secure systems against powerful and evolving adversaries.

References

1. Contagio dump. http://contagiodump.blogspot.com. Accessed 11 Nov 2016
2. The rise of document-based malware. https://www.sophos.com/en-us/security-news-trends/security-trends/the-rise-of-document-based-malware.aspx
3. The rise of machine learning (ml) in cybersecurity. https://www.crowdstrike.com/resources/white-papers/rise-machine-learning-ml-cybersecurity/
4. Mimicus framweork (2017). https://github.com/srndic/mimicus
5. Argyros, G., Stais, I., Jana, S., Keromytis, A.D., Kiayias, A.: Sfadiff: automated evasion attacks and fingerprinting using black-box differential automata learning. In: Proceedings of the 2016 ACM SIGSAC Conference on Computer and Communications Security, pp. 1690–1701. ACM (2016)
6. Argyros, G., Stais, I., Kiayias, A., Keromytis, A.D.: Back in black: towards formal, black box analysis of sanitizers and filters. In: 2016 IEEE Symposium on Security and Privacy (SP), pp. 91–109. IEEE (2016)
7. Ashford, W.: Cyber criminals catching up with nation state attacks. https://www.computerweekly.com/news/252435701/Cyber-criminals-catching-up-with-nation-state-attacks
8. Biggio, B., Corona, I., Maiorca, D., Nelson, B., Šrndić, N., Laskov, P., Giacinto, G., Roli, F.: Evasion attacks against machine learning at test time. In: Blockeel, H., Kersting, K., Nijssen, S., Železný, F. (eds.) ECML PKDD 2013. LNCS (LNAI), vol. 8190, pp. 387–402. Springer, Heidelberg (2013). https://doi.org/10.1007/978-3-642-40994-3_25
9. Bossert, T.P.: It's official: north korea is behind wannacry, December 2017. https://www.wsj.com/articles/its-official-north-korea-is-behind-wannacry-1513642537
10. Chen, Y., Nadji, Y., Kountouras, A., Monrose, F., Perdisci, R., Antonakakis, M., Vasiloglou, N.: Practical attacks against graph-based clustering. arXiv preprint arXiv:1708.09056 (2017)
11. Dang, H., Huang, Y., Chang, E.C.: Evading classifiers by morphing in the dark (2017)
12. Hosseini, H., Xiao, B., Clark, A., Poovendran, R.: Attacking automatic video analysis algorithms: a case study of google cloud video intelligence API. arXiv preprint arXiv:1708.04301 (2017)
13. Hu, W., Tan, Y.: Generating adversarial malware examples for black-box attacks based on gan. arXiv preprint arXiv:1702.05983 (2017)
14. Kantchelian, A., Tygar, J., Joseph, A.: Evasion and hardening of tree ensemble classifiers. In: International Conference on Machine Learning, pp. 2387–2396 (2016)
15. Laskov, P., et al.: Practical evasion of a learning-based classifier: a case study. In: 2014 IEEE Symposium on Security and Privacy (SP), pp. 197–211. IEEE (2014)
16. Li, W.-J., Stolfo, S., Stavrou, A., Androulaki, E., Keromytis, A.D.: A study of malcode-bearing documents. In: M. Hämmerli, B., Sommer, R. (eds.) DIMVA 2007. LNCS, vol. 4579, pp. 231–250. Springer, Heidelberg (2007). https://doi.org/10.1007/978-3-540-73614-1_14

17. MacFarlane, D., Network, I.C.: Why even smaller enterprises should consider nation-state quality cyber defenses, September 2017. https://www.csoonline.com/article/3223866/cyberwarfare/nation-state-quality-cyber-defenses.html

18. Maiorca, D., Corona, I., Giacinto, G.: Looking at the bag is not enough to find the bomb: an evasion of structural methods for malicious PDF files detection. In: Proceedings of the 8th ACM SIGSAC symposium on Information, Computer and Communications Security, pp. 119–130. ACM (2013)

19. Millman, R.: Nation state cyber-attacks on the rise - detect lateral movement quickly, February 2018. https://www.scmagazineuk.com/nation-state-cyber-attacks-on-the-rise-detect-lateral-movement-quickly/article/746561/

20. Riley, M., Robertson, J., Sharpe, A.: The equifax hack has the hallmarks of state-sponsored pros, September 2017. https://www.bloomberg.com/news/features/2017-09-29/the-equifax-hack-has-all-the-hallmarks-of-state-sponsored-pros

21. Rosenberg, I., Shabtai, A., Rokach, L., Elovici, Y.: Generic black-box end-to-end attack against RNNs and other API calls based malware classifiers. arXiv preprint arXiv:1707.05970 (2017)

22. Sethi, T.S., Kantardzic, M.: Data driven exploratory attacks on black box classifiers in adversarial domains. arXiv preprint arXiv:1703.07909 (2017)

23. Sethi, T.S., Kantardzic, M., Ryu, J.W.: 'Security theater': on the vulnerability of classifiers to exploratory attacks. In: Wang, G.A., Chau, M., Chen, H. (eds.) PAISI 2017. LNCS, vol. 10241, pp. 49–63. Springer, Cham (2017). https://doi.org/10.1007/978-3-319-57463-9_4

24. Smutz, C., Stavrou, A.: Malicious PDF detection using metadata and structural features. In: Proceedings of the 28th Annual Computer Security Applications Conference, pp. 239–248. ACM (2012)

25. Smutz, C., Stavrou, A.: When a tree falls: using diversity in ensemble classifiers to identify evasion in malware detectors. In: NDSS (2016)

26. Swearingen, T., Drevo, W., Cyphers, B., Cuesta-Infante, A., Ross, A., Veeramachaneni, K.: ATM: a distributed, collaborative, scalable system for automated machine learning. In: IEEE International Conference on Big Data (2017)

27. Tong, L., Li, B., Hajaj, C., Vorobeychik, Y.: Feature conservation in adversarial classifier evasion: a case study. arXiv preprint arXiv:1708.08327 (2017)

28. Veeramachaneni, K., Arnaldo, I., Korrapati, V., Bassias, C., Li, K.: Ai^2: training a big data machine to defend. In: 2016 IEEE 2nd International Conference on Big Data Security on Cloud (BigDataSecurity), IEEE International Conference on High Performance and Smart Computing (HPSC), and IEEE International Conference on Intelligent Data and Security (IDS), pp. 49–54. IEEE (2016)

29. Wang, B., Gao, J., Qi, Y.: A theoretical framework for robustness of (deep) classifiers under adversarial noise. arXiv preprint arXiv:1612.00334 (2016)

30. Xu, W., Qi, Y., Evans, D.: Automatically evading classifiers. In: Proceedings of the 2016 Network and Distributed Systems Symposium (2016)

Secure Non-interactive User Re-enrollment in Biometrics-Based Identification and Authentication Systems

Ivan De Oliveira Nunes[1,2(✉)], Karim Eldefrawy[1], and Tancrède Lepoint[1]

[1] SRI International, Menlo Park, USA
ivanoliv@uci.edu
[2] University of California Irvine, Irvine, USA
{karim.eldefrawy,tancrede.lepoint}@sri.com

Abstract. Recent years have witnessed an increase in demand for biometrics based identification, authentication and access control (BIA) systems, which offer convenience, ease of use, and (in some cases) improved
security. In contrast to other methods, such as passwords or pins, BIA
systems face new unique challenges; chiefly among them is ensuring
long-term confidentiality of biometric data stored in backends, as such
data has to be secured for the lifetime of an individual. Cryptographic
approaches such as Fuzzy Extractors (FE) and Fuzzy Vaults (FV) have
been developed to address this challenge. FE/FV do not require storing
any biometric data in backends, and instead generate and store helper
data that enables BIA when a new biometric reading is supplied. Security of FE/FV ensures that an adversary obtaining such helper data
cannot (efficiently) learn the biometric. Relying on such cryptographic
approaches raises the following question: *what happens when helper data
is lost or destroyed (e.g., due to a failure, or malicious activity), or when
new helper data has to be generated (e.g., in response to a breach or
to update the system)*? Requiring a large number of users to physically
re-enroll is impractical, and the literature falls short of addressing this
problem. In this paper we develop SNUSE, a secure computation based
approach for non-interactive re-enrollment of a large number of users in
BIA systems. We prototype SNUSE to illustrate its feasibility, and evaluate
its performance and accuracy on two biometric modalities, fingerprints
and iris scans. Our results show that thousands of users can be securely
re-enrolled in seconds without affecting the accuracy of the system.

1 Introduction

Current Biometrics-based Identification and Authentication (BIA) systems[1] [1,
2] typically store in the backend a reference Biometric Templates (BTs) of a
users' biometrics, such as fingerprint minutiae points and/or iris code. If the

---

[1] We omit explicitly mentioning access control, we assume it implicitly when authenticating an individual and then granting access based on the authenticated identity.

© Springer International Publishing AG, part of Springer Nature 2018
I. Dinur et al. (Eds.): CSCML 2018, LNCS 10879, pp. 162–180, 2018.
https://doi.org/10.1007/978-3-319-94147-9_13

backend is compromised or breached, sensitive biometric data of a large number of users may be leaked and enable adversaries to impersonate users and circumvent BIA systems. For example, in 2015, the Office of Personnel Management was compromised and led to the leakage of 5.6 million fingerprints of federal workers that applied for security clearances in the United States of America [3].

In order to protect BTs, typical solutions use secure elements or encryption. Secure elements are applicable when matching is performed against a few biometrics (e.g., the Touch ID technology on iPhones) but do not scale to a large number of biometrics. Encrypting BTs also brings challenges: (1) BTs have to be decrypted to match against the newly supplied biometric readings, (2) the decryption keys associated with the encrypted templates have to be stored somewhere close by (logically and maybe even physically) to perform matching, and (3) when the backend is compromised or breached, the encrypted templates may be leaked. As reference biometrics templates have to be protected for the lifetime of individuals, it is important to select algorithms and key sizes that remain secure for several decades (say, 40–50 years), which remains a challenging task.

A third cryptographic approach to address the above shortcomings and to construct secure BIA systems has been proposed in the form of Fuzzy Vaults (FV) [4] and Fuzzy Extractors (FE) [5]. FE and FV alleviate the need to store BTs in the system's backend; they enable to perform matching during the normal operation of a system from some Helper Data (HD), extracted during the user's initial enrollment into the system. The HD securely encodes[2] a secret, or cryptographic key, that cannot be retrieved unless the same (noisy) biometric used to generate the HD is provided as input. Thus, the system can determine if a new reading of a biometric corresponds to the user being authenticated. The security of the approach stems from the fact that HDs do not convey any information about the underlying biometric; and this guarantee can be information-theoretic/statistical or computational depending on the details of the FE/FV scheme.

To deploy this third cryptographic approach at scale, one has to address the challenge of re-enrollment: in the lifetime of a cryptographically secure BIA system, it will be often necessary to re-generate HDs because of breaches or to perform maintenance or updates. For example, the secret key may need to be revoked and replaced by a fresh one when the HD is leaked, damaged, or corrupted. A second example is when access control is enforced via encryption, e.g., via attribute-based encryption [6]; in this case, changing a given user's permissions implies changing the user's cryptographic keys. In both examples, existing schemes require physical presence of the user to refresh cryptographic material, which is laborious, slow, costly, and hence impractical. This problem is currently not addressed in the literature.

Our Contributions. In this work we introduce SNUSE (Secure Non-interactive User at Scale Enrollment), a new approach for secure user re-enrollment that does not require user involvement, nor storing BTs, in the clear or in encrypted

---

[2] We use the term "encode" very loosely here as helper data may not actually encode the secret, it may only enable constructing it when the biometric is also present.

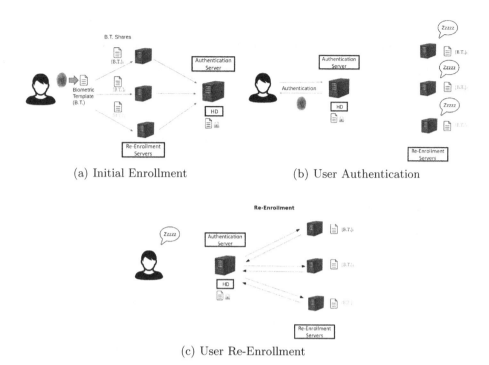

(a) Initial Enrollment   (b) User Authentication

(c) User Re-Enrollment

Fig. 1. Initial user enrollment, user authentication, and user re-enrollment in SNUSE. During regular authentication, the user interacts with the authentication server only, which stores the HD of all the users and enables recovering the user's secret/key when the correct biometric is supplied. When re-enrollment is required, the authentication server communicates with the re-enrollment servers and uses MPC to compute new HD, which encodes a new secret/key, from the secret-shared BT. We emphasize that regular authentication does not require involvement of the re-enrollment servers and, conversely, the re-enrollment phase does not require user involvement.

form, in any single backend server. Instead, SNUSE uses secret sharing to distribute the original biometric templates (BTs) among several offline components, and performs the re-enrollment in a secure distributed manner by computing the required FE/FV HD generation algorithms using efficient secure multiparty computation (MPC). This approach ensures that at no point during the system's operation (after the initial enrollment) are original reference BTs reconstructed in the clear. The components storing shares of BTs can remain offline and inaccessible through the network during normal operation; when re-enrollment is required, the components are brought online and connected to the system for only a brief period (e.g., seconds). Figure 1 illustrates the initial enrollment, subsequent authentication, and re-enrollment phases in SNUSE. To the best of our knowledge, this is the first approach amending BIA systems to enable performing non-interactive user re-enrollment without storing the user's biometric template in clear, or encrypted, in any single component or server. A detailed comparison with related work can be found in Sect. 4. Due to space constraints, we defer the details of the implementation, security analysis, and experimental results of a SNUSE prototype using Fuzzy Vaults to the extended version of this manuscript[3]. Our experimental results demonstrate a high detection accuracy with over 90% Genuine Acceptance Rate (GAR) and less than 5% False Acceptance Rate (FAR). We also show that re-enrollment of thousands of users using standard computing servers only takes a few seconds.

2 Background

We overview here the building blocks used in SNUSE, and introduce the notation that will be used in the rest of the paper.

2.1 Biometrics-Based Authentication

During user enrollment into a BIA system, a reference Biometric Template (BT), composed of features uniquely identifying an individual, is sampled and stored at an Authentication Server (AS). Later, when a user tries to authenticate, a biometric reading is collected and the same feature extraction process is applied to generate a second BT. This new BT is compared to the stored one, and if their similarity exceeds a given threshold, the user is successfully authenticated, otherwise, user authentication fails. A feature extraction procedure, applied to a biometric sample of a user U, results in a BT that can be represented as:

$$BT_U = \{p_1, \ldots, p_M\}, \tag{1}$$

where each p_i is a data point representing a unique detail of U's biometric. For example, fingerprint BTs include location and orientation of *minutiae*, i.e., of regions in the image where fingerprint lines merge and/or split. Such minutiae points are encoded as:

$$p_i = (x_i, y_i, \theta_i), \tag{2}$$

where, x_i is the x coordinate, y_i is y coordinate, and θ_i the orientation of the minutiae i extracted from U's fingerprint. Analogous methods can be proposed to encode other biometrics such as iris scans.

Traditional BIA systems store BTs of thousands (or even hundreds of thousands) of users in clear form. The reason for this is that each time a biometric is sampled by a sensor, it is slightly different due to noise. Standard mechanisms for secure storage of passwords (e.g., salted-hashing) cannot match two noisy readings of the same biometric because they are not exactly the same. Unfortunately, the advantages of biometrics come with a high risk; if the leaked biometric's modality is stable, leakage of a user's biometric at any point in time affects security of *all* authentication systems using this biometric for years. Consequently, protecting the confidentiality of biometric data is of utmost importance.

---

[3] Available at: http://sprout.ics.uci.edu/people/ivan/pubs/2018_snuse.pdf.

As discussed in Sect. 1, Fuzzy Extractors (FE) and Fuzzy Vaults (FV) are cryptographic solutions that use an input BT to (i) generate Helper Data (HD), which encodes a secret/key k; and (ii) ensure that the HD does not reveal anything about the BT or k, unless prompted with (a noisy version of) the same biometric used to generate the HD. Note that k can be an arbitrary secret and not necessarily a symmetric key. In current approaches, the secret k stored in the HD cannot be refreshed without requiring the physical (or remote) presence of user for the re-enrollment process in which a new biometric sample is collected and new HD is computed for the new secret. This manual approach does not scale as re-enrolling thousands (or more) users is impractical; this paper proposes a MPC-based solution that enable large-scale automatic re-enrollments.

2.2 Secret Sharing

In K-out-of-N secret sharing [7] a dealer distributes a secret among N parties such that subsets of K or more parties can recover it; knowing up to $K - 1$ shares leaks no information about it. In SNUSE, we will generate N shares of a biometric template and store them on N re-enrollment servers. Given a secret X, let $[X]_j$ denote the j-th share and denote the generation of N shares of X by:

$$\{[X]_1, \ldots, [X]_N\} \leftarrow X. \tag{3}$$

Denote the reconstruction of secret X from K shares by:

$$X \leftarrow \{[X]_1, \ldots, [X]_K\}. \tag{4}$$

In Shamir's (K, N) secret sharing scheme [7] over a finite field $\mathbb{F}$, one randomly generates the coefficients of a polynomial P of degree $d = K - 1$. The independent term a_0 is then set to the secret X, and one can generate N secret shares $\Phi = \{(i, P(i))\}_{i=1}^N$. Since P has degree $K - 1$, K points in Φ are enough to interpolate P and reconstruct its coefficients, including the secret $a_0 = X$. A set of $L < K$ shares does not reveal any information about X because there's an infinite number of polynomials of degree K that can be constructed from these points.

Summary of Assumptions and Guarantees: Shamir's secret sharing is information theoretic secure. In a K-out-of-N scheme, it is guaranteed that less than K shares of the secret leak no information about such secret. Conversely, K or more shares can be used to reconstruct the secret.

2.3 Secure Multi-party Computation (MPC)

MPC protocols enable mutually distrusting parties to jointly compute a function f of their private inputs while revealing no information (other than the output of f) about their inputs to the other parties [8].

In standard algebraic MPC protocols, each party usually generates shares of its input (using, for instance, Shamir's secret sharing scheme) and distribute one share to each party. A key observation is that if one is able to compute both addition and multiplication on the shares, such that the resulting shares can be combined into the correct result of the operations, one can implement any function f from these two basic operations. Different schemes were proposed to compute addition and multiplication over private inputs [9–13]. Nevertheless, most of them share the following common characteristics in the computation of these operations:

- **Addition** of secret shares can be computed locally. To that purpose each party computes addition of its own secret shares. The N local results, once combined, yield the result of an addition of the actual secret(s).
- **Multiplication** of secret shares requires communication. Even though different schemes exist, most require parties to broadcast an intermediate (blinded) result during the computation of multiplication, such that individual shares of the multiplication result can be correctly computed.

We do not specify details of the multiplication sub-protocols and we refer the reader to [11,12] for further details. The takeaway is that multiplication requires communication between parties, and in practice the overhead to multiply is usually orders of magnitude higher than the overhead of addition. In our design we take advantage of the unique characteristics of a biometric enrollment system to reduce the number of multiplications to a minimum (sometimes even eliminating it), allowing cost-effective scalable MPC-based user re-enrollment.

Summary of Assumptions and Guarantees: MPC assumptions and guarantees vary depending on the specific MPC scheme used. In this work we focus on the honest-but-curious (HBC) threat model with honest majority, in which corrupted parties might collaborate to learn private inputs of other parties, but they do not deviate from the protocol specification. The scheme remains secure if the number of colluding parties is smaller than half of the total number of parties.

2.4 Fuzzy Vault Scheme

A Fuzzy Vault (FV) scheme [4] is designed to work with a BT represented as an unordered set of points as shown in Eq. (1). FVs allow using a user's biometric template BT_U to hide a secret k. The secret can be, for instance, some private data or a cryptographic key.

A FV scheme consists of two algorithms, a generation part (FV_{GEN}) and a secret reconstruction part (FV_{OPEN}), which can be informally defined as follows:

- FV_{GEN}: receives as input the secret k and a biometric template BT_U. It uses BT_U and k to generate a Helper Data HD. This HD encodes the secret k without revealing information about k and BT_U:

$$HD = FV_{GEN}(BT_U, k) \tag{5}$$

- FV_{OPEN}: receives as input the HD and BT'_U. It retrieves k from HD if, and only if, BT'_U is a template extracted from the same biometric as extracted in the input BT_U to FV_{GEN} when generating HD. In other words, given a distance function D and a threshold w:

$$FV_{OPEN}(BT'_U, HD) = \begin{cases} k & \text{if } D(BT_U, BT'_U) \leq w \\ \perp & \text{otherwise} \end{cases}. \tag{6}$$

The threshold w is a security parameter that allows control of the trade-off between minimizing false acceptance (revealing k to the wrong user) and false rejection (refusing to reveal k to the rightful user).

FV_{GEN} algorithm starts by selecting a polynomial P of degree d defined over a field $GF(2^\tau)$ and encoding (or splitting) the secret k into the $d+1$ coefficients $a_0, \ldots, a_d$ of P. The resulting polynomial is defined as:

$$P_k(x) = \sum_{i=0}^{d} a_i x^i \tag{7}$$

where the coefficients $\{a_0, \ldots, a_d\}$ are generated from k and can be used by anyone to reconstruct k. Since P is defined over $GF(2^\tau)$, each coefficient can encode τ bits; this implies that the maximum size of the secret k that can be encoded is $(d+1) \times \tau$. After encoding k as $P_k(x)$, each of the M data points in BT_U is evaluated with the polynomial $P_k(x)$ generating a set of points in a two-dimensional space:

$$L_P = \{(p_1, P_k(p_1)), \ldots, (p_M, P_k(p_M))\}. \tag{8}$$

Note that the field must be large enough to encode a data point from BT_U as a single field element. The resulting set L_P contains only points in the plane that belong to the polynomial $P_k(x)$. In addition to L_P, a set of chaff points L_S of size $S \gg M$ is generated by randomly selecting pairs (r_x, r_y), where r_x and $r_y \in GF(2^\tau)$ resulting in:

$$L_S = \{(r_{x1}, r_{y1}), \ldots, (r_{xS}, r_{yS})\} \tag{9}$$

Finally, L_P and L_S are shuffled together using a random permutation π and the result is included in the HD:

$$\pi(L_P + L_S) \in HD \tag{10}$$

The HD also includes the set of public parameters $\Phi = \{\mathbb{F}, d, M, H(k)\}$, where $\mathbb{F}$ is the field in which $P_k(x)$ is defined and d is its degree, M is the size of BT_U, i.e., the number of points in the HD that belong to $P_k(x)$, and $H(k)$ is a cryptographic hash of the secret k allowing one to verify if the correct secret was reconstructed using FV_{OPEN}.

The key idea behind security of the FV scheme is that with $d+1$ distinct points $(p_i, P_k(p_i))$, one can interpolate $P_k(x)$, retrieve its coefficients and thus recover k. However, finding which $d + 1$ points to interpolate out of the $M + S$ in HD is hard if $M + S$ is sufficiently larger than d.

When attempting to reconstruct k from the HD using a new biometric reading BT'_U, the FV_{OPEN} algorithm will use a distance function (which must be defined according to the biometric type) to select, out of the $M + S$ points in the HD, the M points that are the closest matches to the points in BT'_U. If, out of the M selected points, at least $d+1$ points are points that belong to the original L_P, then the algorithm will be able to interpolate the correct polynomial and recover k. To verify that k was correctly recovered, the algorithm hashes the result and compares it to $H(k)$, which was published together with the HD. If less than $d + 1$ correct points are among the M points selected via distance matching, no interpolation with combinations of $d+1$ points out of M will yield a match in the hash, because $P_k(x)$ will not be interpolated correctly. Therefore, FV_{OPEN} will reject BT'_U.

Note that the FV scheme does not rely on the order of the elements in BT_U and BT'_U and does not require all points to be present in both templates. Instead, $d+1$ data points in BT'_U must be close enough to points in BT_U. In that sense, the polynomial degree d acts as a security parameter that allows calibration of the scheme to reduce false acceptance by increasing the required number of matching data points.

Summary of Assumptions and Guarantees: *The security of FV relies on the infeasibility of the polynomial reconstruction problem* [14]. *The scheme's security relies on the inability to distinguish the statistical distribution of minutiae from that of chaff points. The degree d of the polynomial used to encode k determines the minimal number coincidental minutiae that are necessary to reveal k.*

3   The SNUSE Approach

Figure 1 illustrates the operations of SNUSE during (a) initial user enrollment, (b) regular user authentication, and (c) non-interactive user re-enrollment. SNUSE involves three types of parties: a User (U), an Authentication Server (AS), and n Re-Enrollment Servers $(\{RES_1, \ldots, RES_n\})$.

During the initial enrollment of U into the system, U's biometric template BT_U is secret shared into n shares $([BT_U]_1, \ldots, [BT_U]_n)$; each share is distributed to one of the n Re-Enrollment Servers $(\{RES_1, \ldots, RES_n\})$. The RESs then jointly generate U's secret k_U and use an MPC protocol to compute FV helper data, HD_{k_U}, from their shares $([BT_U]_1, \ldots, [BT_U]_n)$, thus securely "locking" k using the user's BT_U.

Regular user authentication happens between U and AS. Since AS only stores HD_{k_U} (which reveals nothing about BT_U), the FV_{OPEN} algorithm must be used to retrieve k. Now, if HD_{k_U} was correctly computed using a secure FV_{GEN} algorithm, the only way to retrieve k from HD_{k_U} is by providing a second biometric template BT'_U which is close enough to the original BT_U used in computation of FV_{GEN}. Therefore, $FV_{OPEN}(BT'_U, HD_{k_U})$ will successfully retrieve k if and only if $BT'_U \approx BT_U$, i.e., BT'_U is a noisy version of the same biometric used to generate the HD. After this stage, k can be used in standard cryptographic primitives, e.g., decrypt a user's files or messages, or grant them access to resources based on the recovered secret.

Whenever k needs to be replaced with a fresh secret k' (this process may happen periodically within an organization to guarantee freshness of users' cryptographic keys), RESs are brought online and connected to the system, and AS issues a request to the RESs to compute a new HD for U. The RESs will securely generate a new k' and (as in the enrollment protocol) use U's secret shares $[BT_U]_1, \ldots, [BT_U]_n$, stored during U's enrollment, to compute a new $HD_{k'_U}$. This way, users' cryptographic material can be refreshed and brand new HD can be constructed without bringing the user in to re-sample her biometric and without storing her biometric template in clear.

At a first sight, a simple approach for generating a fresh k' would be to multiply the y-coordinates in the FV by a random σ, which would result in a fresh random key $k' = k \cdot \sigma$ encoded in the FV. However, this approach does not work for several reasons. First, as discussed before, k might not be an independent random byte-stream, such as a symmetric key; it could, for instance, include a set of user permissions or an asymmetric private-key (sk') associated with the user's public-key. In the latter case, the generation of fresh $k' = sk'$ implies deriving the corresponding new public-key (pk'). SNUSE can handle these cases while simple multiplication by randomness can not. Second, while this approach using randomization might work for a classic FV with symmetric keys, because the key is encoded as coefficients of a polynomial, it does not necessarily apply to other FV/FE constructions; SNUSE on the other hand is a generic approach (as any computable function can be computed using MPC), that could be implemented with other FV/FEs. Third, even in the case where the scheme uses a classic FV to encode a random symmetric key, multiplying by randomness σ will update the encoded secret but will prevent the reconstruction of the secret, i.e., $k' = FV_{OPEN}(BT'_U, HD)$ cannot be computed; this is because FV_{OPEN} must verify the hash of each candidate recovered key with the stored $H(k)$ to decide if the correct key was reconstructed. However, $H(k')$ cannot be updated in the same way, because for any reasonable hash function, $H(\sigma \cdot k) \neq \sigma \times H(k)$.

Throughout the rest of this section we describe the steps of SNUSE in more details. We have implemented SNUSE with fingerprints and iris scans, and evaluated SNUSE performance in terms of computation and storage requirements. Our evaluation shows that SNUSE can re-enroll thousands of users in seconds; the storage requirements for thousands of users is a few MBytes. Though SNUSE does not affect the accuracy of the underlying biometric feature extraction tool, we also provide accuracy results for completeness. For implementation and evaluation details, we defer the reader to the extended version of this paper.

Remark 1. All message exchanged in the following protocols are assumed to be through secure authenticated channels, such as standard TLS. Establishing such secure channels is omitted from the protocols for the sake of clarity.

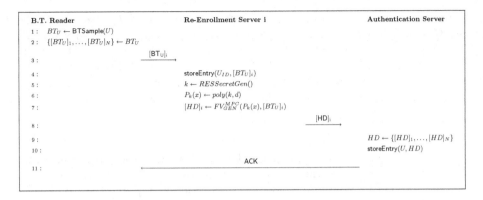

Fig. 2. User enrollment protocol in SNUSE

3.1 Initial User Enrollment

The initial user enrollment, presented in Fig. 2, is the only protocol in SNUSE that involves all parties, i.e., U, AS, and RES_i $\forall i \in [1, N]$. This protocol is executed only once for each user, all interactions after the initial enrollment are performed either between U and AS (authentication), or between AS and $RESs$ (re-enrollment).

The protocol starts with U using a trusted enrollment device (e.g., fingerprint sensor, iris reader), referred to as B.T. Reader, to measure and extract U's biometric template BT_U (Fig. 2, line 1). BT_U is then split into N secret shares, where N is the number of $RESs$. Each share $[BT_U]_i$ transmitted to, and stored by, the respective RES_i (Fig. 2, lines 2–4). Note that in Fig. 2 we only depict one RES_i, however, in reality each of the N $RESs$ receives and stores its share $[BT_U]_i$.

Once each share $[BT_U]_i$ is stored on the corresponding RES_i, the RESs agree on the new authentication material k for the user, by using $RESSecretGen$ (Fig. 2, line 5). $RESSecretGen$ may be implemented by simply generating k at one of the RES_i and transmitting the value of k to the other $RESs$ through secure channels; alternatively, group key agreement protocols could be utilized depending on the structure of k. Each RES then encodes k as a polynomial of degree d, denoted as $P_k(x)$. The polynomial $P_k(x)$ is the same polynomial described in the standard FV scheme (see Sect. 2). The difference is that, instead of generating the HD with BT_U, as in the standard FV scheme, the protocol uses $FV_{GEN}^{MPC}(P_k(x), [BT_U]_i)$, which uses MPC to compute a share of the HD ($[HD]_i$) from the secret shared $[BT_U]_i$. Each of the $RESs$ computes the same function with its own share. Finally, each share $[HD]_i$ is then sent to AS, which reconstructs the actual HD from the received shares and acknowledges the completion of the enrollment protocol. This is depicted in Fig. 2, lines 9–11. Finally, AS stores an entry for newly generated HD, associated with user U.

Note that, during user enrollment protocol, BT_U is only visible in clear to the trusted sensor device that reads and then secret shares the biometric. Each RES_i

only sees its own share which leaks no information about BT_U itself. AS only sees HD, which can not be used to reconstruct BT_U by the security of FV. Therefore, confidentiality of the biometric is ensured during user enrollment. In fact, there is no single server from which BT_U can be retrieved in clear. BT_U only exists in clear ephemerally at the B.T. Reader and that must happen anyway because B.T. Reader is the sensor device used to sample the user's biometric.

3.2   User Authentication

A consequence of correct computation of HD using MPC in the enrollment phase is that the user authentication protocol consists of simply using standard local computation of FV_{OPEN} with a new biometric reading BT'_U and the stored HD. The RESs do not participate in user authentication, but only in the enrollment and re-enrollment protocols.

The authentication protocol, shown in Fig. 3, starts with a user supplying its ID (U_{ID}) and biometric sample to the B.T. Reader. A biometric template BT'_U is generated from the new sample and kept locally. U_{ID} is sent to AS which fetches U's HD from its database based on the supplied U_{ID}, and sends the associated HD as a reply. BT'_U is then used to extract k from HD using the FV_{OPEN} algorithm. Note that here, and similar to user enrollment, U's BT only exists in clear in the B.T. Reader.

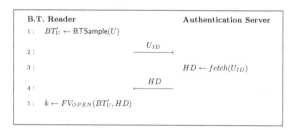

Fig. 3. User authentication protocol in SNUSE

After retrieving the secret k, the user can authenticate to the server using, for example, standard challenge-response mechanisms based on the secret k. Suppose, for instance, that k is actually part of an asymmetric encryption scheme (sk, pk), where $k = sk$ and pk is a public key, known to AS, associated to the secret key sk. AS can then authenticate the user by sending a challenge $Enc_{pk}(nonce)$, where $nonce \leftarrow_\$ \{0, 1\}^n$. If the user is able to retrieve k from the FV, it can then use $k = sk$ to compute $nonce \leftarrow Dec_{sk}(Enc_{pk}(nonce))$ and send the result back to AS, proving its claimed identity.

3.3   Non-interactive User Re-enrollment

Non-interactive user re-enrollment works by having the RESs compute a new HD based on a fresh secret k'. Since the shares of the biometric template are stored during the initial enrollment, this step does not need user involvement, even though the biometric template does not exist in clear neither at RESs nor AS.

In this protocol, AS sends a request for re-enrollment for each RES_i, containing the ID of the user. The RESs will then jointly generate a new k' for the user and encode it as a polynomial of degree d. Each RES_i then uses U_{ID} to fetch the secret share $[BT_U]_i$ associated with U_{ID} and uses FV_{GEN}^{MPC} to compute $P_{k'}(x)$ on the secret share $[BT_U]_i$. The result is a secret share $[HD]_i$ for a brand new HD which encodes the freshly generated k' under the same biometric template BT_U. This process is depicted on Fig. 4. Finally, AS receives all N shares $[HD]_i$ and compute the new HD, which can, from this point on, be used for authenticating user U with the protocol in Fig. 3. Notice that, during the execution of the re-enrollment protocol, BT_U is not reconstructed in clear at any point. This is only possible because of the computation of HD using MPC over the secret shares. Otherwise, this process would require either (i) user involvement to collect another biometric reading or (ii) storing the biometric template in clear in the backend servers.

Authentication Server		Re-Enrollment Server i
	$\xrightarrow{\quad U_{ID} \quad}$	
1 :		
2 :		$k' \leftarrow RESSecretGen()$
3 :		$P_{k'}(x) \leftarrow poly(k', d)$
4 :		$[BT_U]_i \leftarrow fetch(U_{ID})$
5 :		$[HD]_i \leftarrow FV_{GEN}^{MPC}(P_{k'}(x), [BT_U]_i)$
6 :	$\xleftarrow{\quad [HD]_i \quad}$	
7 :	$HD \leftarrow \{[HD]_1, \ldots, [HD]_N\}$	

Fig. 4. Re-enrollment protocol in SNUSE

3.4   Using MPC to Generate the HD

The fundamental part of SNUSE that allows non-interactive user re-enrollment (without storing BT_U in clear) is the ability to compute the HD from the secret shares $\{[BT_U]_1, \ldots, [BT_U]_N\}$ of BT_U. In the protocols of Figs. 2 and 4, this is represented by the computation of the function $FV_{GEN}^{MPC}(P_k(x), [BT_U]_i)$, resulting in a secret share $[HD]_i$ that can be interpolated to reconstruct the actual HD.

In this section, we discuss how FV_{GEN} is computed from the secret shares. We start by outlining the basic operations needed to compute FV_{GEN}, and then describe details of how each is performed using secret shared data. The standard local computation of FV_{GEN} algorithm, involves three basic types of operations:

1. Evaluation of the polynomial $P_k(x)$, that encodes k, on each of the M data points ($\{p_1, \ldots, p_M\}$) that compose BT_U to generate the list of points in the polynomial P_k:

$$P = \{(p_1, P_k(p_1)), \ldots, (p_i, P_k(p_i)), \ldots, (p_M, P_k(p_M))\} \tag{11}$$

2. Generation of a list S composed of s random chaff points (r_x, r_y), to be shuffled together with the polynomial points:

$$S = \{(r_{x,1}, r_{y,1}), \ldots, (r_{x,i}, r_{y,i}), \ldots, (r_{x,s}, r_{y,s})\} \tag{12}$$

3. Random permutation π to shuffle the elements of lists R and P together generating the HD:

$$HD = \pi(P \cup R) \tag{13}$$

Steps 2 and 3 are relatively easy to compute when compared step 1. For the random chaff point generation (Step 2), each RES_j computes a random share $[(r_{x,i}, r_{y,i})]_j$. When the randomly generated shares are merged together they will result in random chaff points.

For Step 3, all M RESs agree on a single random permutation π, and all of them permute their secret shares according to this same randomly chosen permutation. Note that the permutation π is kept secret from AS, because knowing π would allow an adversary who takes control of AS to separate chaff points from the points in the polynomial by computing π, allowing reconstruction of BT_U. Nevertheless, even though AS does not know which permutation was used, because each RES use the same permutation to compute $\pi(P \cup R)$ on their shares, each share will be matched to its correct set of shares during the reconstruction of the HD at AS.

Since we are able to generate random chaff points and compute a permutation on the secret shares (which results in a permutation on the reconstructed HD), the remaining task for FV_{GEN} is to compute the polynomial P_k using MPC on each of the secret shares. We discuss the classic approach to compute P_k using MPC and then we introduce our optimized version that takes advantage of pre-computation of secret exponents before the secret sharing phase.

In the classical approach, assuming that BT_U is composed of M data points, each secret share $[BT_U]_j$ corresponding to RES_j would be a vector in the form:

$$[BT_U]_j = \{[p_1]_j, \ldots, [p_M]_j\} \tag{14}$$

The polynomial P_k can be generically defined as:

$$P_k(x) = \sum_{i=0}^{d} a_i x^i \tag{15}$$

where:

$$\{a_0, \ldots, a_d\} \leftarrow k \tag{16}$$

denoting that the coefficients of the polynomial encode k. Therefore, computing $P_k(x)$ implies computing exponentiation up to degree d on the secret shared variables, multiplication of the resulting values by the respective constants $a_0, \ldots, a_d$, and addition on the resulting terms $a_i x^i$ for all $i \in [1, \ldots, d]$.

As discussed in Sect. 2, the bulk of the overhead on MPC comes from multiplication of secret shares, because addition (and consequently multiplication by a constant) can be computed locally by adding the secret shares. Multiplication, on the other hand, requires communication, since the parties must publish (broadcast) intermediate results to all other parties that hold secret shares. Since each multiplication involves network delays this is usually the major source of overhead in the MPC evaluation.

The computation of x^d, with no optimization, requires d multiplication operations, i.e., computing $\prod^d x$. In such approach, computing all terms in Eq. (15) would take $\sum_{i=0}^{d} i$. Therefore, the number of communication rounds to compute the HD shares would be:

$$T = M * \sum_{i=0}^{d} i \tag{17}$$

In terms of asymptotic complexity this *naive* approach yields $\Theta(d \times log(d))$ multiplications, where d is the polynomial degree. Such number of multiplications can be trivially reduced to d if we take into account that $x^n = x \cdot x^{n-1}$. This implies that the result of a lower order polynomial term can be used as an intermediate result for the computation of the subsequent term, reducing the number of communication rounds to compute the HD to:

$$T = M * d \tag{18}$$

resulting in linear asymptotic complexity of $\Theta(d)$ for the number of required multiplications.

To make the process more efficient, we take a different approach. We bring the number of necessary multiplications to zero by pre-computing the powers of x before distributing the shares to the RESs. From the standard BT_U, which is a vector of M data points in the format $\{p_1, \ldots, p_M\}$, we pre-compute the x^i exponents for all $i \in [1, d]$ and secret share each of the pre-computed exponents for each data point. Therefore, a secret share $[BT_U]_j$ of a biometric template with pre-computed exponents becomes a $d \times M$ matrix in the format:

$$[BT_U]_j = \begin{pmatrix} [(p_1)^1]_j & \cdots & [(p_M)^1]_j \\ \vdots & & \vdots \\ [(p_1)^i]_j & & [(p_M)^i]_j \\ \vdots & & \vdots \\ [(p_1)^d]_j & \cdots & [(p_M)^d]_j \end{pmatrix} \tag{19}$$

Each column in $[BT_U]_j$ represents a data point p_i and each line an exponentiation of such data point. For example, line 3, column 4, would contain a secret share of the fourth data point in BT_U raised to the cubic power: $[(p_4)^3]_j$.

In this approach, a secret share is included for each of the exponents of each data point. Thus, the evaluation of the polynomial requires no multiplications of secret shares because all exponents are now individual secret shares in the matrix $[BT_U]_j$. Therefore, the computation of $[HD]_j$ to be done locally. Specifically, let $[BT_U]_j(x, y)$ denote the element in line x and column y of the matrix[4]. Then $[y]_j = P_k([p_i]_j)$ can be computed locally for every i as:

$$P_k([p_i]_j) = a_0 + \sum_{k=1}^{d} a_k \times [BT_U]_j(k, i) \tag{20}$$

where $\{a_0, \ldots, a_d\}$ are the polynomial coefficients defined in Eq. (15).

This eliminates the need for network communication making the scheme much faster. Such approach is feasible because in practice $d \approx 10$ and because one single entity (B.T. Reader) possesses all data points for enrollment. This is a necessary condition for user enrollment in biometric authentication systems (the BT must be read by a given sensor). We take advantage of that to improve efficiency of SNUSE by pre-computing the exponents and including them in the secret shares of the biometric templates. As detailed in the experiments available in the extended version of this paper, by pre-computing the exponentiations, SNUSE achieves high performance in terms of processing time with reasonable storage requirements that are comfortably within the capacity of modern computers.

Algorithm 1. FV_{GEN}^{MPC} computation on RES_j

---

inputs : Secret share matrix $[BT_U]_j$ (Eq. (19)); fresh secret k; random
         permutation π; number of chaff points s; and polynomial degree d.
output: $[HD]_j$
1   $\{a_0, \ldots, a_d\} \leftarrow$ EncodeAsPolynomialCoeffs(k, d);
2   $L_p \leftarrow$ emptyList()
3   **forall the** $i \in [1, 2, \ldots, M]$ **do**
4 | $pi_x \leftarrow [BT_U]_j(1, i)$
5 | $pi_y \leftarrow a_0 + \sum_{k=1}^{d} a_k \times [BT_U]_j(k, i)$ /* MPC */
6 | $L_p.append([pi_x, pi_y])$
7   **end**
8   $L_s \leftarrow$ emptyList()
9   **forall the** $i \in [1, 2, \ldots, s]$ **do**
10 | $r_x \leftarrow rand_{GF(2^\tau)}()$
11 | $r_y \leftarrow rand_{GF(2^\tau)}()$
12 | $L_s.append([r_x, r_y])$
13   **end**
14   $L \leftarrow concat(L_p, L_s)$
15   $[HD]_j \leftarrow permute(L, \pi)$
16   **return** $[HD]_j$;

---

[4] In our notation the first row/column of a matrix is indexed by 1 and not 0.

Algorithm 1 synthesizes what is discussed in this section with a formalization of the method to compute a share of a HD using MPC on the matrix $[BT_U]_j$ of Eq. (19). AS receives all shares $[HD]_j \ \forall j \in [1, N]$ and use them to reconstruct $HD \leftarrow \{[HD]_1, \ldots, [HD]_N\}$. Note that the MPC evaluation in line 5 of Algorithm 1 only involves additions and multiplication by constants. Therefore, Algorithm 1 can be computed locally at RES_j not requiring communication.

3.5   Secret (k) Confidentiality Discussion

SNUSE is designed to provide non-interactive re-enrollment, allowing one to refresh the stored secret k without interaction and without compromising the confidentiality of the BT. One may argue that the attack surface for an attacker interested in stealing the user's secret k, instead of the BT, will increase because now all RESs need to ephemerally store K at some point in time to enable computation of the new vault. This restriction can be addressed by generating k in a single separate server and using MPC with k as a secret share as well. We consider this optional in SNUSE design, as our focus is to protect the BT itself. Having such feature would add one communication round to the re-enrollment process, because the polynomial coefficients that encode k need to be multiplied by BT shares. For more details on the security analysis and implementation of SNUSE we refer the reader to the extended version of this paper.

4   Related Work

A study [15] of security and privacy challenges facing biometrics, especially iris based ones, investigated suitability and viability of relying on them as the sole method for identification and authentication. The results of the study in terms of accuracy and entropy of extracted keys were both positive and encouraging. The first Fuzzy Vault (FV) scheme was developed in [4]. It was later implemented and tested with actual fingerprint biometrics in [16]. Subsequently, Fuzzy Extractors (FE) were formalized in [17], and further applied to biometrics in [18]. Most FE schemes provided statistical or information-theoretic security, until the scheme of [5] was developed; this computational FE scheme relies on hardness of the Learning with Errors (LWE) problem.

Secure two/multi-party computation (2PC/MPC) has been an active area of research for the past three decades [9, 19–21]. Recent models and practical schemes [22] provide a trade off between security and privacy guarantees on one hand, and required computation and communication on the other. For example, the covert model [23] accounts for settings where the involved parties are less likely to cheat if they get caught with a high probability (e.g., 0.5) and the work in [24] proposes protocols in which a malicious adversary may learn a single (arbitrary) bit of additional information about the honest party's input. Generic 2/MPC protocols can be utilized as is in cryptographically secured BIA systems but may incur higher overhead. If performance of generic protocols is unsatisfying, one can design function specific secure protocols for the generate function of FE/FV; several function-specific two and multiparty protocols for pattern matching were also developed in [25, 26].

5   Conclusion and Future Work

We study secure re-enrollment in cryptographically secured biometrics-based identification and authentication (BIA) systems. We argue that addressing this issue is paramount for real-life deployments of such systems and ensuring long-term confidentiality of biometrics. We develop a new approach for Secure Non-interactive Users at Scale Enrollment (SNUSE) and prototype it. Our experimental results (available in extended version of this paper) show a high BIA detection accuracy and efficient re-enrollment for thousands of users, e.g., in a few seconds, using standard computing servers. As interesting future work we highlight (1) the implementation of SNUSE with other FE/FV schemes, e.g., computational and reusable ones; and (2) adding support to other biometrics in addition to fingerprints and iris, and other types of devices, e.g., smart-phones.

Acknowledgement. This work was funded by the US Department of Homeland Security (DHS) Science and Technology (S&T) Directorate under contract no. HSHQDC-16-C-00034. The views and conclusions contained herein are the authors' and should not be interpreted as necessarily representing the official policies or endorsements, either expressed or implied, of DHS or the US government.

References

1. Jain, A.K., Bolle, R.M., Pankanti, S.: Biometrics: Personal Identification in Networked Society, vol. 479. Springer, New York (2006). https://doi.org/10.1007/978-0-387-32659-7
2. Ratha, N.K., Connell, J.H., Bolle, R.M.: Enhancing security and privacy in biometrics-based authentication systems. IBM Syst. J. **40**(3), 614–634 (2001)
3. Wikipedia: Office of Personnel Management data breach. https://en.wikipedia.org/wiki/Office_of_Personnel_Management_data_breach. Accessed 5 Dec 2017
4. Juels, A., Sudan, M.: A fuzzy vault scheme. Des. Codes Cryptogr. **38**(2), 237–257 (2006). https://doi.org/10.1007/s10623-005-6343-z
5. Fuller, B., Meng, X., Reyzin, L.: Computational fuzzy extractors. In: Sako, K., Sarkar, P. (eds.) ASIACRYPT 2013. LNCS, vol. 8269, pp. 174–193. Springer, Heidelberg (2013). https://doi.org/10.1007/978-3-642-42033-7_10
6. Goyal, V., Pandey, O., Sahai, A., Waters, B.: Attribute-based encryption for fine-grained access control of encrypted data. In: Proceedings of the 13th ACM Conference on Computer and Communications Security, pp. 89–98. ACM (2006)
7. Shamir, A.: How to share a secret. Commun. ACM **22**(11), 612–613 (1979)
8. Yao, A.C.: Protocols for secure computations. In: 23rd Annual Symposium on Foundations of Computer Science, SFCS '08, pp. 160–164. IEEE (1982)
9. Goldreich, O., Micali, S., Wigderson, A.: How to play any mental game. In: Proceedings of the Nineteenth Annual ACM Symposium on Theory of Computing, pp. 218–229. ACM (1987)

10. Damgård, I., Pastro, V., Smart, N., Zakarias, S.: Multiparty computation from somewhat homomorphic encryption. In: Safavi-Naini, R., Canetti, R. (eds.) CRYPTO 2012. LNCS, vol. 7417, pp. 643–662. Springer, Heidelberg (2012). https://doi.org/10.1007/978-3-642-32009-5_38

11. Rabin, T., Ben-Or, M.: Verifiable secret sharing and multiparty protocols with honest majority. In: Proceedings of the Twenty-First Annual ACM Symposium on Theory of Computing, pp. 73–85. ACM (1989)

12. Beaver, D.: Efficient multiparty protocols using circuit randomization. In: Feigenbaum, J. (ed.) CRYPTO 1991. LNCS, vol. 576, pp. 420–432. Springer, Heidelberg (1992). https://doi.org/10.1007/3-540-46766-1_34

13. Cramer, R., Damgård, I., Maurer, U.: General secure multi-party computation from any linear secret-sharing scheme. In: Preneel, B. (ed.) EUROCRYPT 2000. LNCS, vol. 1807, pp. 316–334. Springer, Heidelberg (2000). https://doi.org/10.1007/3-540-45539-6_22

14. Kiayias, A., Yung, M.: Cryptographic hardness based on the decoding of Reed-Solomon codes. In: Widmayer, P., Eidenbenz, S., Triguero, F., Morales, R., Conejo, R., Hennessy, M. (eds.) ICALP 2002. LNCS, vol. 2380, pp. 232–243. Springer, Heidelberg (2002). https://doi.org/10.1007/3-540-45465-9_21

15. Itkis, G., Chandar, V., Fuller, B.W., Campbell, J.P., Cunningham, R.K.: Iris biometric security challenges and possible solutions: for your eyes only? Using the iris as a key. IEEE Sig. Process. Mag. 32(5), 42–53 (2015)

16. Nandakumar, K., Jain, A.K., Pankanti, S.: Fingerprint-based fuzzy vault: implementation and performance. IEEE Trans. Inf. Forensics Secur. 2(4), 744–757 (2007)

17. Dodis, Y., Reyzin, L., Smith, A.: Fuzzy extractors: how to generate strong keys from biometrics and other noisy data. In: Cachin, C., Camenisch, J.L. (eds.) EUROCRYPT 2004. LNCS, vol. 3027, pp. 523–540. Springer, Heidelberg (2004). https://doi.org/10.1007/978-3-540-24676-3_31

18. Boyen, X., Dodis, Y., Katz, J., Ostrovsky, R., Smith, A.: Secure remote authentication using biometric data. In: Cramer, R. (ed.) EUROCRYPT 2005. LNCS, vol. 3494, pp. 147–163. Springer, Heidelberg (2005). https://doi.org/10.1007/11426639_9

19. Yao, A.C.-C.: How to generate and exchange secrets. In: Proceedings of the 27th Annual Symposium on Foundations of Computer Science. SFCS 1986, pp. 162–167. IEEE Computer Society, Washington, DC (1986). https://doi.org/10.1109/SFCS.1986.25

20. Ben-Or, M., Goldwasser, S., Wigderson, A.: Completeness theorems for non-cryptographic fault-tolerant distributed computation. In: STOC 1988. ACM (1988)

21. Damgård, I., Keller, M., Larraia, E., Pastro, V., Scholl, P., Smart, N.P.: Practical covertly secure MPC for dishonest majority – or: breaking the SPDZ limits. In: Crampton, J., Jajodia, S., Mayes, K. (eds.) ESORICS 2013. LNCS, vol. 8134, pp. 1–18. Springer, Heidelberg (2013). https://doi.org/10.1007/978-3-642-40203-6_1

22. Archer, D.W., Bogdanov, D., Pinkas, B., Pullonen, P.: Maturity and performance of programmable secure computation. In: IEEE S & P (2016)

23. Aumann, Y., Lindell, Y.: Security against covert adversaries: efficient protocols for realistic adversaries. J. Cryptol. 23(2), 281–343 (2010). https://doi.org/10.1007/s00145-009-9040-7

24. Huang, Y., Katz, J., Evans, D.: Quid-Pro-Quo-tocols: strengthening semi-honest protocols with dual execution. In: IEEE S & P 2012. IEEE Computer Society (2012)

25. Hazay, C., Lindell, Y.: Efficient protocols for set intersection and pattern matching with security against malicious and covert adversaries. In: Canetti, R. (ed.) TCC 2008. LNCS, vol. 4948, pp. 155–175. Springer, Heidelberg (2008). https://doi.org/10.1007/978-3-540-78524-8_10
26. Baron, J., El Defrawy, K., Minkovich, K., Ostrovsky, R., Tressler, E.: 5PM: secure pattern matching. In: Visconti, I., De Prisco, R. (eds.) SCN 2012. LNCS, vol. 7485, pp. 222–240. Springer, Heidelberg (2012). https://doi.org/10.1007/978-3-642-32928-9_13

Brief Announcement: Image Authentication Using Hyperspectral Layers

Guy Leshem$^{(\boxtimes)}$ and Menachem Domb$^{(\boxtimes)}$

Department of Computer Science, Ashkelon Academic College, Ashkelon, Israel
leshemg@cs.bgu.ac.il, menachem.domb@gmail.com

Abstract. Access control systems using face recognition, are widely implemented. This technic lacks the ability to bypass it. To avoid it, an authentication process is required. In this paper we propose a new security image-signature, which authenticates the given image. The proposed signature is generated from the corresponding hyperspectral image layers. The process extracts unique patterns from the hyperspectral layers, these are collected to build a unique biometric signature for the related person. Experiments show the potential of enhancing image authentication using the proposed signature.

Keywords: Face recognition · Hyperspectral image · Inner layer
Multiple identification algorithms

1 Introduction

Common face recognition algorithms use features extracted from an image to identify a person. Other algorithms use several images to build a 3D model. However, all can be bypassed. To resolve this issue, we propose an enhanced identification process, which combines current face recognition along with an image signature. The signature is semi arbitrary, and as such, it provides a significant addition to complement the image authentication. This combination is reliable and difficult to break as it captures distinctive unique patterns, while preserving the ability to recover the signature.

2 Literature Review

The use of hyperspectral methods in face recognition have been used to improve traditional face recognition. Pan et al. [1] explored hyperspectral face recognition in the near-infrared spectral range. Denes et al. [2] used visible bands to test spectral asymmetry. Chang et al. [3] fused hyperspectral images in the visible spectrum to validate the improvement of image fusion to face recognition. Cho et al. [4] used an automatic selection framework for the optimal alignment method, to improve the performance of face recognition. Ghasemzadeh and Demirel [5] introduced a three-dimension discrete wavelet transform based feature extraction for the classification hyperspectral.

© Springer International Publishing AG, part of Springer Nature 2018
I. Dinur et al. (Eds.): CSCML 2018, LNCS 10879, pp. 181–183, 2018.
https://doi.org/10.1007/978-3-319-94147-9_14

3   The Proposed System

A spectral camera, generates multiple and separated layers of the same image. Each layer is analyzed separately to generate a unique binary signature string, which then contributes to the image signature. The image-signature generation process: (1) Obtains a hyperspectral image; (2) Separates the image into layers, where each layer is a 2D binary image; (3) Analyzes layer features, selects function and generates a unique binary string and; (4) Concatenates the string to previous strings and composes the image-signature. The image signature is relatively strong, because it is captured from different locations of the layer, the extract function is one of a list per layer, the extracted string represents a detected pattern, unknown to the public, and the person's image itself has already been confirmed.

4   Experiment of the Proposed Process

Below is an outline of the experiment we conducted.

1. Obtain the image layers: Run the **converting** function (MATLAB software) to separate the different five layers.
2. First layer analysis using the CascadeObjectDetector function, which is characterized by its high recognition rate. The output bit string of this stage is: 10111100101000100001101100111001111011000110110000101011010
3. Second layer: use the Local Binary Patterns algorithm, which divides the image into squares, count the pixels in it. Then it builds a binary string for each square. The algorithm can identify features, such as the line of the lips, the eyes and its inner components. The output of this stage is 10101110101111000100, and 1001001100001000000.
4. Third layer: used the Circular Hough Transform algorithm to find circles in an image. It was directed to find pupils, provide the min/max size of a pupil and finde the darkest circle in the eye. The output of this stage is: 10000111
5. Forth layer: a combination of the angles between the eyes, and the distance between the two eyes. The eye detection is done using Viola Jones. The output is: 100110010101
6. Fifth layer performed by **bwarea** function. It measures the area of all the white pixels in the image by a schema of their area. The output is: 1100110010001000010110
7. Composition of image-signature is done by concatenating the above strings: 1001010010100101010100110011010010100001100110010001000101101000101000001100010110010101110

We used 30 face hyperspectral images. For the signature composition we selected 5 layers. We ran the system with a mix of classified and unclassified images. In each test we compared an un-classified image with a classified image. For simplicity, we skipped the basic image comparison and focused on generating the image signatures and then compared the signatures of the two images. Below are the results of three test cases:

Case 1: First signature: 1111110100111100010111001001001011100101110000 01100011011110000100100110 0 0
Second signature: 1000111001110110110001101010101001110101001011011111 0000
Case 2: First signature: 1110000001101010010010011101001001100001000000 11011001110111000001 1 1
Second signature: 1011110010100010000110110011100111101100011011000010 1 01110101101100110011000111000110001000010000 0 100100
Case 3: First signature: 1100011100111011011000110101010100111010100101 10 111110000
Second signature: 1100011100111011011000110101010100111010100101101 1 1111 0000

5 Conclusions and Future Work

Person identification is required for access control and similar needs. In this work we deal with the reliability of standard image processing used for person identification. To enhance its authentication, we propose the use of image-signature in addition to the existing standard image identification. To generate the image signature, we require a hyperspectral image, which is a multi-layer image. The proposed process, accepts the hyperspectral image, analyzes it and generates a unique signature. To verify a face image, we require the signature together with face image. In future work we plan to improve the image signature process and reliability, further test various cases and apply this approach to other areas.

References

1. Zhihong, P., Healey, G., Prasad, M., Tromberg, B.: Face recognition in hyperspectral images. IEEE Trans. Pattern Anal. Mach. Intell. **25**(12), 1552–1560 (2003). https://doi.org/10.1109/tpami.2003.1251148
2. Denes, L., Metes, P., Liu, Y.: Hyperspectral face database. Technical report CMU-RI-TR-02-25, Robot. Institute Carnegie Mellon University, Pittsburgh, PA (2002)
3. Chang, H., Harishwaran, H., Yi, M., Koschan, A., Abidi, B., Abidi, M.: An indoor and outdoor, multimodal, multispectral and multi-illuminant database for face recognition. Paper presented at the 2006 Conference on Computer Vision and Pattern Recognition Workshop (CVPRW 2006), New York, 17–22 June 2006
4. Cho, W., Jang, J., Koschan, A., Abidi, M.A., Paik, J.: Hyperspectral face recognition using improved inter-channel alignment based on qualitative prediction models. Opt. Expr. **24**(24), 27637–27662 (2016)
5. Ghasemzadeh, A., Demirel, H.: Hyperspectral face recognition using 3D discrete wavelet transform. Paper presented at the 2016 Sixth International Conference on Image Processing Theory, Tools and Applications (IPTA), Finland, 12–15 December 2016

Brief Announcement: Graph-Based and Probabilistic Discrete Models Used in Detection of Malicious Attacks

Sergey Frenkel[✉] and Victor Zakharov

Federal Research Center "Computer Science and Control"
Russian Academy of Sciences, Moscow, Russia
fsergei51@gmail.com, VZakharov@ipiran.ru

Abstract. Design of program secure systems is connected with choice of mathematical models of the systems. A widely-used approach to malware detection (or classification as "benign-malicious") is based on the system calls traces similarity measurement. Presently both the set-theoretical metrics (for example, Jaccard similarity, the Edit (Levenshtein) distance (ED) [1]) between the traces of system calls and the Markov chain based models of attack effect are used. Jaccard similarity is used when the traces are considered as a non-ordering set. The Edit Distance, namely, the minimal number of edit operations (delete, insert and substitute of a single symbol) required to convert one sequence to the other, is used as it reflects the traces ordering and semantics. However, the time and space complexity of the edit distance between two strings requires quadratic (in symbol numbers) complexity [1]. The traces can also be represented as a system calls graphs [2], the nodes of which are the system calls (or the items of the q-grams [1]). That is, we can consider the traces description by the ordered string as a partial case of the graph representation, for which it is possible to use the same similarity metrics with the same computational complexity.

This work demonstrates a framework for combining both graph-based and probabilistic models enabling both the analysis of the system robustness to malicious attacks and malicious codes recognition and detection.

1 Introduction

Design of program secure systems is connected with choice of mathematical models of the systems. A widely-used approach to malware detection (or classification as "benign-malicious") is based on the system calls traces similarity measurement. Presently both the set-theoretical metrics (for example, Jaccard similarity, the Edit (Levenshtein) distance (ED) [1]) between the traces of system calls and the Markov models of attack effect are used. Jaccard similarity is used when the traces are considered as a non-ordering set. The Edit Distance, namely, the minimal number of edit operations (delete, insert and substitute of a single symbol) required to convert one sequence to the other, is used as it reflects the traces ordering and semantics. However, the time and space complexity of the edit distance between two strings requires quadratic (in symbol

© Springer International Publishing AG, part of Springer Nature 2018
I. Dinur et al. (Eds.): CSCML 2018, LNCS 10879, pp. 184–187, 2018.
https://doi.org/10.1007/978-3-319-94147-9_15

numbers) complexity [1]. The traces can also be represented as a system calls graphs [2], the nodes of which are the system calls (or the items of the q-grams [1]). That is, we can consider the traces description by the ordered string as a partial case of the graph representation.

The Markov models of the malicious behavior detection [2] have more appropriate complexity, but they do not reflect the semantics of the traces.

In this Announcement, we demonstrate a framework for combining both graph-based and probabilistic models enabling both the analysis of the system robustness to malicious attacks and malicious codes recognition and detection, which allow overcome partially these drawbacks (more details there are in [3, 4]).

2   Automaton Based Malicious Attack Model

Known presently approaches to modeling the effect of malicious attacks based on the use of metrics similarity traces, implicitly suggest the presence of some differences in the traces of benign and malicious codes. We consider an application program model as a Finite State Machine (FSM), corresponding to the program system call graph mentioned above. This FSM can be built either from a program source or from system calls sequences [2–4].

Malicious Behavior Model: $\{(a_i, a_j) \rightarrow (a_i, a_k)\}$, where $\{(a_i, a_j), (a_i, a_k)\}$ are the inter-state transitions in the FSM, represented the program with normal behavior, which was changed to the transition in the states a_k due to an attack. No new states arise.

Let the system calls level (parametrized by the system calls parameters, e.g. Windows API) trace of a program which may be subjected to an attack is represented by an Mealy automaton (FSM) $s_{t+1} = \delta(x_{t+1}, s_t)$, where δ is the FSM transition function, x_t and s_t are input and state variables correspondingly. A way of the FSM building from a system calls sequences see, e.g., in [3, 4]. Vector x (corresponding to the system calls parameters) are independent random input variables. The malicious behavior is described as the Markov chain (MC) $Z_t = \{M_t, F_t, t \geq 0\}$, where $\{M_t, t \geq 0\}$ is MC corresponding to the FSM with a set S of n states functioning without effect of any attack (altering the flow graph, corresponding to the transition function δ) under independent random inputs, and $\{F_t, t \geq 0\}$ is the MC based on the same FSM but exposed by some altering transition. Then Z_t is MC with state space $S^2 = S \times S$ of pairs (a_i, a_j), $a_i, a_j \in S$. The size of the MC is $n(n-1) + 2$ with two absorbing states (See details in [3]).

We characterize the security regarding a malicious attack as the probability of event that the trajectories (states, transitions and outputs) of M_t and F_t will be coincided after the termination the attack causing a flow graph deviation, before a clock t when outputs of both FSMs (underlying these MCs) become mismatched.

3   Examples of Joint Use of the Markov Model and Similarity Metric Based Models

Let us consider a segment of API system calls trace [2] represented as a system call graph in the Fig. 1(a) or in more abstract, as Fig. 1(b), and an attack provokes execution

NtCreateSection(SectionHandle = C,..., FileHandle = B) instead of NtCreateFile(File-Handle = A,..., ObjectAttributes = "Sample.exe"). We assume in this example that in the fragment the "Exit" state is formed normally if either "NtQueryInformationFile" is executed (assuming that after receiving information about the file object we finish the sub-task), or *NtMapViewofSection*. Using a methodology described in [3] we can receive an FSM, representing the transitions of the system calls corresponding to the call graph. Then, in accordance with definition of malicious behavior by an attack mentioned above, this attack altering the program flow graph as in Fig. 1(b) (arrow to node SC2) is described in terms of this automaton malicious transformation (see [3]). Then, we can estimate the probability of this attack manifestation. Obviously, this probability depends on the probability of whether it is produced in the normal functioning "FileHandle = B" or "FileHandle = A". That is the FSM input data play a role in the abstraction of the program behavior by affecting the branching choice probability, namely, the probability that input data provide the choice just a given branch. For example, let's this probability is equal to 0.4, which is the probability that result of node 2 Fig. 1(b) activates the exit from the module. Then, using technique [3], the probabilities that the output values (say, output of the node 4 (SC3) in Fig. 1(b)) of the program has already manifested itself to the given clock as corrupted, what means the attack detection, can be obtained by the solution of Kolmogorov-Chapman equation for above Markov chain Z_t, and we get that this probability is about 0.3, that is this attack has some chances to be manifested and detected. In contrast to the Markov Chain based models [2] mentioned above, our model can indicate explicitly the specific malicious transformation of the system calls sequence, and, therefore, the probability computed can be connected with a similarity metric (Jaccard, ED). Indeed, since the probability considered above determines the frequency of possible appearances of traces corruption as a result of the attacks, in order to fully characterize the security of the software and hardware being developed to malicious codes, we should relate these probabilities to the probabilities of correct traces classification as malicious, which is determined by similarity/distance (in a given metric) between the considered (suspicious) trace and corresponding clusters of malicious program traces. Let us consider two API traces fragments:

X = LoadLibrary NtCreateFile NtQueryInformationFile
Y = LoadLibrary NtCreateSection NtmapViewOfSystem (the right branch).
(the left branch of the Fig. 1(a))

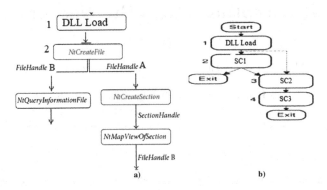

Fig. 1. System call graph of a trace fragment (a) and its abstraction (b).

The Jaccard similarity J(X, Y) of their 3-g representation (see [4]), which is the probability that the string will be assigned to the same cluster [1], that is, that the malicious trace will be identified as a normal, is equal 17/50. Taking into account above probability of the result of the malicious attack manifestation, and obvious independence of these two events (the attack manifestation and classification), we may define the probability of undetection of this attack as $0.3 \times 17/50$.

Thus, due to using the Markov Chain-based model which allows (in contrast to [2]) to specify possible malicious system calls traces transformation, we can use together two classes of the malicious attacks detection models with quadratic of the MC states, number of which less essentially than the trace length [4], while the Jaccard (or ED) metric should be calculated only for the calls, affected by the attack.

Acknowledgements. Research partially supported by the Russian Foundation for Basic Research under grants RFBR 18-07-00669 and 18-07-00576.

References

1. Leskovec, J., Rajaraman, A., Ullman, J.: Mining of Massive Datasets. Cambridge University Press, Cambridge (2014)
2. Maggi, F., Matteucci, M., Zanero, S.: Detecting intrusions through system call sequence and argument analysis. Trans. Dependable Secure Comput. 7(4), 381–396 (2010)
3. Frenkel, S., Zakharov, V., Basok, B.: Technical report of FRC "Computer Science and Control" of RAS, Moscow, Russia (2017). http://www.ipiran.ru/publications/Tech_report.pdf
4. Frenkel, S., Zakharov, V.: Technical report of FRC "Computer Science and Control" of RAS, Moscow, Russia (2018). http://www.ipiran.ru/publications/Report FR_Zakh.pdf

Intercepting a Stealthy Network

Mai Ben Adar Bessos[1(✉)] and Amir Herzberg[1,2]

[1] Bar-Ilan University, 5290002 Ramat Gan, Israel
mai.bessos@gmail.com, amir.herzberg@gmail.com
[2] University of Connecticut, Storrs, CT 06269-2157, USA

Abstract. We investigate an understudied threat: *networks of stealthy routers (S-Routers)*, communicating across a restricted area. S-Routers use short-range, low-energy communication, detectable only by nearby devices.

We examine algorithms to intercept S-Routers, using one or more mobile devices, called *Interceptors*. We focus on *Destination-Search* scenarios, in which the goal of the Interceptors is to find a (single) destination S-Router, by detecting transmissions along one or more paths from a given (single) source S-Router. We evaluate the algorithms analytically and experimentally (simulations), including against a parametric, optimized S-Routers algorithm.

Our main result is an Interceptors algorithm which bounds the expected time until the destination is found to $O\left(\frac{N}{\hat{B}}\log^2(N)\right)$, where N is the number of S-Routers and $\hat{B}$ is the average rate of transmission.

1 Introduction

Stealthy wireless communication channels have been widely deployed and studied, already since World War I, and mainly for (human) intelligence. Stealthy channels involve a stealthy *source*, communicating to a remote *destination*. Advanced stealthy transmission and encoding methods were developed to hide the transmissions and location of the source, while ensuring reliable communication to the remote destination. Counter-intelligence efforts involved deployment of intercepting-devices (*Interceptors*), deploying advanced techniques to detect the communication and locate the stealthy source. See details in [5–7,10,15,20] (additional related topics are surveyed in the draft of the full version of this work [3]).

Recent advances in Wireless Sensor Networks (WSNs) introduce a new variant of stealthy communication: a *stealthy network*. In a stealthy network, communication is relayed along a *path* consisting of stealthy devices, which we refer to as S-Routers. The S-Routers are covert devices, 'hidden' within a restricted-access area; the source and destination are simply (special) S-Routers. Since adjacent S-Routers are physically near, they can use low-energy, short-range, communication. Energy savings are important; however, it is even more beneficial that such low-energy communication can be hard to detect and localize by

© Springer International Publishing AG, part of Springer Nature 2018
I. Dinur et al. (Eds.): CSCML 2018, LNCS 10879, pp. 188–205, 2018.
https://doi.org/10.1007/978-3-319-94147-9_16

remote Interceptors. On the other hand, S-Router may not be able to deploy the most stealthy techniques, due to size, cost and energy considerations. As a result, an Interceptor would often succeed in detecting and locating a nearby transmitting S-Router.

Prevention of stealthy communication, and interception of S-Routers, is important for many scenarios, including commercial (prevention of industrial espionage) and personal/political, as well as the 'classical' military, counter-intelligence and counter-terrorism. Stealthy networks have been studied and deployed since roughly 2006 [17–19].

The proliferation of Wireless Sensor Networks (WSN), based on low-cost, miniature networked devices [2,8,12,14], may facilitate extensive use of stealthy networks, including commercial and private privacy-intrusive applications. Examples include outdoor or indoor eavesdropping [4,16], industrial espionage, and a command-and-control channel for physical attacks e.g. against communication or energy infrastructure.

In spite of the wide-ranging implications of stealthy networks, this is the first work which presents defense mechanisms to efficiently intercept the stealthy network, using a set of mobile Interceptors, which can detect an S-Router transmitting nearby. We present, analyze and experimentally evaluate algorithms for ensuring efficient search by a set of Interceptors, to intercept and locate the destination of a stealthy communication network (e.g. a base station of a WSN).

Model. To study the problem analytically, we introduce a model for evaluating stealthy routing algorithms, as well as stealthy network interception algorithms.

The model is a round-based process between two parties: S-Routers and Interceptors, both operating on the plane (two-dimensions space). Our focus is on a protocol for the Interceptors party, which defines the operation of all Interceptors, and whose goal is to find the destination S-Router.

We focus on *Destination-Search* scenarios, where the S-Routers route all information from one special S-Router, called the *source*, to another special S-Router, called the *destination*, and denoted by D; the goal of the Interceptors is to find D. A single round of the process is illustrated in Fig. 1.

We assume a single, known *source* of the stealthy communication, which, WLOG, we fix at the center of coordinates, i.e., $(0,0)$. There are N S-Routers, one of which is the destination D; the other S-Routers relay information from the source $(0,0)$ to D. S-Routers may transmit up to one data unit at each round. Out model assumes that an Interceptor exposes an S-Router if the S-Router transmits within the interception range of the Interceptor.

S-Routers transmit data at a certain average rate, denoted $\hat{B}$. Intuitively, it seems easier to find D when the rate is high; indeed, we show that when the rate $\hat{B}$ is 1, the Interceptors can use a more optimized algorithm which improves their performance. Note that $\hat{B} = 1$ means that the S-Routers have to transmit data at every round.

Algorithms are measured by their impact on the *lifetime*, which is the number of rounds that pass before the destination is exposed by Interceptors; S-Routers

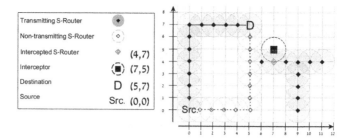

Fig. 1. Illustration of one round, with a single Interceptor searching for destination D, located at $(7,5)$ (in 2D, i.e., $\mathbb{R}^2$). Two routes of S-Routers connect the source $(0,0)$ to D (at $(7,5)$); the first (via $(0,7)$) transmits at this round (black diamonds), the other not (white diamonds). Several other S-Routers (from $(6,4)$ to $(11,4)$ and to $(9,0)$) transmit 'dummy' messages, to divert the Interceptor from the real routes, and lead it to two 'dead ends': $(11,4)$ and $(9,0)$.

attempt to maximize the lifetime, while Interceptors attempt to minimize it. The Interceptors may utilize up to M separate agents simultaneously.

In the particular scenario we study, Interceptors do not 'disable' non-destination S-Routers, even if they 'know' their location. Our model may be extended to consider scenarios/models where the Interceptors can disable an S-Router, e.g. by installing a nearby jamming device or by physical elimination. However, we prefer, in this work, to focus on the scenario where S-Routers are not disabled, both for simplicity and since such scenarios may be important. In particular, in the expected case where there is a large network of multiple S-Router paths, sources and destinations, the Interceptors may prefer not to disable an S-Router, since this may alert the S-Routers network and foil detection of the rest of the stealthy network. Disabling S-Routers before exposing end-points may trigger defensive and evasive reactions by S-Routers e.g. temporarily shutting down nearby communication, alternating communication routes and even activating nearby disconnected S-Routers in order to mislead Interceptors.

Video illustration. A video illustration of the algorithms presented in this work is available online in [1].

Contributions.

- We introduce the threat of stealthy networks, with a flexible model facilitating analysis of algorithms.
- We study two approaches for Interceptors: network graph search and spatial search, and explore advantages and disadvantages of each approach.
- We present a Divide and Conquer (D&C) Algorithm: a spatial search algorithm for Interceptors which ensures that the expected lifetime is in $O(\frac{N}{M\hat{B}} \cdot \log^2(N))$.
- We introduce a parametric algorithm for S-Routers, and use it to experimentally evaluate the Interceptors algorithms presented in this work, via simulations.

2   The 2D Stealthy Network Model

In this section, we present a model for studying problems involving S-Routers, whose goal is to route data without discovery of D, and Interceptors, whose goal is to find D.

The model is round-based; i.e., it operates in consecutive, discrete rounds $t \in \mathbb{N}$ of equal duration, starting from $t = 1$. At each round, the model invokes two algorithms, Π_I for the Interceptors and Π_S for the S-Routers. We model both as centralized algorithms; see [3] for discussion on this simplification. To end a round, the model invokes a third algorithm, Π_E, which models the *environment*.

The environment algorithm Π_E determines the results of the actions of the Interceptors and S-Routers, including the inputs for next round. We believe that the modeling of the environment by an algorithm Π_E gives significant flexibility to the model. For example, in the scope of this work we study two environments, one of which enforces that S-Routers transmit continuously throughout the process, i.e., where $\hat{B} = 1$, and another which does not enforce $\hat{B} = 1$. Other variants of the problem may be modeled by other environments. For example, our model does not restrict the movement of Interceptors from one round to the next; but only a minor change in the environment is required to limit the movement of each Interceptor.

The initial input to all three algorithms are the number of Interceptors, denoted M, and the number of S-Routers, denoted N.

We next define the Interceptors algorithm Π_I. The input for Π_I combines the observations of all Interceptors into a list of points from which transmissions were detected at the previous round (up to one point per Interceptor), and the output is the joint list of locations at the following round. That is, the output for each Interceptor is an encoding of the location, in $\mathbb{R}^2$, for the corresponding Interceptor. The algorithm also has a state as input and output, modeled as binary string.

Next, we define the S-Routers algorithm Π_S. In this work, S-Routers do not move; we denote the location of S-Router $i \in \{0, ...N\}$ by $S_L(i) \in \mathbb{R}$, where 0 is the destination and $1 ... N$ are the (other) S-Routers. Furthermore, in this work, the model does not include any input to the S-Routers, in particular, S-Routers do not have any way to detect Interceptors, and S-Routers are never impacted in any way; hence, their entire behavior can be determined initially. Their 'behavior' only consists of a *transmission schedule*, $S_T(t) \subset \{0, ..., N\}$, identifying the S-Routers that transmit at round $t \in \mathbb{N}$, and by the (fixed) location of the destination D. The S-Routers algorithm Π_S implements S_T; details omitted.

In most of the paper, we refer directly to S_L, S_T and D, instead of to Π_S. One of the S-Routers serves as the source, which is always at $(0,0)$; i.e. $(0,0) \in \cup S_L$. The set of S-Routers must ensure connectivity from source $(0,0)$ to destination; see definition of connectivity below.

We next model the environment, Π_E. Π_E models the behavior of the environment, as a (probabilistic) algorithm, allowing analysis of different stealthy-network scenarios and goals. The inputs to Π_E are the outputs of Π_I, and the locations of the S-Routers that transmit at the current round. The outputs are the interceptions to be provided, in the next step, to Π_I. In addition, the environment has a state of its own as input and output. Upon termination, the environment also outputs the average rate of transmission; $\hat{B}$, as defined in Definition 2.

The values returned at the end of the execution of the process are $\hat{B}$, as returned by Π_E, and the *lifetime* of the process, i.e., the number of rounds until the process terminates. If the process never ends, then lifetime is ∞.

2.1   Destination-Search Environments

In this subsection define two environments Π_E used in this work. In the destination-search scenario, the goal of the S-Routers is to maintain a connection between $(0,0)$ and D using the available S-Routers. Therefore, we begin with the notion of connectivity.

Definition 1 (Connectivity). *Two points $p_1, p_2 \in \mathbb{R}^2$ are* connected *iff their Euclidean distance is at most one, i.e., $\|p_1 - p_2\| \leq 1$. Let $Connected(p_1, p_2) = 1$ if p_1, p_2 are connected, and 0 otherwise. A list of points is* connected *if every pair of two consecutive points is connected. Two points $p_1, p_2 \in \mathbb{R}^2$ are* connected via *a list of points P if there is a list of points in P, say $(l_1, ...l_k)|(\forall i)(l_i \in P)$, s.t. the list $(p_1, l_1, \ldots, l_k, p_2)$ is connected.*

The value of lifetime represents the success/reward of the S-Routers. This value is also the cost/penalty for the Interceptors, whose goal is to minimize the time needed for intercepting D. However, measuring performance using lifetime alone may be misleading, as S-Routers are often able to increase it simply by decreasing the number of transmissions.

Therefore, we define an additional criteria, the transmission rate $\hat{B}$, which indicates how often S-Routers transmit. For denoting whether a data unit was transmitted at a specific round $t \in \mathbb{N}$, we use $b(t)$.

Definition 2 (Transmission rate measurement $b(), \hat{B}$). *Let $b(t) = 1$ if data is transmitted from $(0,0)$ to D at round t. That is, $b(t) = 1$ if exists is a list of S-Routers $S = (r_1, r_2 \ldots)$ s.t. the source $(0,0)$ and D are connected via their locations $\{S_L(r_i) : r_i \in S\}$, and all S-Routers in the list transmit at round t i.e. $S \subseteq S_T(t)$. Otherwise, let $b(t) = 0$.*

The average transmission rate of S-Routers, denoted with $\hat{B}$, is $\sum_t b(t)$ divided by the lifetime.

Note that in this work, S-Routers can not buffer messages and deliver them later. Future work may remove this restriction, to allow for delay-tolerant networking by S-Routers. One reason for this (simplifying) restriction, is that each round represents a (potential) physical movement by the Interceptors, and movements are normally much slower than communication.

Initialization. Upon initialization, Π_E is provided with an indication of whether the *continuous transmission constraint* has to be enforced (the constraint is defined in Definition 3).

Termination. The environment Π_E will terminate the execution if either party acts in a way forbidden for that execution or if one of the Interceptors is directly connected to the destination D.

The process terminates when Interceptors visit a point near D, due to the assumption that D is a long-range transmitter, which is expected to be larger cf. to S-Routers and therefore easier to localize.

For simplicity, we also assume that Interceptors are able to expose D even while S-Routers do not transmit. Note that for the algorithms presented in this work, this assumption does not affect their performance asymptotic complexity (with an exception for the Naive Disc Search algorithm).

As previously mentioned, if the process never ends, the lifetime is ∞; however, in this work, Interceptors may always avoid this, since D necessarily may be reached after a finite number of rounds.

Transmission rate constraints. As presented so far, the model allows rounds t in which there no transmission-path from source $(0,0)$ to D (i.e., $b(t) = 0$). However, it seems that in many scenarios, Interceptors will transmit continuously to D. We refer to this as the *continuous transmission constraint/assumption*.

Definition 3 (Continuous Transmission Constraint). *We say that the* Continuous transmission constraint *holds for an execution, if for every round* t *in the execution,* $b(t) = 1$ *holds, i.e.,* $(0,0)$ *and* D *are connected via some set of S-Routers at that round. We say that* Π_E Enforces Continuous Transmission *if* Π_E *terminates the execution, with* $\hat{B} = 0$, *upon a round in which the constraint does not hold.*

3   Introducing Interceptors Algorithms

We found that the design of efficient Interceptors algorithm is more challenging than appears initially, with resulting algorithm being somewhat counter-intuitive. Obviously, we cannot repeat here all the variations we experimented with; however, we present few basic algorithms, which we believe will help the reader understand the problem better, preparing the ground for the more efficient - but less intuitive - algorithms presented in the following sections.

In this section we present three Interceptors algorithms. We begin with an observation that the Interceptors may limit their search to a bounded area, specifically, a *disc*. We then present an algorithm which essentially searches this disc. Afterwards, we present two naive attempts to find D by 'following the path' from the source $(0,0)$ to D, which are reminiscent of graph-search algorithms.

To our disappointment, we did not yet find an efficient graph-search algorithm, that works in the general case. However, it is possible that future work would find better ways to use the graph-search approach. For a more thorough

194 M. Ben Adar Bessos and A. Herzberg

discussion on this topic, and for proofs of the Propositions and Lemma presented in this section, see the draft of the full version of this work [3].

Naive Disc Search Algorithm. We begin with a very simple algorithm that we call *Naive Disc Search*. Basically, the Naive Disc Search algorithm exhaustively searches for D in a bounded disc. We first show in Lemma 1 that it suffices to search for D within a disc, specifically, the disc of radius N whose center is the source $(0,0)$; this simple bound may also used by the more advanced algorithms. The lemma uses the following notation.

Notation: *Disc.* Given a point $c \in \mathbb{R}^2$ and a distance $r \in \mathbb{R}$, let $\textbf{\textit{Disc}}(r,c) = \{p \in \mathbb{R}^2 | \, ||p - c|| \leq r\}$ denote the region of a disc whose center is c and whose radius is r.

Lemma 1. *If D is connected to the source $(0,0)$ via any list of S-Routers $S \subset \cup S_L$, then $D \in Disc(N, (0,0))$.*

Since it suffices to search for D within $Disc(N, (0,0))$, a simple, naive approach is to exhaustively search this disc. More precisely, such algorithm will visit different points within the disc, where each point results in covering a disc of radius 1 centered in that point, until the entire disc was covered - or D found.

The order of visitations may affect the performance of the algorithm. For example, if all S-Routers are located 'densely' around $(0,0)$, as illustrated in Fig. 2(a), Interceptors may sort all points by (increasing) distance from $(0,0)$ then visit them in that order in order to find D efficiently in $O(N)$ rounds. However, if the search is deterministic and known in advance, D may be placed so it is found only by the very last searched points. For example, if Interceptors keep visiting points with increasing distance from $(0,0)$ but S-Routers are located as illustrated in Fig. 2(b), roughly the entire disc $Disc(\frac{N}{2}, (0,0))$ will be covered before D is found. Hence, a random order is slightly preferable for Interceptors.

The Naive Disc Search Algorithm uses a predefined set of points to search, i.e., the covering of a disc of radius N by discs of radius 1. Let $\textbf{\textit{DiscCoverage(N)}}$ denote the set of points that cover $Disc(N, (0,0))$. It is difficult to minimize the size of $DiscCoverage(N)$ [11], but its complexity is necessarily $O(N^2)$, and efficient implementations for $DiscCoverage(N)$ can achieve it [9].

In the The Naive Disc Search Algorithm, the Interceptors search for D by visiting every point in $DiscCoverage(N)$ in random order. At each invocation, the algorithm keeps all previously visited points, and outputs a list of M previously unvisited points to visit next.

Proposition 2. *The expected lifetime of the Naive Disc Search algorithm is in $O(N^2/M)$.*

Naive Graph Search Algorithm. The Naive Disc Search Algorithm does not use the interceptions (detections), which seems wasteful. Surely, we can use interceptions to find D more efficiently. One natural idea is to exploit the fact that D must receive transmissions from the source $(0,0)$; we can try to 'follow' these transmissions, by always searching in the vicinity of one of the points where we intercepted a transmission, plus the source $(0,0)$. This Naive Graph Search Algorithm keeps a set of locations from which a transmission was intercepted (initialized to the source $\{(0,0)\}$), then visits at each round at points that are within distance ≤ 3 from one of the points in the vicinity of previously successful search locations, chosen at random with uniform probability.

Proposition 3. *The expected lifetime of the Naive Graph Search algorithm is in $O(\frac{N^2}{\hat{B} \cdot M})$.*

Uniform Graph Search Algorithm. Since the Naive Graph Search algorithm selects points with uniform probability at each step, points in the vicinity of earlier interceptions have more opportunities for being selected. Intuitively, if newly discovered interceptions will be visited more frequently, the performance of the algorithm may be significantly improved. In order to examine this approach, we have designed the *Uniform Graph Search* algorithm. The algorithm assumes that only a single Interceptor is available, i.e., $M = 1$. The algorithm is defined similarly to the Naive Graph Search Algorithm, with the following modifications:

1. For each point in $DiscCoverage(N)$, initialize a counter to 0.
2. Each time a point is visited by the algorithm, increase its counter by 1.
3. When selecting a point to visit, select points with minimal counter value.

Namely, all points that the algorithm may visit will be roughly visited an equal number of times, and for each interception, the new points and its vicinity will be visited repeatedly, until their associated counter value is no longer minimal. For example, if S-Routers use the network illustrated in Fig. 2(b), and the continuous transmission constraint, as defined in Definition 3, holds, then the Uniform Graph Search will frequently intercept new S-Routers (and eventually D) at the 'front' of the few routes, since data is transmitted through at least one of the routes at each step, and the algorithm will repeatedly visit points near the 'front' after each interception. This scenario is handled far less efficiently by the Naive Disc Search and the Naive Graph Search algorithms.

Unfortunately, in the worst case, the performance of this algorithm is not significantly better (compared to the naive algorithm). Even if the continuous transmission constraint holds, if the network graph includes many separate alternate routes that connect $(0,0)$ and D, the transmission rate in each route may be reduced proportionally (as illustrated in Fig. 2(c)), and interception of new S-Routers will be infrequent.

Proposition 4. *If the continuous transmission constraint holds, the expected lifetime of the Uniform Graph Search is in $\Omega(N^2/log(N))$.*

Note that in the above result, the term used for bounding on the lifetime excludes $\hat{B}$, since the continuous transmission constraint ensures that $\hat{B} = 1$.

In the following sections we present Interceptors algorithms and prove their expected performance is significantly better (assymptocially) compared to the Uniform Graph Search algorithm. However, for many cases, where S-Routers

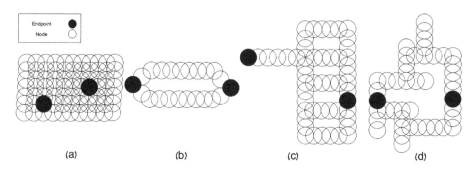

Fig. 2. Examples of different S-Router networks: (a) All S-Routers are located 'densely' around the source $(0,0)$. An exhaustive search around $(0,0)$ may reach D efficiently. (b) Only few separate 'long' alternate routes connect $(0,0)$ and D. If Continuous Transmission Assumption holds, the rate of transmission in at least one of these routes is relatively high, which allows the Uniform Graph Search algorithm to expose S-Routers efficiently. (c) A network which uses numerous separate alternate routes. The transmission rates in different points may vary significantly; in particular, this prevents Uniform Graph Search algorithm from exposing new S-Routers efficiently even if the Continuous Transmission Assumption holds. (d) A network with few paths but numerous 'dead ends'. Even if a graph-search algorithm can efficiently cope with routes which transmit slowly, it is difficult to discern such routes from actual 'dead ends'. Since most 'walks' in the network lead to a 'dead end' and S-Routers may make 'dead end' routes appear exactly like other routes, it is also difficult to avoid them

is small enough, the Uniform Graph Search algorithm outperforms all other algorithms, as illustrated in Fig. 6. A more detailed performance comparison is given in Sect. 5.

4   Divide and Conquer Interceptors Algorithm

In this section we present an algorithm for the Interceptors, the *Divide And Conquer Algorithm*, which bounds the expected lifetime to $O(\frac{N}{B \cdot M} \log^2(N))$ for M Interceptors. Counter-intuitively, and in contrast to the less efficient graph search algorithms of the previous section, this algorithm does *not* try to 'search' the graph of S-Routers from $(0,0)$ to D. Instead, this algorithm takes a 'divide and conquer' method, to find the destination D 'directly' - without exposing the entire path to it.

We begin with few a preliminaries in Sect. 4.1, then describe the algorithm in Sect. 4.2. For a thorough analysis of the algorithm, proofs, and additional topics, see the draft of the full version of this work [3].

4.1   Preliminaries: Ranges and Walls

We begin this section with few additional topological concepts which are used in this section.

First, given a location $l \in \mathbb{R}^2$, let $Range(l)$ denote its *range*, i.e., the set of points whose communication would be intercepted by an Interceptor located at location l. Formally, **$Range$**$(l) = \{x \in \mathbb{R}^2 | Connected(l, x)\}$. The range notation extends to a set of points L, namely we denote $Range(L) = \bigcup_{l \in L} Range(l)$.

The Divide And Conquer Algorithm uses the fact that D must be within $Disc(N, (0, 0))$, as shown in Lemma 1. The algorithm partitions $Disc(N, (0, 0))$ into smaller regions, then examines these regions by visiting points on their boundaries. If the boundaries 'separate' between $(0, 0)$ and D, and considering the S-Routers transmit from $(0, 0)$ to D, it follows that these transmissions must 'cross' one or more of the boundaries which are visited by the algorithm. If the points of a boundary are sufficiently-close, then the algorithm may intercept transmissions from at least one of these points.

We define two topological notions which are important in this algorithm: a $\sqrt{3}$-spaced *wall* and *closed wall*.

Definition 4 (Wall, closed wall, and In/Out regions). *An $\sqrt{3}$-spaced wall is a list of points $L = \{l_1, l_2, \ldots, l_k\} \in \left(\mathbb{R}^2\right)^k$ such that the distance between every two consecutive points l_i, l_{i+1} is at most $\sqrt{3}$. A $\sqrt{3}$-spaced closed wall L (abbreviated to closed wall), is a wall where the distance between l_1 and l_k is at most $\sqrt{3}$. We denote the outer region by **Out**(L), and the internal region, excluding $Range(L)$ itself, by **In**(L).*

In the definition above, to define the inner and outer region, we use basic topological notions such as boundary and region, which are quite intuitive and standard; precise definitions can be found, e.g., in [13].

We focus on $\sqrt{3}$-spaced walls and closed walls, since a $\sqrt{3}$-spaced closed wall separates between (points in) its *internal region*, $In(L)$, and (points in) its *outer region*, $Out(L)$. The formal statement is given in the next Lemma, and illustrated in Fig. 3. Therefore, we write *walls* and *closed walls*, always referring to $\sqrt{3}$-spaced walls and closed walls (an x-spaced closed wall with $x > \sqrt{3}$ may fail to provide the separation property referred to in the Lemma).

Lemma 5. *Given a ($\sqrt{3}$-spaced) closed wall L, no pair of points $p_{In} \in In(L)$ and $p_{Out} \in Out(L)$ are connected. Namely, for all $p_{In} \in In(L)$ and $p_{Out} \in Out(L)$ holds $\|p_{In} - p_{Out}\| > 1$.*

The Divide And Conquer Algorithm generates closed walls, then instructs the Interceptors to visit them in a random order. In order to calculate the probability of interception when visiting a point in a closed wall that 'separates' $(0, 0)$ and D (such as the closed wall illustrated in Fig. 3).

Definition 5 (Separating closed walls). *We use separating closed wall to refer to a closed wall that contains D but excludes $(0, 0)$, namely a closed wall L for which $D \in In(L) \cup Range(L)$ and $(0, 0) \in Out(L) \cup Range(L)$ hold.*

Proposition 6. *Let L be a separating closed wall, and let $t \in \mathbb{N}$ be a round s.t. $b(t) = 1$. There exist $v \in L$ and $x \in S_T(t)$ s.t. $Connected(S_L(x), v)$.*

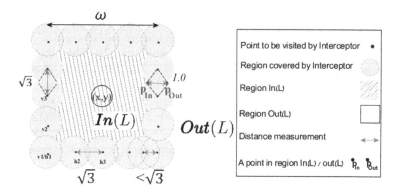

Fig. 3. The closed wall $L = \{v_1, v_2, \ldots\} \cup \{h_1, h_2, \ldots\} \cup \ldots$ separates the plane into the the inner and outer regions $In(L)$, $Out(L)$ and its range $Range(L)$. Note that all points in L rest on the boundary of the same square. Any point from region $In(L)$ is not connected to any point in region $Out(L)$. This property is used by our algorithms: Interceptors inspect, randomly, the points in closed walls separating D (inside) from the source $(0,0)$.

Finally, we define *leading square walls*, which are the 'smallest' separating square walls; by dividing these walls, the algorithm 'zooms in' on D.

Definition 6 (Leading square walls). *Let* $\mathcal{L} = \{L_1, L_2, \ldots\}$, *where* L_i *is a subset of the 'watched' points, be the set of all separating square walls in the 'watched' points. We refer to a separating square wall* $L \in \mathcal{L}$ *as a* leading square wall *if no other separating square wall is contained in L i.e.* $\forall L' \in \mathcal{L} : L' \not\subseteq (In(L) \cup Range(L))$.

4.2 Divide And Conquer Algorithm

We now present the Divide And Conquer Algorithm. The algorithm visits at each round M distinct points out of a set of 'watched' points. These 'watched' points are placed as a closed wall around *squares* containing D; we begin with very large squares and repeatedly divide them into smaller squares, until we find D. Let us first present our notation for a square.

Notation. (Square). Given a point $(x, y) \in \mathbb{R}^2$, and a length $w \in \mathbb{R}$, let ***Square***$(w, (x, y)) = [-\frac{w}{2} + x, \frac{w}{2} + x] \times [-\frac{w}{2} + y, \frac{w}{2} + y]$ denote the region of a square whose center is (x, y) and each of its edges are of length w. For example, Fig. 3 gives a visualization of the closed wall L, where all points in L rest on the boundary of the square $Square(w, (x, y))$.

The algorithm searches for D in $Square(2N, (0, 0))$ (which contains $Disc(N, (0, 0))$ and D in particular). The algorithm partitions $Square(2N, (0, 0))$ into smaller squares and places Interceptors at several random points along walls on their boundaries. That is, a closed wall is kept per square, s.t. one of these 'watched' closed walls contains the destination D; we refer to these square-shaped closed walls as *square wall*.

When a square wall shows signs of possibly containing D, namely when a transmission was intercepted from one of it's points, the algorithm further divides the corresponding square into four quarters, and repeats the process for these smaller squares, until finding D. For efficiency, the total size of walls of 'watched' squares should be small; to find D, the regions must contain it. The algorithm carefully ensures both properties.

It is crucial to randomize the location of the squares, to foil S-Router placements that exploit predictable locations of square walls. S-Routers may lead the algorithm into 'watching' many additional regions that do not contain D, due to S-Routers that deliberately expose themselves at specific locations. Hence, we first select a random point o from $Square(2N, (0, 0))$. From the beginning, we 'watch' the four $2N \times 2N$ squares shown in Fig. 4(a), s.t. o is a shared corner. From Lemma 1, we can assume that D is within one of these four squares.

At each round, we put the Interceptors at distinct random points in the walls of 'watched' squares. We try to detect the transmissions by S-Routers crossing these walls, from source $(0, 0)$ to D; this identifies 'suspect' squares, worthy of further decomposing into four sub-squares. From Lemma 1, it suffices, however, to put Interceptors at points which are within $Disc(N, (0, 0))$; see our 'focus' on $Disc(N, (0, 0))$ in Fig. 4(b).

If we use only large squares, e.g., the four large, $2N \times 2N$ squares shown in Fig. 4(a), then it is quite possible that no path from $(0, 0)$ to D will cross their walls at all - since D will be within the same large square. Indeed, in Fig. 4(a), we see that D and $(0, 0)$ are both in the lower-right large square (shown more closely in Fig. 4(b)). To ensure that each path of S-Routers from $(0, 0)$ to D will cross one of the 'watched walls', we divide, from the very beginning, each of the 'watched squares' containing $(0, 0)$ into its four sub-squares, until reaching squares small enough to ensure localization of D. This is not *that* wasteful: the additional length of all these initial sub-squares is less than the length of the initial $2N \times 2N$ squares.

Assuming D is in $Disc(N, (0, 0))$, at least one of the initial square walls includes D and excludes $(0, 0)$ i.e. the algorithm begins with at least one leading square wall; this ensures Proposition 6 holds, and at least one point in one of the initial square walls will allow the Interceptor to detect a transmission. Points to search are selected at random with uniform probability from all square walls, such that the probability of selecting a point from any certain square wall is proportional to its size. For each successful search, on top of the square walls associated with that search point, four smaller square walls that encircle the same region are added, in an attempt to decrease the size of the leading square wall. Sufficiently small squares are covered by a single visit.

Theorem 7. *The expected lifetime of Divide And Conquer Algorithm is* $O(\frac{N}{B \cdot M} \cdot \log^2(N))$.

Theorem 7 holds since the algorithm has to divide a leading square wall at most $\lceil log_2(N) - 1 \rceil$, until D is found, and since the expected number of transmitted data units until the algorithm divides a leading square wall is $O(\frac{N}{M} log(N))$.

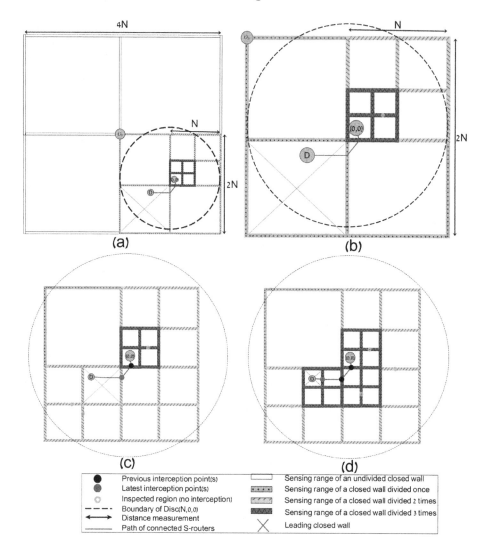

Fig. 4. Illustration of the operation Divide And Conquer Algorithm. The algorithm begins by visiting points on square walls generated upon initialization, and with each interception it generates additional square walls which potentially include D. (a) Illustrates the initialization of the algorithm, where four square walls of size $2N \times 2N$ are generated with random offset such that the entire searched region $Disc(N, (0, 0))$ is contained by them. (b) Illustrates the next initialization step. The square wall that includes $(0, 0)$ is repeatedly divided for $log(N)$ times. Note that one of the square walls includes D, but excludes $(0, 0)$ (c) Illustrates the first detection of a transmission while visiting a point in one of the square walls (the red circle). The two square walls adjacent to the point of detection(s) are divided into four. After this division, the smallest square wall which includes the destination and excludes $(0, 0)$ (the leading square wall) is smaller. (d) Illustrates the second and third transmission detection. The second detection occurs at the same point, which leads to additional division. The third transmission detection (the red circle) will lead to another division (excluded from this illustration), and then to the detection of D. (Color figure online)

As previously mentioned, a more thorough discussion and analysis of the algorithm is given in the draft of the full version of this work [3]. In particular, we present a modified version of the Divide And Conquer Algorithm, referred to as D&CCTA, which achieves better performance if the continuous assumption holds (the difference in performance is illustrated in Fig. 6).

5   Evaluation and Results

There is always a challenge in evaluating the practical performance of a defensive mechanism, whose results depend completely on the behaviour of the adversary - in our case, the S-Routers. Our approach was to use a set of S-Routers algorithms, each one 'optimized' for per each Interceptors algorithm. Of course, we do not really know how to produce the 'best' S-Router algorithm (in general, or for a particular Interceptors algorithm). Instead, we tried our best to develop good S-Routers algorithms, in two steps.

In the first step, we developed a parametric heuristic S-Routers algorithm, the *Parametric Segmented Network* algorithm, based on our analysis of different Interceptors algorithms, and on 'trial and error', using a simulation and visualization environment we developed for this purpose. We will make this tool freely available in [1], to allow further research and reproducibility.

In the second step, we used genetic programming to optimize the parameters of Parametric Segmented Network algorithm for each of the Interceptors algorithms, and then compared the results of the different Interceptors algorithms - each running against the 'best' parameters (of the parametrized S-Routers algorithm). We begin with a description of the Parametric Segmented Network algorithm, then present simulation results.

The Parametric Segmented Network Algorithm. The algorithm receives as input parameters for selecting the number of segments, the number of S-Routers that compose each segment, the number of parallel paths in each segment, and what portion of these paths lead to a 'dead end'.

Figure 5 illustrates a network composed of $N = 128$ S-Routers which are separated into three segments. The segments are composed of $0.35N, 0.3N, 0.3N$ S-Routers (from left to right), where $N = 128$. D is the leftmost S-Router, while $(0, 0)$ is the rightmost S-Router. Parallel paths of the same segment begin and end with a joint path (perpendicular to the parallel paths), and adjacent segments share one S-Router that connects the joint paths. The remaining $0.05N$ S-Routers that are disconnected transmit continuously, in order to mislead non-graph search algorithms (such as the Divide And Conquer Algorithm). The leftmost segment transmits into one additional 'dead end' parallel route.

The Parametric Segmented Network algorithm satisfies the continuous-transmission constraint. At each round, every segment transmits data through one of its paths, and S-Routers of joint (perpendicular) paths transmit continuously. In Fig. 5, the top path in each segment transmits.

More specific details on the implementation of the Parametric Segmented Network algorithm are given in the full version of this paper in [3].

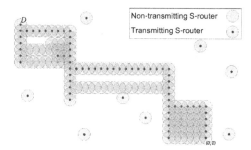

Fig. 5. The Parametric Segmented Network algorithm as visualized by the simulation, for 128 S-Routers that are separated to three segments, at a specific round. At each step, one of the parallel paths in each segment transmits. The S-Routers connecting adjacent segments transmit continuously. Hence, data may flow from the rightmost S-Router in the network, $(0,0)$ to the leftmost S-Router, D.

Results. We now compare results obtained by simulations. In Fig. 6, S-Routers use the Parametric Segmented Network algorithm. In order to limit the dimension of optimization, the number of segments was limited to four. The parameters for the algorithm were optimized separately for each scenario i.e. for each Interceptor algorithm and each N pairing. After the selection of the best parameters for each scenario, we ran the process enough times to ensure the evaluation of the parameter set is accurate. The results are displayed in Fig. 6. A confidence interval of 99% is used. All Interceptors algorithms utilize a single Interceptor. Due to the high computational costs, we used the genetic algorithm only for $N \leq 128$. For larger N values, the parameter sets were selected according to the conclusions found from smaller values.

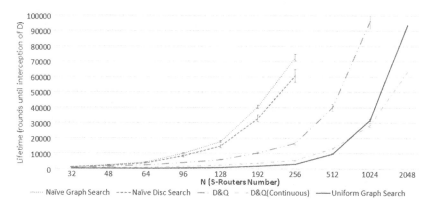

Fig. 6. Performance comparison of different Interceptors algorithms. Performance is estimated against Parametric Segmented Network S-Routers algorithm - optimized for each Interceptor algorithm (using a single Interceptor) and each N value separately. Confidence interval is displayed as vertical lines.

The genetic algorithm consistently keeps a population size of 500. At each epoch, the algorithm applies standard roulette selection. Crossovers are done until for 75% of the population is replaced, and then mutation is applied to 20% of the population where on average two elements of each mutated parameter set change. Changes are done by adding a Gaussian distributed random value. Since the S-Routers algorithms are making extensive use of randomization, the value of a given parameter set was the average lifetime when running the process repeatedly for 15 times. We allocated 24 h of CPU time for each scenario. Running a larger scenario takes significantly more time; therefore, the number of generations varied from several hundreds (for $N \leq 64$) to 48 (for $N = 128$).

Discussion. After examining the parameter sets that resulted from optimization, we discovered that certain properties consistently maximize the outcome for S-Routers. S-Routers that were disconnected from $(0,0)$ and 'dead end' paths were not useful, and optimized solutions did not express them.

When S-Routers face the Uniform Graph Search algorithm, only two segments are needed. The segment adjacent to $(0,0)$ is composed of single route, with $0.1N$ to $0.4N$ of the S-Routers, which is always intercepted in its entirety by the algorithm. The segment adjacent to D uses the remaining S-Routers for composing parallel paths. The number of parallel paths increases with N. Additionally, these routes are spread with distance of three from each other, in order to minimize the number of distinct 'watched' points.

A result we did not expect was that when S-Routers face the Divide and Conquer algorithm, the lifetime is maximized when only a single, long route was used, i.e., all segments express a single transmission route. In hindsight, we understood this; if using multiple routes than the Interceptors algorithms will eventually generate square walls that are too small to intersect all parallel paths of the same segment. As a consequence, these square walls have a lower probability for being divided, the total number of points that are 'watched' by the Interceptors algorithms decreases, and their performance improves (lower outcome). This may motivate an S-Routers algorithm that combines the graph-search approach with the divide-and-conquer approaches.

6   Conclusions and Extensions

Stealthy networks, comprised of hard-to-locate devices, are becoming a part of reality; we use the term S-Routers for such devices, who can relay information, to form large networks. Stealthy networks will be used for different applications; many of the applications may represent threats to privacy of individuals and organizations. Hence, it is important to develop efficient countermeasures. Due to the small size of the devices and their use of short-range communication, we envision the use of mobile devices, Interceptors, to localize the S-Routers.

In this work, we investigated algorithmic issues related the interception of stealthy networks. Our focus was on developing efficient algorithms for Interceptors, to expose the destination of stealthy network; we believe that such algorithms may be deployed as part of the design of countermeasures to stealthy networks.

There are many directions for improvements, extensions/variations, and further research. For example, if Interceptors may predict the S-Routers' transmission schedule, they may be able to accelerate their search significantly.

Improvements may also be possible for the analysis. The current results provide an upper bound for the expected lifetime in the studied environment, but a lower bound is yet to be found. While it is relatively simple to prove that S-Routers may ensure the expected lifetime is bounded from by $O(N)$, developing additional algorithms for S-Routers may be required in order to find the exact bounds, or at least for narrowing the gap between $O(N)$ and $O(\frac{N}{MB}log^2(N))$.

The presented model is general enough, to allow investigation of several related problems, including (1) multiple sources and/or destinations, (2) allowing S-Routers to buffer data, (3) introducing mobility, and much more.

Finally, note that the current model does not support decentralized algorithms. We expect that in some practical scenarios, S-Routers may have to risk exposure in order to coordinate. An extension to the model is necessary for studying such scenarios.

Acknowledgments. This work is supported by the Israeli Ministry of Science and Technology.

References

1. Herzberg, A., Ben Adar Bessos, M.: Intercepting a stealthy network - simulation demonstration (2018). https://sites.google.com/view/stealthynetinterception/home
2. Baisch, A.T., Ozcan, O., Goldberg, B., Wood, R.J.: High speed locomotion for a quadrupedal microrobot. Int. J. Robot. Res. **33**(8), 1063–1082 (2014)
3. Herzberg, A., Ben Adar Bessos, M.: Intercepting a stealthy network. vixra.org/abs/1712.0510
4. Bobic, I.: Ted cruz wants police to 'patrol and secure' U.S. Muslim communities after brussels, March 2016. www.huffingtonpost.com. Accessed 21 Nov 2017
5. Bash, B.A., Goeckel, D., Towsley, D.: Hiding information in noise: fundamental limits of covert wireless communication. IEEE Commun. Mag. **53**(12), 26–31 (2015)
6. Che, P.H., Bakshi, M., Jaggi, S.: Reliable deniable communication: hiding messages in noise. In: 2013 IEEE International Symposium on Information Theory Proceedings (ISIT), pp. 2945–2949. IEEE (2013)
7. Chen, O., Meadows, C., Trivedi, G.: Stealthy protocols: metrics and open problems. In: Gibson-Robinson, T., Hopcroft, P., Lazić, R. (eds.) Concurrency, Security, and Puzzles. LNCS, vol. 10160, pp. 1–17. Springer, Cham (2017). https://doi.org/10.1007/978-3-319-51046-0_1
8. Chen, X., Purohit, A., Pan, S., Ruiz, C., Han, J., Sun, Z., Mokaya, F., Tague, P., Zhang, P.: Design experiences in minimalistic flying sensor node platform through sensorfly. ACM Trans. Sensor Netw. (TOSN) **13**(4), 33 (2017)
9. Das, G.K., Das, S., Nandy, S.C., Sinha, B.P.: Efficient algorithm for placing a given number of base stations to cover a convex region. J. Parallel Distrib. Comput. **66**(11), 1353–1358 (2006)

10. Hu, J., Yan, S., Zhou, X., Shu, F., Wang, J.: Covert communication in wireless relay networks (2017). CoRR abs/1704.04946
11. Kershner, R.: The number of circles covering a set. Am. J. Math. **61**(3), 665–671 (1939)
12. MacGregor, A.: Russian scientists create cockroach spy robot. thestack.com. Accessed 2 May 2018
13. Munkres, J.R.: Topology. Prentice Hall, Upper Saddle River (2000)
14. Rubenstein, M., Ahler, C., Nagpal, R.: Kilobot: a low cost scalable robot system for collective behaviors. In: 2012 IEEE International Conference on Robotics and Automation (ICRA), pp. 3293–3298. IEEE (2012)
15. Shabsigh, G.: Covert Communications in the RF Band of Primary Wireless Networks. Ph.D. thesis, University of Kansas (2017)
16. Sidahmed, M.: NYPD's muslim surveillance violated regulations as recently as 2015: report, August 2016. www.theguardian.com. Accessed 21 Nov 2017
17. Bokareva, T., Hu, W., Kanhere, S.S., Jha, S.: Wireless sensor networks for battlefield surveillance. In: Proceedings of the Land Warfare Conference, pp. 1–8 (2006)
18. He, T., Krishnamurthy, S., Luo, L., Yan, T., Gu, L., Stoleru, R., Zhou, G., Cao, Q., Vicaire, P., Stankovic, J.A., Abdelzaher, T.F., Hui, J., Krogh, B.: VigilNet: an integrated sensor network system for energy-efficient surveillance. ACM Trans. Sensor Netw. (TOSN) **2**(1), 1–38 (2006)
19. He, T., Vicaire, P., Cao, Q., Yan, T., Zhou, G., Gu, L., Luo, L., Stoleru, R., Stankovic, J.A., Abdelzaher, T.F.: Achieving long-term surveillance in vigilnet. Technical report. Department of Computer Science, Virginia Univ Charlottesville (2006)
20. Wang, L., Wornell, G.W., Zheng, L.: Fundamental limits of communication with low probability of detection. IEEE Trans. Inf. Theory **62**(6), 3493–3503 (2016)

Privacy in e-Shopping Transactions: Exploring and Addressing the Trade-Offs

Jesus Diaz[1]([⊠]), Seung Geol Choi[2], David Arroyo[3], Angelos D. Keromytis[4], Francisco B. Rodriguez[3], and Moti Yung[5]

[1] Blue Indico - BEEVA, Madrid, Spain
jesus.diaz@beeva.com, jesus.diaz.vico@gmail.com
[2] United States Naval Academy, Annapolis, USA
choi@usna.edu
[3] Universidad Autónoma de Madrid, Madrid, Spain
{david.arroyo,f.rodriguez}@uam.es
[4] Georgia Institute of Technology, Atlanta, USA
angelos@gatech.edu
[5] Columbia University, New York, USA
moti@cs.columbia.edu

Abstract. The huge growth of e-shopping has brought convenience to customers, increased revenue to merchants and financial entities and evolved to possess a rich set of functionalities and requirements (e.g., regulatory ones). However, enhancing customer privacy remains to be a challenging problem; while it is easy to create a simple system with privacy, this typically causes *loss of functions*.

In this work, we look into current e-shopping infrastructures and aim at enhancing customer privacy while retaining important features and requiring the system to *maintain the topology and transaction flow of established e-shopping systems* that are currently operational. Thus, we apply what we call the "utility, privacy, and then utility again" paradigm: we start from the state of the art of e-shopping (utility); then we add privacy enhancing mechanisms, reducing its functionality in order to tighten privacy to the fullest (privacy); and finally, we incorporate tools which add back lost features, carefully relaxing privacy this time (utility again).

We also implemented and tested our design, verifying its reasonable added costs.

1 Introduction

Privacy vs. Utility: The Case of Group Signatures. The evolution of privacy primitives in various specific domains often centers around the notion of balancing privacy needs and utility requirements. Consider the notion of "digital signature" [22,39] whose initial realization as a public key infrastructure [37] mandated that a key owner be certified with its identity and its public verification key: a certification authority (CA) signs a record (called certificate) identifying the user and its signature public verification key.

© Springer International Publishing AG, part of Springer Nature 2018
I. Dinur et al. (Eds.): CSCML 2018, LNCS 10879, pp. 206–226, 2018.
https://doi.org/10.1007/978-3-319-94147-9_17

Later on, it was suggested that CA's sign anonymous certificates which only identify the keys (for example, a bulk of keys from a group of users is sent to the CA via a mix-net and the CA signs and publish the certificates on a bulletin board: only the owner of a key can sign anonymously with its certified key. Alternatively the CA blindly signs certificates). This brings digital signing to the domain of anonymous yet certified action (i.e., the action/ message is known to originate from the group that was certified).

However, it was noted quite early that under the mask of anonymity users can abuse their power and sign undesired messages, where no one can find the abuser. Therefore, primitives like group signature [14] or traceable signature [29] were designed, assuring that the anonymity property of a signed message usually stays, but there are *authorities which can unmask abusers, or unmask certain message signatures* in order to keep balance between anonymity of well behaving signers while protecting the community against unacceptable message signing practices.

Privacy by Design for Systems in Production? While privacy by design principles mandate that privacy enhancing mechanisms be taken into account already at the design stage of any system, for well established processes and infrastructures this is not possible. Moreover, trying to re-engineer an existing system from scratch, now including privacy tools by design, must nevertheless be constrained at every step by maintaining the same main processes and information flows. Otherwise, there exists a too high risk of rejection due to the unacceptable chain-effect changes its adoption would imply.

Utility, Privacy, and then Utility Again. The above development on group signatures shows that even in one of the simplest case of anonymity vs. basic message authenticity, there is already certain advantage in providing partial anonymity to perform in a desirable environment which balances various needs. Additionally, the described case of privacy by design for already deployed systems calls out for variants of this methodology. Extrapolating from the above staged methodology that gave us the primitives of group signature and traceable signature, we follow a methodology that can be viewed as "utility, privacy, and then utility again": First translating a primitive to an idealized anonymous primitive, but then identifying lost utility which complete anonymity prevents: and, in turn, relaxing privacy for additional utility.

Application to e-Shopping. We put forward our approach for this methodology through to the involved case of the real world (compound) process of e-shopping, where we find numerous trade-offs which we unveil and discuss (based on utility needed in various steps of the system). We begin by modelling the e-shopping ecosystem, identifying its entities, main processes and added-value mechanisms; then, we implement a fully anonymous system that keeping the entities and main processes, at the cost of losing the added-value parts; finally, we recover them by giving end-users the option to act fully anonymously or pseudonymously. Importantly, our methodology allows us to maintain the main processes of current e-shopping systems, making it easier to come up with a proposal compatible with the existing complex e-commerce ecosystem.

Note that we have not aimed solely at a theoretical exercise. We demonstrate feasibility of our approach by an exemplifying implementation which demonstrates that we keep a large portion of the utility of the original systems (without anonymity) for a reasonable added performance cost (with anonymity). The achieved practicality of a privacy-respectful system in a real-world context is of relevance, specially considering the latest regulations towards privacy, such as the European GDPR (General Data Protection Regulation[1]) and PSD2 (Payment Services Directive[2].)

1.1   Related Work

The most prolific related area are anonymous payments, e-cash [13] being its main representative, which has seen a huge boost since Bitcoin [34]. While Bitcoin itself does not provide robust privacy, more advanced proposals address this [5,12,24,32][3]. Still, they address only the payment process, and are typically not concerned with additional functionality, except [24], which adds support for regulatory concerns. Some traditional e-cash proposals also incorporate utility to some extent, mainly through tracing (after the payment has been done) [11,18,35] or some kind of spending limitation [35,41]. Privacy respectful payment systems out of the e-cash domain also exist, such as [28], built on mix networks to prevent linking customers and merchants, and [43], which uses discounts based on the (always pseudonymous) users' history. Private purchase systems have been constructed preventing merchants from learning what digital goods customers buy [38], but are not suitable for physical goods; [42] works by interleaving proxies that remove identifiable information about customers. Some works focus specifically on privacy respectful user profiling [17,36,44], mostly for affinity programs, although some approaches are also applicable to fraud prevention [17]. Anonymous delivery systems of physical goods have also been proposed [3,42], covering a crucial phase that has received much less attention. Finally, solutions related to the completion phase (feedback, complaints, etc.) have been basically ignored, although this phase have been shown to allow de-anonymization attacks [33]. Underlying most of these proposals are, often, cryptographic primitives such as oblivious transfer [2] or anonymous credentials [9,15], which are of natural interest in this domain as core building blocks.

The above proposals focus on the two steps of the methodology above (i.e., the *"utility, privacy"* stages), with a few limited exceptions [17,24,35,41], thus restricting the extended utility recovered by our last stage of *"utility again."* Moreover, none covers all the e-shopping core processes, reducing the privacy of the composed overall system to that of the weakest link [20]. Some proposals

---

[1]   https://www.eugdpr.org/.

[2]   https://ec.europa.eu/info/law/payment-services-psd-2-directive-eu-2015-2366_en.

[3]   As well as many proposals in non-academic forums. See, for instance, https://z.cash/ (a modified implementation of Zerocash) and https://cryptonote.org/.

introduce extensive changes into the infrastructure and processes [28] or require modifications that conflict with regulations or practical concerns, like requiring the outsourcing of information that would probably be proprietary in many scenarios [17,44]. Therefore, at present, the utility-privacy trade-off is leaning towards utility in the industry and towards full privacy in the literature.

1.2   Organization

After some preliminaries in Sect. 2, we sketch in Sect. 3 how we apply privacy to the traditional system. We analyze this system to show its shortcomings and recover utility in Sect. 4. We conclude in Sect. 5. For lack of space, we omit formal security definitions and proofs and a detailed analysis on the experiments performed with our prototype. We refer to the full version of this paper for the details [21].

2   Preliminaries

Notation. For an algorithm A, let $A(x_1, \ldots, x_n; r)$ denote the output of A on inputs $x_1, \ldots x_n$ and random coins r; in addition, $y \leftarrow A(x_1, \ldots, x_n)$ means choosing r uniformly at random and setting $y \leftarrow A(x_1, \ldots x_n; r)$. For a set S, let $x \leftarrow S$ denote choosing x uniformly at random from S. We let $\langle O_A, O_B \rangle \leftarrow P(I_C)[A(I_A), B(I_B)]$ denote a two-party process P between parties A and B, where O_A (resp. O_B) is the output to party A (resp. B), I_C is the common input, and I_A (resp. I_B) is A's (resp. B's) private input; when party B does not have output, we sometimes write $O_A \leftarrow P(I_C)[A(I_A), B(I_B)]$. When a single party algorithm P uses a public key pk, we may write $O \leftarrow P_{pk}(I)$ (although we omit it if it is clear from the context). For readability, we assume that if any internal step fails, the overall process fails and stops.

Basic Cryptographic Primitives. We assume readers are familiar with public-key encryption [22,39], digital signature and commitment schemes [8], and zero-knowledge proofs of knowledge (ZK-PoKs) [26]. Let (EGen, Enc, Dec) denote a public-key encryption scheme, and (SGen, Sign, SVer) denote a digital signature scheme. For readability, we assume that it is possible to extract the signed message from the corresponding signature. We let $com_m \leftarrow Com(m; r_m)$ denote a commitment to a message m, where the sender uses uniform random coins r_m; the sender can open the commitment by sending (m, r_m) to the receiver. We use $\pi \leftarrow \mathtt{ProveZK}_L(x; w)$ and $\mathtt{VerifyZK}_L(x, \pi)$ to refer to creating non-interactive proof π showing that the statement x is in language L (which we sometimes omit if obvious from the context) with the witness w, and to verifying the statement x based on the proof π.

Group Signatures. Group signatures [10,14,29–31] provide anonymity. A public key is set up with respect to a group consisting of multiple members. Any member of the group can create a signature ϱ revealing no more information about the signer than the fact that a member of the group created ϱ. Group signatures also provide accountability: the group manager (GM) can open signatures and identify the actual signer.

- $(pk_G, sk_G) \leftarrow$ GS.Setup(1^k) sets up a key pair; GM holds sk_G.
- $\langle mk_i, \ell' \rangle \leftarrow$ GS.Join$(pk_G)[M(s_i), GM(\ell, sk_G)]$ allows member M with secret s_i to join group G, generating the private member key mk_i and updating the Group Membership List ℓ to ℓ'.
- $\varrho \leftarrow$ GS.Sign$_{mk_i}(msg)$ issues a group signature ϱ.
- GS.Ver$_{pk_G}(\varrho, msg)$ verifies whether ϱ is a valid group signature.
- $i \leftarrow$ GS.Open$_{pk_G}(sk_G, \varrho)$ returns the identity i having issued the signature ϱ.
- $\pi \leftarrow$ GS.Claim$_{mk_i}(\varrho)$ creates a claim π of the ownership of ϱ.
- GS.ClaimVer$_{pk_G}(\pi, \varrho)$ verifies if π is a valid claim over ϱ.

Traceable Signatures. Traceable signatures [29] are essentially group signatures with additional support of tracing (when we use the previous group signature operations, but with a traceable signature scheme, we use the prefix TS instead of GS).

- $t_i \leftarrow$ TS.Reveal$_{sk_G}(i)$. The GM outputs the tracing trapdoor of identity i.
- $b \leftarrow$ TS.Trace(t_i, ϱ). Given the tracing trapdoor t_i, this algorithm checks if ϱ is issued by the identity i and outputs a boolean value b reflecting the check.

Partially Blind Signatures. A blind signature scheme [13] allows a user U to have a signer S blindly sign the user's message m. Partially blind signatures [1], besides the blinded message m, also allow including a common public message in the signature.

- $(pk_S, sk_S) \leftarrow$ PBS.KeyGen(1^k) sets up a key pair.
- $(\tilde{m}, \pi) \leftarrow$ PBS.Blind$_{pk_S}(m, r)$. Run by a user U, it blinds the message m using a secret value r. It produces the blinded message $\tilde{m}$ and a correctness proof π of $\tilde{m}$.
- $\tilde{\varrho} \leftarrow$ PBS.Sign$_{sk_S}(cm, \tilde{m}, \pi)$. Signer S verifies proof π and issues a partially blind signature $\tilde{\varrho}$ on $(cm, \tilde{m})$, where cm is the common message.
- $\varrho \leftarrow$ PBS.Unblind$_{pk_S}(\tilde{\varrho}, \tilde{m}, r)$. Run by the user U, who verifies $\tilde{\varrho}$ and then uses the secret value r to produce a final partially blind signature ϱ.
- PBS.Ver$_{pk_S}(\varrho, cm, m)$ checks if ϱ is valid.

3   System with a High Level of Privacy and Less Functionalities

Following the approach of "utility, privacy, and then utility again", we first overview the existing e-shopping system (*utility*) and then add privacy enhancing mechanisms, relaxing its functionality in order to achieve a high level of privacy (*privacy*). In the next section, we add other important features, carefully relaxing privacy (*utility again*).

The General e-Shopping Process. Assuming users have already registered in the system, we may consider four phases: purchase, checkout, delivery and completion (see Fig. 1). The involved parties are customers (C), merchants (M), the payment system (PS), financial entities processing and executing transactions (that we bundle in our abstraction as FN) and delivery companies (DC). PS basically connects merchants and FN, providing advanced services. First, in the *purchase phase*, C picks the products he wants to buy from M and any coupons he may be eligible for (task in which PS may be involved). In the *checkout phase*, the payment and delivery information specified by C are routed to PS, probably through M, and processed and executed by FN. During checkout, M, PS and FN may apply fraud prevention mechanisms and update C's purchase history. Subsequently, in the *delivery phase*, and for physical goods, DC delivers them to C. Finally, in the *completion phase*, C verifies that everything is correct, maybe initiating a complaint and/or leaving feedback.

Fig. 1. The overall process of a traditional e-shopping.

Many aspects in this process enter in conflict with privacy (e.g., coupons, fraud prevention and physical delivery), but they are necessary to foster industry acceptance.

3.1 Privacy Goal

We assume that merchants can act maliciously, but PS, FN and DC are semi-honest. Informally, we aim at achieving customer privacy satisfying the following properties:

- *Hide the* identity of a customer *and reveal it only if necessary:* The identity of a customer is sometimes sensitive information, and we want to hide it from other parties as much as possible. In the overall e-shopping process merchants, PS, and DC don't really need the identity of the customer in order for the transaction to go through. However, FN must know the identity to withdraw the actual amount of money from the customer's account and to comply with current regulations.

– *Hide the* payment information *and reveal it only if necessary:* The information about the credit card number (or other auxiliary payment information) that a customer uses during the transaction is quite sensitive and thereby needs to be protected. In the overall e-shopping process, like the case of the customer identity, observe that only FN must know this information to complete the financial transaction.

– *Hide the* product information *and reveal it only if necessary:* The information about which product a customer buys can also be sensitive. However, note that PS and FN don't really need to know what the customer is buying in order for the transaction to go through, but the merchants and DC must handle the actual product.

3.2 Approach for Privacy-Enhancements

In the full version of this paper, we describe in detail the privacy enhanced system. Below, we highlight our approach towards privacy and sketch the system in Fig. 2.

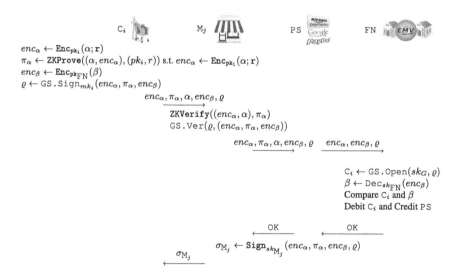

Fig. 2. The overall process of the system. Here, α and β are the product and purchase information respectively. α has been obtained previously by C_i, browsing M_j's web anonymously.

Controlling the Information of Customer Identity. We use the following privacy-enhancing mechanisms to control the information of customer identity.

– *Sender anonymous channel from customers:* Customers use sender-anonymous channels such as Tor [23] for their communications.

– *Customer group signatures on transaction data:* The transaction data on the customer side is authenticated by the customer's group signature. In our context, FN takes the role of the group manager. Thus, if a merchant M verifies the group signature included by a customer in a transaction, M is confident that the customer has an account with FN. Moreover, due to the group signatures, the customer's identity is hidden from other parties based on. However, since FN takes the role of the group manager, it can identify the customer by opening the signature if required, but it is otherwise not requested to take any active role with respect to managing the group or processing group signatures. Note that the group manager must be a trusted entity concerning the group management tasks, although this trust can be reduced with threshold techniques like those in [6].

Controlling the Payment Information. Customers encrypt their payment information with FN's public key. Thus, only FN can check if the identity in the payment information matches the one extracted from the customer's group signature.

Controlling the Product Information. The customer encrypts the information about the product he wants to purchase using a key-private public key encryption scheme (e.g., ElGamal encryption) [4]; he generates a key pair and uses the public key to encrypt the product information. The key pair can be used repeatedly since the scheme is key-private[4], and the public encryption key is never sent to other parties. The main purpose of doing this is for logging. Once FN logs the transactions, the customer can check the product information in each transaction by simply decrypting the related ciphertext.

Obviously, the encryption doesn't reveal any product information to other parties. Yet, merchants must obtain this data to proceed. To handle it, customers send the product information both in plaintext and ciphertext, and then prove consistency using a ZK proof. When this step is cleared, only the ciphertext part is transferred to other entities.

Note that this system satisfies all our privacy goals. However, it reduces utility, as is not compatible with many features required by the industry (or by regulation), specifically, marketing and fraud prevention tools, or extensions like customer support, subscriptions or taxation [20].

4 Privacy-Enhanced System with Richer Functionality

Next, we add important functionalities, in particular marketing and antifraud mechanisms, to the system described in Sect. 3, carefully relaxing privacy (*utility again*).

---

[4] Key-privacy security requires that an eavesdropper in possession of a ciphertext not be able to tell which specific key, out of a set of known public keys, is the one under which the ciphertext was created, meaning the receiver is anonymous from the point of view of the adversary.

Adding Marketing Tools: Utility vs Privacy. We would like the payment system PS (or merchants) to use marketing tools (e.g., coupons) so as to incentivize customers to purchase more products and thereby increase their revenue. For clarity of exposition, we will consider adding a feature of coupons and discuss the consequential privacy loss; other marketing features essentially follow the same framework.

When we try to add this feature to the system, PS *must at least have access to the amount of money each customer has spent so far;* otherwise, it's impossible for the coupons to be issued for more loyal customers. Obviously, revealing this information is a privacy loss. However, this trade-off between utility and privacy seems to be unavoidable, if the system is to be practically efficient, ruling out the use of fully-homomorphic encryptions [25] or functional encryptions [7], which are potentially promising but, as of now, prohibitively expensive to address our problem. The main question is as follows:

- Can we reveal *nothing more than* the purchase history of encrypted products?
- Can we provide the customers with an option to *control the leakage of this history?* In other words, can we give the customers an option to *exclude some or all of their purchase activities* from the history?

We address both of the above questions affirmatively. In order to do so, we first allow *each customer to use a pseudonym selectively.* That is, the payment system can aggregate the customer's purchase history of encrypted products only if the customer uses his pseudonym when buying a product. If the customer wants to exclude some purchase activity from this history, he can proceed with the transaction anonymously.

Still, there are a couple of issues to be addressed. First, we would like the system to work in *a single work flow* whether a customer chooses to go *pseudonymously or anonymously.* More importantly, we want a customer to be able to *use coupons even if he buys a product anonymously.* We will show below how we address these issues, when we introduce the notion of a checkout-credential.

Adding Antifraud Mechanisms: Utility vs Privacy. Merchants need to be protected against fraudulent or risky transactions, e.g. transactions that are likely to end up in non-payments, or that are probably the result of stolen credit cards and similar cases. This is typically done by having the PS send a risk estimation value to merchants, who can also apply their own *filters* based on the specifics of the transaction (number of items, price, etc.). At this point, we have an utility-privacy trade-off. In particular, if the risk estimation is too specific and identifying, it will hinder the system from supporting anonymous transactions. We believe that this trade-off is inherent, and in this paper, we treat the specificity of risk estimation to be given as an appropriately-chosen system parameter, depending on the volume of the overall transactions and only mildly degrading the quality of anonymity in anonymous transactions. The main question we ask is:

Can we relax anonymity of transactions but only to reveal *the risk estimation?*

As with the marketing tools, we use the checkout-credential for implementing this.

4.1   Our Approach

Checkout Credentials. We want to allow customers to perform unlinkable (anonymous) purchases, and we also need to provide merchants with the fraud estimation of a transaction based on each customer's previous transactions. This goal is achieved in a privacy-respectul manner through the checkout-credential retrieval process.

The checkout-credential retrieval process is carried out before the actual checkout, and it is executed between PS and the customer. The resulting checkout-credential is the means used by PS to aggregate the available information related to each pseudonym and provide the marketing and antifraud information for merchants without violating each customer's privacy. Figure 3 shows the augmented information flow of the *purchase* and *checkout* phases in our system. Delivery and completion are not depicted in Fig. 3 since, as we show in the following description, they are quite straightforward and do not suffer further modifications (with respect to the system in Sect. 3) besides integrating them with the new *purchase* and *checkout* processes. Specifically, note that while we have partitioned the main processes in multiple sub-processes, the overall flow is still the same. That is, *purchase → checkout → delivery → completion.* Finally, note also that the parties involved in each process are maintained compared to current systems.

Basically, a checkout-credential is a partially blind signature, requested by a customer and issued by PS, where the common message includes *aggregated*

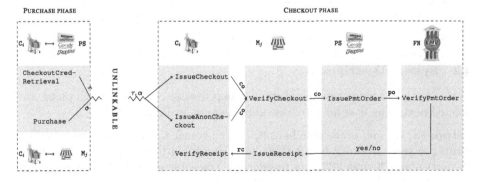

Fig. 3. System process flow. Here, τ is the checkout-credential and α is the product information.

data related to fraud and marketing and the blinded message is a *commitment to the customer key*. During checkout, a customer proves to merchants in ZK that he knows the committed key embedded in the checkout credential. *Since it was blindly signed,* PS *and merchants cannot establish a link* beyond what the aggregated common information allows.

At this point, when the customer decides to perform a pseudonymous checkout (in this case, the pseudonym is also shown during checkout), PS will be able to link the current checkout to the previous ones and update the customer's history (updating his eligibility to promotions and risk estimation). If he chooses an anonymous checkout, PS will not be able to link this transaction with others.

Protection Against Fraudulent Anonymous Transactions. There is an additional issue. An attacker may execute a large volume of pseudonymous transactions honestly, making its pseudonym have a low risk-estimate value, and then perform a fraudulent anonymous transaction. Note in this case, the checkout-credential will contain low risk estimate and the transaction will likely go through, but problematically, *because of unlinkability of this fraudulent transaction,* PS *cannot reflect this fraud into the pseudonym's transaction history.* Moreover, taking advantage of this, the attacker can repeatedly perform fraudulent anonymous transactions with low risk estimate. However, in this variant of our system, we use traceable signatures. Thus, if an anonymous transaction proves to be fraudulent a posteriori, FN can open the signature and give PS the tracing trapdoor associated with the token (i.e., the traceable signature). Given this trapdoor, PS can update the risk estimation even for anonymous checkouts.

Note that customers are offered a trade-off. When customers always checkout anonymously, they have no previous record and receive worse promotions and fraud estimates. When they always checkout pseudonymously, they get better offers and probably better fraud estimates, in exchange of low privacy. But there are also intermediate options. In all cases, they can take advantage of any coupons they are eligible for and receive fraud estimates based on previous pseudonymous purchases.

However, we emphasize that our system is natively compatible with many antifraud techniques in the industry without needing to resort to tracing and which are also applicable with anonymous checkouts and do not reduce privacy (see [21]).

4.2   System Description

In this section, we describe our system. The processes composing each phase are defined next. The flow for purchase and checkout is depicted in Fig. 3.

Setup. FN, PS, and every merchant M_j and customer C_i run their corresponding setup processes in order to get their keys, according to the processes in Fig. 4. In particular, FN runs FNSetup to generate traceable signature and encryption keys. PS runs PSSetup to generate a key pair for partially blind signatures. M_j runs MSetup to generate signing keys. C_i and FN interact in order to generate key pairs for C_i, running CSetup. C_i contacts FN, creates an account and joins a group

G, obtaining a membership key mk_i using a secret s_i. In this case, C_i also sets up a pseudonym P_i, known to FN. The pseudonym P_i is a traceable signature on a random message created using his membership key mk_i; we let $P_i.r$ denote the random message and $P_i.\varrho$ the traceable signature on $P_i.r$. During the process, FN updates its membership database ℓ into ℓ'.

$$
\begin{aligned}
&\text{FNSetup}(1^k): &&\text{MSetup}(1^k): \\
&\quad (pk_G, sk_G) \leftarrow \text{TS.Setup}(1^k) &&\quad (pk_{M_j}, sk_{M_j}) \leftarrow \text{SGen}(1^k) \\
&\quad (pk_{FN}, sk_{FN}) \leftarrow \text{EGen}(1^k) &&\quad \text{PK}_{M_j} \leftarrow pk_{M_j}; \text{SK}_{M_j} \leftarrow sk_{M_j} \\
&\quad \text{PK}_{FN} \leftarrow (pk_{FN}, pk_G) \\
&\quad \text{SK}_{FN} \leftarrow (sk_{FN}, sk_G) \\
& && \text{CSetup}(pk_G)[C_i(s_i), \text{FN}(sk_G, \ell)]: \\
& && \quad \langle mk_i, \ell' \rangle \leftarrow \text{TS.Join}(pk_G)[C_i(s_i), \text{FN}(\ell, sk_G)] \\
&\text{PSSetup}(1^k): && \quad (pk_i, sk_i) \leftarrow \text{EGen}(1^k) \\
&\quad (pk_{PS}, sk_{PS}) \leftarrow \text{SGen}(1^k) && \quad C_i \text{ chooses } r \leftarrow \{0,1\}^* \\
&\quad (pk_{PBS}, sk_{PBS}) \leftarrow \text{PBS.KeyGen}(1^k) && \quad C_i \text{ computes } \varrho \leftarrow \text{TS.Sign}_{mk_i}(r; r_{P_i}) \\
&\quad \text{PK}_{PS} \leftarrow (pk_{PS}, pk_{PBS}) && \quad C_i \text{ sends } P_i = (r, \varrho) \text{ to FN} \\
&\quad \text{SK}_{PS} \leftarrow (sk_{PS}, sk_{PBS}) && \quad \text{SK}_{C_i} \leftarrow (P_i, mk_i, r_{P_i}, pk_i, sk_i)
\end{aligned}
$$

Fig. 4. Full system setup processes.

Checkout-Credential Retrieval and Purchase. The purchase phase includes the `Purchase` and `CheckoutCredRetrieval` processes. The purpose of this phase is for C_i to obtain a description of the products to buy from M_j and a credential authorizing him to proceed to checkout, including information necessary to apply marketing and antifraud tools.

During `CheckoutCredRetrieval`, C_i interacts pseudonymously with PS. The protocol starts by having the customer C_i send his pseudonym P_i. Then, PS retrieves the information of how loyal P_i is (i.e., rk), whether (and how) P_i is eligible for promotion (i.e., pr), and the deadline of the checkout-credential to be issued (i.e., dl), sending back (rk, pr, dl) to C_i. C_i chooses a subset pr' from the eligible promotions pr. Finally, C_i will have PS create a partially blind signature such that its common message is (rk, pr', dl) and its blinded message is a commitment , to his membership key mk_i. We stress that the private member key mk_i of the customer C_i links the pseudonym (i.e., $P_i.\varrho \leftarrow \text{TS.Sign}_{mk_i}(P_i.r)$) and the blinded message (i.e., $\text{com} \leftarrow \text{Com}(mk_i; r_{com})$). The customer is supposed to create a ZK-PoK ϕ showing this link. Upon successful execution, the checkout-credential is set to τ. We use $\tau.\text{rk}, \tau.\text{pr}, \tau.\text{dl}, \tau.,, \tau.\varrho$ to denote the risk factor, promotion, deadline, commitment to the member key, and the resulting blind signature respectively. Refer to Fig. 5 for pictorial description. A checkout-credential issued with the process in Fig. 5 would be verified during checkout using the `VerifyCheckoutCred` process, defined as follows:

$$\text{VerifyCheckoutCred}_{\text{PK}_{PS}}(\tau): \textbf{return } \text{PBS.Ver}_{pk_{PBS}}(\tau.\varrho, (\tau.\text{pr}, \tau.\text{rk}, \tau.\text{dl}), \tau.\text{com})$$

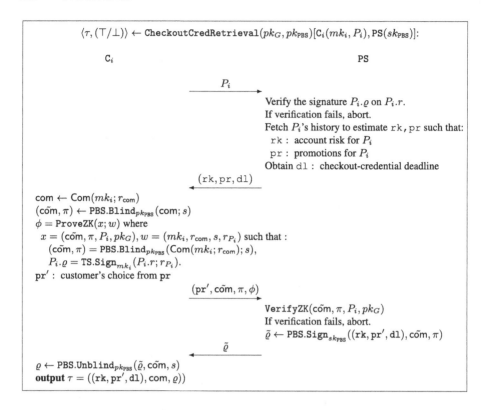

Fig. 5. The CheckoutCredRetrieval process.

Concurrently, C_i obtains through the Purchase process a product description of the items he wants to buy. Note that this can be done just by having C_i browse M_j's website using sender anonymous channels:

$$\alpha \leftarrow \text{Purchase}[C_i, M_j] : \textbf{return} \text{ product description from } M_j\text{'s website}$$

Finally, with both the product description α and the checkout-credential τ, C_i can initiate the checkout phase.

Checkout. After receiving the checkout-credential τ and having obtained a product description, C_i decides whether to perform an anonymous (IssueAnonCheckout) or pseudonymous (IssueCheckout) checkout process. Let α be the product information with the product name, merchant, etc.; also, let \$ be the price of the product and let β be the customer's payment information containing a random number uniquely identifying each transaction. The checkout process is formed as follows (refer to Fig. 6 for a detailed description of the algorithms). Note that the information flow is equivalent to that in Fig. 2, but here we include additional cryptographic tokens.

Step 1: Client issues a checkout object. A customer C_i enters the checkout phase by creating a checkout object co, executing Issue(Anon)Checkout using the

co ← IssueCheckout($\text{SK}_{C_i}, \tau, \alpha, \$, \beta$):
  Parse $\text{SK}_{C_i} = (P_i, mk_i, r_{P_i}, pk_i, sk_i)$
  $\text{enc}_\alpha \leftarrow \text{Enc}_{pk_i}(\alpha; r_\alpha)$
  $\text{enc}_\beta \leftarrow \text{Enc}_{pk_{\text{PN}}}(\beta)$
  $\varrho \leftarrow \text{TS.Sign}_{mk_i}((\$, \text{enc}_\alpha, \text{enc}_\beta); r_{gs})$
  $\psi \leftarrow \text{ProveZK}(x, w)$ with
    $x = (P_i, \tau.\text{com}, \$, \alpha, \text{enc}_\alpha, \text{enc}_\beta, \varrho)$
    $w = (mk_i, r_{P_i}, r_\alpha, r_{\text{com}}, r_{gs})$
    such that
    $P_i.\varrho = \text{TS.Sign}_{mk_i}(P_i.r; r_{P_i})$
    $\text{enc}_\alpha = \text{Enc}_{pk_i}(\alpha; r_\alpha)$
    $\tau.\text{com} = \text{Com}(mk_i; r_{\text{com}})$
    $\varrho = \text{TS.Sign}_{mk_i}((\$, \text{enc}_\alpha, \text{enc}_\beta); r_{gs})$
  co ← $(P_i, \tau, \$, \alpha, \text{enc}_\alpha, \text{enc}_\beta, \varrho, \psi)$
  return co

co ← IssueAnonCheckout($\text{SK}_{C_i}, \tau, \alpha, \$, \beta$):
  Parse $\text{SK}_{C_i} = (P_i, mk_i, r_{P_i}, pk_i, sk_i)$
  $\text{enc}_\alpha \leftarrow \text{Enc}_{pk_i}(\alpha; r_\alpha)$
  $\text{enc}_\beta \leftarrow \text{Enc}_{pk_{\text{PN}}}(\beta)$
  $\varrho \leftarrow \text{TS.Sign}_{mk_i}((\$, \text{enc}_\alpha, \text{enc}_\beta); r_{gs})$
  $\psi \leftarrow \text{ProveZK}(x, w)$ with
    $x = (\tau.\text{com}, \$, \alpha, \text{enc}_\alpha, \text{enc}_\beta, \varrho)$
    $w = (mk_i, r_\alpha, r_{\text{com}}, r_{gs})$
    such that
    $\text{enc}_\alpha = \text{Enc}_{pk_i}(\alpha; r_\alpha)$
    $\tau.\text{com} = \text{Com}(mk_i; r_{\text{com}})$
    $\varrho = \text{TS.Sign}_{mk_i}((\$, \text{enc}_\alpha, \text{enc}_\beta); r_{gs})$
  co ← $(\tau, \$, \alpha, \text{enc}_\alpha, \text{enc}_\beta, \varrho, \psi)$
  return co

$\top/\bot$ ← VerifyCheckout(co):
  Parse co into $([P_i,]\tau, \$, \alpha, \text{enc}_\alpha, \text{enc}_\beta, \varrho, \psi)$
  $\text{VerifyCheckoutCred}_{pk_{\text{PS}}}(\tau)$
  Check if $(\tau.\text{rk}, \tau.\text{pr}, \tau.\text{dl})$ is acceptable
  Check if τ is unique within $\tau.\text{dl}$
  $\text{TS.Ver}_{pk_G}((\$, \text{enc}_\alpha, \text{enc}_\beta), \varrho)$
  $\text{VerifyZK}(([P_i,]\tau.\text{com}, \$, \alpha, \text{enc}_\alpha, \text{enc}_\beta, \varrho), \psi)$
  If all the checks above pass, return 1
  Otherwise return 0

po ← IssuePmtOrder(co):
  VerifyCheckout(co)
  If verification fails, return 0
  Parse co into $([P_i,]\tau, \$, \alpha, \text{enc}_\alpha, \text{enc}_\beta, \varrho, \psi)$
  If P_i is present, update P_i's history
  po ← $(\$, \text{enc}_\alpha, \text{enc}_\beta, \varrho)$
  return po

rc ← IssueReceipt(SK_{M_j}, co):
  rc ← $\text{Sign}_{sk_{M_j}}(\text{co})$
  return rc

$\top/\bot$ ← VerifyPmtOrder(SK_{PN}, po):
  Parse po into $(\$, \text{enc}_\alpha, \text{enc}_\beta, \varrho)$
  $\text{TS.Ver}_{pk_G}((\$, \text{enc}_\alpha, \text{enc}_\beta), \varrho)$
  $\beta = \text{Dec}_{sk_{\text{PN}}}(\text{enc}_\beta)$
  Check if β has not been used before
  Check if $\text{TS.Open}_{sk_G}(\varrho)$ equals C_i in β
  Verify the other billing info in β
  If all the checks above pass, return 1
  Otherwise return 0

$\top/\bot$ ← VerifyReceipt(rc, co):
  Find identity M_j from α in co
  return $\text{SVer}_{pk_{M_j}}(\text{co}, \text{rc})$

Fig. 6. Checkout algorithms.

checkout-credential τ obtained during checkout-credential retrieval. In either procedure, C_i generates a traceable signature ϱ on $(\$, \text{enc}_\alpha, \text{enc}_\beta)$, where enc_α is an encryption of the product information α, and enc_β is an encryption of the payment information β, and $\$$ is the price of the product. Then, C_i generates a ZK proof ψ showing that the checkout-credential and the traceable signature (and the pseudonym for IssueCheckout) use the same mk_i. In summary, we have co = $([P_i,]\tau, \$, \alpha, \text{enc}_\alpha, \text{enc}_\beta, \varrho, \psi)$.

Step 2: Merchant processes checkout co. When M_j receives the checkout object co (which includes the product information α in the clear, as well as encrypted), verifies it with VerifyCheckout. If verification succeeds, M_j passes co to PS. Note that τ needs to be checked for uniqueness to prevent replay attacks. However, a used credential τ only needs to be stored up to $\tau.dl$. It is also possible for M_j to include additional antifraud information, like an Address Verification Service value[5] (see [21]).

Step 3: PS *issues a payment order po.* On receiving co from M_j, PS verifies co, runs IssuePmtOrder and issues a payment order po with the minimum information required by FN for processing the payment that is, $po = (\$, enc_\alpha, enc_\beta, \varrho)$.

Step 4–5: Payment confirmations. Given the payment order po, FN verifies it by running VerifyPmtOrder. If the verification succeeds, FN processes the order and notifies PS of the completion; PS in turn sends the confirmation back to M_j.

Step 6: M_j *issues a receipt.* M_j receives the confirmation from PS and runs IssueReceipt, issuing rc, a signature on co. Finally, C_i verifies rc with VerifyReceipt.

Delivery. Once C_i receives rc, he can use it to prove in ZK that he actually payed for some transaction co, and initiate additional processes, like having DC deliver the goods through APOD [3]. This proof is obtained with the processes in Fig. 7. In the showing process, if C_i received a receipt rc, he shows rc along with the corresponding checkout object co; then, using his membership key mk_i, he claims ownership of a traceable signature contained in co. Even if he did not receive a receipt, he can prove ownership of ϱ to FN (using ShowReceiptZK too). Since FN is semi-honest, C_i may ask FN to cancel the associated payment (or *force* PS and M_j to reissue the receipt).

$\pi \leftarrow$ ShowReceiptZK(SK_{C_i}, rc, co):	$\top/\bot \leftarrow$ VerifyReceiptZK(rc, co, π):
Parse co $= ([P_i,]\tau, \$, \alpha, enc_\alpha, enc_\beta, \varrho, \psi)$	Parse co $= ([P_i,]\tau, \$, \alpha, enc_\alpha, enc_\beta, \varrho, \psi)$
$\pi \leftarrow$ TS.Claim$_{mk_i}(\varrho)$	VerifyReceipt(rc, co)
return π	TS.ClaimVer$_{pk_G}(\pi, \varrho)$
	If all the checks pass, return 1
	Otherwise return 0

Fig. 7. Full system processes for claiming rc in Zero-Knowledge.

In order to interconnect with APOD, C_i proves M_j being the owner of rc (through ShowReceiptZK). Then, M_j issues the credential cred required by APOD as in [3]. Note however that the incorporation of APOD incurs in additional costs and the need for further cryptographic tokens for merchants (who could delegate this task to PS). A less anonymous delivery method, but probably good enough for many contexts, could be using Post Office boxes (or equivalent delivery methods) [20].

Completion. When C_i receives the goods, the completion phase may take place. In this phase, C_i may leave feedback or initiate a claim, for which he needs to prove having purchased the associated items. For this purpose, C_i can again make use of the ShowReceiptZK and VerifyReceiptZK processes, defined in Fig. 7.

---

[5] https://en.wikipedia.org/wiki/Address_Verification_System.

4.3 Security

We assume that customers and merchants can act maliciously. PS is assumed to be semi-honest during checkout-credential retrieval, but malicious otherwise. FN is semi-honest.

Here, for lack of space, we informally describe the security properties of our system. We give formal security definitions and proofs in the full version [21].

Privacy. The system possesses the following privacy properties.

- *Customer anonymity.* If a customer executes the checkout process anonymously, no coalition of merchants, PS, and other customers should be able to determine the identity or pseudonym of the customer from the checkout process beyond what the common message in the checkout credential reveals.
- *Transaction privacy against merchants and* PS. No coalition of merchants, PS and other customers should be able to determine the payment information associated to the checkout process.
- *Transaction privacy against* FN. The financial network FN should not be able to determine the detail of a customer's transaction beyond what is necessary, i.e., the customer identity and the amount of payment; in particular, M_j's identity and the product information should be hidden from FN.
- *Unlinkable checkout-credential retrieval and checkout.* If a customer runs an anonymous checkout, no coalition of merchants, PS, and other customers should be able to link the customer or his pseudonym to the corresponding checkout-credential retrieval procedure beyond what the common message in the credential reveals.

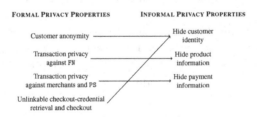

Fig. 8. Mapping between informal properties in Sect. 3.1 and formal properties in this section.

Note that this properties map to the properties in Sect. 3.1, with some additional conditions (see Fig. 8 for a pictorial representation). It is also worth noting that there are indirect connections between them. For instance, *Transaction privacy against* FN and *Transaction privacy against merchants and* PS undoubtedly improves resistance against differential privacy attacks aimed at deanonymizing customers (hence, affecting the *Customer anonymity*). However, as stated in the conclusion, a detailed analysis of these aspects is out of the scope of this work and is left for future work.

Robustness. The system also ensures the following robustness properties.

– *Checkout-credential unforgeability.* A customer should not be able to forge a valid checkout-credential with a risk factor, promotions or deadline set by his own choice.
– *Checkout unforgeability.* When C_i receives a checkout-credential from PS, it cannot be used by C_j $(i \neq j)$ to create a valid co, even if they collude.
– *Fraudulent transaction traceability.* When C_i performs a fraudulent transaction, FN and PS can trace the pseudonym used by C_i even if the transaction is anonymous.
– *Receipt unforgeability.* No coalition of customers, merchants (other than the target M_j), and PS should be able to forge a valid receipt that looks originating from M_j.
– *Receipt claimability.* For any valid receipt issued to an uncorrupted customer, no other customer should succeed in claiming ownership of the receipt.

4.4   Outline of the Methodology and Experiments Summary

We achieve a privacy-enhanced e-shopping system by applying the *utility, privacy and utility again* methodology as follows:

– *(Utility, privacy)* Following [20], we first identify the core components of the existing e-shopping system as follows:
  • The participating parties: users, merchants, payment systems, financial network, and delivery companies.
  • The basic e-shopping processes: purchase, checkout, delivery, completion.
  • Added-value tools: marketing and fraud prevention.
  When applying the privacy-enhancing mechanisms, we minimize the modification of these core functionalities. In particular, we change neither the participating parties nor the actual transaction flow. However, we add full anonymity at the cost of marketing and fraud prevention tools.
– *(Utility again)* In this stage, we add the following important real-world features:
  • Marketing tools such as targeted coupons.
  • Fraud preventions measures, allowing to include unpayment risk estimations.
  When providing these important utility features, we carefully relax privacy. In particular, each customer is associated with a pseudonym, and fraud prevention and marketing tools are applied by aggregating certain pieces of transaction history based on the pseudonym. Yet, we allow customers to act anonymously in each transaction, ensuring *privacy is not reduced beyond what this aggregation implies.*

Finally, we have implemented a prototype of our system. Here, for lack of space, we do not include a full report on our results, which will be made available in the full version [21]. As as a summary, we point out that in an unoptimized version of our prototype, we achieve between 1–3 full-cycle purchases per second. For comparison, other similar systems (e.g., Magento) report between 0.17 and 0.7 purchases per second[6]. It is important to note that we have simplified some parts of the process, such as payments (simulated through a database modification). This, however, is likely to be a relatively negligible operation within the overall process: e.g. VISA processed 141 billion transactions in 2016[7], which makes roughly 4500 transactions per second. Concerning the sizes of the groups of customers in the group signature schemes, we note that this is a highly configurable aspect. For instance, groups can be set based on geographies, based on sign up time, or other heuristics. As for the impact on performance of the sizes of the groups, we refer to [19], which we used to implement our prototype and offers some statistics about the group sizes and throughput of the main operations.

5   Conclusion

We have put forth our proposal for reaching a balance between privacy and utility in e-shopping. This is a complex scenario, where the diverse set of functionalities required by the industry makes it hard to provide them in a privacy respectful manner [20]. Moreover, the restriction of maintaining a similar system topology, limits the application of traditional privacy by design principles. With respect to the related work, our proposal integrates all core components of e-shopping (purchase, checkout, delivery and completion) and the advanced functionality in industry systems (marketing and fraud prevention). To the best of our knowledge this is an unsolved problem [20, 40].

Note that our system provides a basic infrastructure for building privacy respectful systems requiring user profiling. Specifically, users pseudonymously obtain customized credentials based on their history, and then anonymously prove possession of those credentials unlinkably to the pseudonymous phase. We have also implemented a prototype of our system, showing its practicability and low added costs. We refer to the full paper for further details on experiments, formal security proofs and possible extensions [21].

Nevertheless, further work is necessary. We include aggregated antifraud and promotions information that is publicly accessible from the checkout-credential. Hence, an open problem is reducing the impact of this leak for reidentification.

Finally, we used a *"utility, privacy, and then utility again"* methodology for designing our system. This strategy is can be applied to transition from policy to engineering in privacy protection in already deployed systems [16]. In other words, our work contributes to build up the Business, Legal, and Technical framework [27] demanded to reconcile economic interests, citizens' rights, and users' needs in today's scenario.

---

[6] https://magento.com/sites/default/files/White%20Paper%20-%20Magento%202.0
%20Performance%20and%20Scalability%2003.31.16.pdf.

[7] https://usa.visa.com/dam/VCOM/global/about-visa/documents/visa-facts-
figures-jan-2017.pdf.

Acknowledgements. The work of Jesus Diaz was done in part while visiting the Network Security Lab at Columbia University. The work of Seung Geol Choi was supported in part by the Office of Naval Research under Grant Number N0001415WX01232. The work of David Arroyo was supported by projects S2013/ICE-3095-CM (CIBERDINE) and MINECO DPI2015-65833-P of the Spanish Government. The work of Francisco B. Rodriguez was supported by projects MINECO TIN2014-54580-R and TIN2017-84452-R of the Spanish Government. The work of Moti Yung was done in part while visiting the Simons Institute for Theory of Computing, UC Berkeley.

References

1. Abe, M., Fujisaki, E.: How to date blind signatures. In: Kim, K., Matsumoto, T. (eds.) ASIACRYPT 1996. LNCS, vol. 1163, pp. 244–251. Springer, Heidelberg (1996). https://doi.org/10.1007/BFb0034851
2. Aiello, B., Ishai, Y., Reingold, O.: Priced oblivious transfer: how to sell digital goods. In: Pfitzmann, B. (ed.) EUROCRYPT 2001. LNCS, vol. 2045, pp. 119–135. Springer, Heidelberg (2001). https://doi.org/10.1007/3-540-44987-6_8
3. Androulaki, E., Bellovin, S.M.: APOD: anonymous physical object delivery. In: Privacy Enhancing Technologies, pp. 202–215 (2009)
4. Bellare, M., Boldyreva, A., Desai, A., Pointcheval, D.: Key-privacy in public-key encryption. In: Boyd, C. (ed.) ASIACRYPT 2001. LNCS, vol. 2248, pp. 566–582. Springer, Heidelberg (2001). https://doi.org/10.1007/3-540-45682-1_33
5. Ben-Sasson, E., Chiesa, A., Garman, C., Green, M., Miers, I., Tromer, E., Virza, M.: Zerocash: decentralized anonymous payments from bitcoin. In: 2014 IEEE Symposium on Security and Privacy, pp. 459–474 (2014)
6. Benjumea, V., Choi, S.G., Lopez, J., Yung, M.: Fair traceable multi-group signatures. In: Tsudik, G. (ed.) FC 2008. LNCS, vol. 5143, pp. 231–246. Springer, Heidelberg (2008). https://doi.org/10.1007/978-3-540-85230-8_21
7. Boneh, D., Sahai, A., Waters, B.: Functional encryption: definitions and challenges. In: Ishai, Y. (ed.) TCC 2011. LNCS, vol. 6597, pp. 253–273. Springer, Heidelberg (2011). https://doi.org/10.1007/978-3-642-19571-6_16
8. Brassard, G., Chaum, D., Crépeau, C.: Minimum disclosure proofs of knowledge. J. Comput. Syst. Sci. **37**(2), 156–189 (1988)
9. Camenisch, J., Dubovitskaya, M., Neven, G.: Oblivious transfer with access control. In: ACM CCS, CCS 2009, pp. 131–140. ACM (2009)
10. Camenisch, J., Lysyanskaya, A.: Dynamic accumulators and application to efficient revocation of anonymous credentials. In: Yung, M. (ed.) CRYPTO 2002. LNCS, vol. 2442, pp. 61–76. Springer, Heidelberg (2002). https://doi.org/10.1007/3-540-45708-9_5
11. Camenisch, J., Piveteau, J.-M., Stadler, M.: An efficient fair payment system. In: ACM Conference on Computer and Communications Security, pp. 88–94 (1996)
12. Campanelli, M., Gennaro, R., Goldfeder, S., Nizzardo, L.: Zero-knowledge contingent payments revisited: attacks and payments for services. In: Proceedings of the 2017 ACM SIGSAC CCS, pp. 229–243 (2017)
13. Chaum, D.: Blind signatures for untraceable payments. In: Chaum, D., Rivest, R.L., Sherman, A.T. (eds.) Advances in Cryptology, pp. 199–203. Springer, Boston (1983). https://doi.org/10.1007/978-1-4757-0602-4_18
14. Chaum, D., van Heyst, E.: Group signatures. In: Davies, D.W. (ed.) EUROCRYPT 1991. LNCS, vol. 547, pp. 257–265. Springer, Heidelberg (1991). https://doi.org/10.1007/3-540-46416-6_22

15. Coull, S.E., Green, M., Hohenberger, S.: Access controls for oblivious and anonymous systems. ACM Trans. Inf. Syst. Secur. **14**, 10:1–10:28 (2011). http://doi.acm.org/10.1145/1952982.1952992
16. Danezis, G., Domingo-Ferrer, J., Hansen, M., Hoepman, J.-H., Le Metayer, D., Tirtea, R., Schiffner, S.: Privacy and data protection by design-from policy to engineering. Technical report, ENISA (2014)
17. Danezis, G., Kohlweiss, M., Livshits, B., Rial, A.: Private client-side profiling with random forests and hidden Markov models. In: Fischer-Hübner, S., Wright, M. (eds.) PETS 2012. LNCS, vol. 7384, pp. 18–37. Springer, Heidelberg (2012). https://doi.org/10.1007/978-3-642-31680-7_2
18. Davida, G., Frankel, Y., Tsiounis, Y., Yung, M.: Anonymity control in E-cash systems. In: Hirschfeld, R. (ed.) FC 1997. LNCS, vol. 1318, pp. 1–16. Springer, Heidelberg (1997). https://doi.org/10.1007/3-540-63594-7_63
19. Diaz, J., Arroyo, D., de Borja Rodríguez, F.: libgroupsig: an extensible C library for group signatures. IACR Cryptology ePrint Archive, 2015:1146 (2015)
20. Diaz, J., Choi, S.G., Arroyo, D., Keromytis, A.D., Rodriguez, F.B., Yung, M.: Privacy threats in e-Shopping (position paper). In: Garcia-Alfaro, J., Navarro-Arribas, G., Aldini, A., Martinelli, F., Suri, N. (eds.) DPM/QASA -2015. LNCS, vol. 9481, pp. 217–225. Springer, Cham (2016). https://doi.org/10.1007/978-3-319-29883-2_14
21. Diaz, J., Choi, S.G., Arroyo, D., Keromytis, A.D., Rodriguez, F.B., Yung, M.: A methodology for retrofitting privacy and its application to e-Shopping transactions (2018, to appear)
22. Diffie, W., Hellman, M.E.: New directions in cryptography. IEEE Trans. Inf. Theor. **22**(6), 644–654 (1976)
23. Dingledine, R., Mathewson, N., Syverson, P.: Tor: the second-generation onion router. In: USENIX Security Symposium, SSYM 2004, Berkeley, CA, USA, pp. 21–21. (2004)
24. Garman, C., Green, M., Miers, I.: Accountable privacy for decentralized anonymous payments. IACR Cryptology ePrint Archive, 2016:61 (2016)
25. Gentry, C.: Fully homomorphic encryption using ideal lattices. In: Mitzenmacher, M. (ed.) 41st ACM STOC, pp. 169–178. ACM Press, May/June 2009
26. Goldwasser, S., Micali, S., Rackoff, C.: The knowledge complexity of interactive proof systems. SIAM J. Comput. **18**(1), 186–208 (1989)
27. Greenwood, D., Stopczynski, A., Sweatt, B., Hardjono, T., Pentland, A.: The new deal on data: a framework for institutional controls. In: Privacy, Big Data, and the Public Good: Frameworks for Engagement, p. 192 (2014)
28. Jacobson, M., M'Raïhi, D.: Mix-based electronic payments. In: Tavares, S., Meijer, H. (eds.) SAC 1998. LNCS, vol. 1556, pp. 157–173. Springer, Heidelberg (1999). https://doi.org/10.1007/3-540-48892-8_13
29. Kiayias, A., Tsiounis, Y., Yung, M.: Traceable signatures. In: Cachin, C., Camenisch, J.L. (eds.) EUROCRYPT 2004. LNCS, vol. 3027, pp. 571–589. Springer, Heidelberg (2004). https://doi.org/10.1007/978-3-540-24676-3_34
30. Libert, B., Peters, T., Yung, M.: Group signatures with almost-for-free revocation. In: Safavi-Naini, R., Canetti, R. (eds.) CRYPTO 2012. LNCS, vol. 7417, pp. 571–589. Springer, Heidelberg (2012). https://doi.org/10.1007/978-3-642-32009-5_34
31. Libert, B., Yung, M.: Fully forward-secure group signatures. In: Naccache, D. (ed.) Cryptography and Security: From Theory to Applications. LNCS, vol. 6805, pp. 156–184. Springer, Heidelberg (2012). https://doi.org/10.1007/978-3-642-28368-0_13

32. Miers, I., Garman, C., Green, M., Rubin, A.D.: Zerocoin: anonymous distributed e-cash from bitcoin. In: 2013 IEEE Symposium on Security and Privacy (2013)
33. Minkus, T., Ross, K.W.: I know what you're buying: privacy breaches on eBay. In: De Cristofaro, E., Murdoch, S.J. (eds.) PETS 2014. LNCS, vol. 8555, pp. 164–183. Springer, Cham (2014). https://doi.org/10.1007/978-3-319-08506-7_9
34. Nakamoto, S.: Bitcoin: a peer-to-peer electronic cash system (2009). http://www.bitcoin.org/bitcoin.pdf
35. Nakanishi, T., Haruna, N., Sugiyama, Y.: Unlinkable electronic coupon protocol with anonymity control. ISW 1999. LNCS, vol. 1729, pp. 37–46. Springer, Heidelberg (1999). https://doi.org/10.1007/3-540-47790-X_4
36. Partridge, K., Pathak, M.A., Uzun, E., Wang, C.: PiCoDa: privacy-preserving smart coupon delivery architecture (2012)
37. ITU-T Recommendation. X.509. Information technology - open systems interconnection - the directory: authentication framework, June 1997
38. Rial, A., Kohlweiss, M., Preneel, B.: Universally composable adaptive priced oblivious transfer. In: Shacham, H., Waters, B. (eds.) Pairing 2009. LNCS, vol. 5671, pp. 231–247. Springer, Heidelberg (2009). https://doi.org/10.1007/978-3-642-03298-1_15
39. Rivest, R.L., Shamir, A., Adleman, L.M.: A method for obtaining digital signatures and public-key cryptosystems. Commun. ACM **21**(2), 120–126 (1978)
40. Ruiz-Martinez, A.: Towards a web payment framework: State-of-the-art and challenges. Electron. Commer. Res. Appl. **14**, 345–350 (2015)
41. Sander, T., Ta-Shma, A.: Flow control: a new approach for anonymity control in electronic cash systems. In: Franklin, M. (ed.) FC 1999. LNCS, vol. 1648, pp. 46–61. Springer, Heidelberg (1999). https://doi.org/10.1007/3-540-48390-X_4
42. Stolfo, S., Yemini, Y., Shaykin, L.: Electronic purchase of goods over a communications network including physical delivery while securing private and personal information of the purchasing party. US Patent App. 11/476,304, 2 November 2006
43. Tan, C., Zhou, J.: An electronic payment scheme allowing special rates for anonymous regular customers. In: DEXA Workshops, pp. 428–434 (2002)
44. Toubiana, V., Narayanan, A., Boneh, D., Nissenbaum, H., Barocas, S.: Adnostic: privacy preserving targeted advertising. In: NDSS (2010)

Detection in the Dark – Exploiting XSS Vulnerability in C&C Panels to Detect Malwares

Shay Nachum[1]([⊠]), Assaf Schuster[1]([⊠]), and Opher Etzion[2]([⊠])

[1] Technion – Israel Institute of Technology, 32000 Haifa, Israel
nachus@technion.ac.il, assaf@cs.technion.ac.il
[2] Yezreel Valley College, 19300 Yezreel Valley, Israel
ophere@yvc.ac.il

Abstract. Numerous defense techniques exist for preventing and detecting malware on end stations and servers (endpoints). Although these techniques are widely deployed on enterprise networks, many types of malware manage to stay under the radar, executing their malicious actions time and again. Therefore, a more creative and effective solution is necessary, especially as classic threat detection techniques do not utilize all stages of the attack kill chain in their attempt to detect malicious behavior on endpoints.

In this paper, we propose a novel approach for detecting malware. Our approach uses offensive and defensive techniques for detecting active malware attacks by exploiting the vulnerabilities of their command and control panels and manipulating significant values in the operating systems of endpoints – in order to attack these panels and utilize trusted communications between them and the infected machine.

Keywords: XSS · C&C · Detection

1 Introduction

Over the past decade, numerous cyber attacks have made headlines, each attack more serious than the previous ones: larger in scope, greater in sophistication and more advanced in data exfiltration. These attacks have been possible thanks to malware programs (e.g., Bot, Trojan and POS) that are incorporated with new and advanced stealth techniques that keep them undetected. Although a variety of threat-detection techniques for combating such malware have been developed over the years, the attackers continue to find new and creative methods for getting around these techniques. In other words, classic prevention and detection defense techniques, such as firewall-based software, signatures and rules, antivirus software, and intrusion detection systems (IDS), do not suffice in the battle against cyber attacks. While they may be able to protect against known attacks, classic defense systems are to no avail when protecting against new ones that lack a fingerprint.

Moreover, malware technologies seems to always be one step ahead of IDS, security information and event management (SIEM) technologies [1, 2], as the latter focus on the technology used by the attacker, are based on predefined rules for detecting breaches,

© Springer International Publishing AG, part of Springer Nature 2018
I. Dinur et al. (Eds.): CSCML 2018, LNCS 10879, pp. 227–242, 2018.
https://doi.org/10.1007/978-3-319-94147-9_18

have low flexibility to noise, and have restricted visibility due to limited access to data. There is, therefore, a genuine demand for a more creative and adaptive method that will provide *a last line of defense* against malware.

This paper proposes a novel counter-attack method for automatically detecting malware infection at endpoints, using both offensive and defensive techniques. We leveraged the idea that each piece of malicious code has vulnerable security holes that can be utilized for detecting and preventing malicious infection of endpoints. Using controlled experiments, we investigated and analyzed several types of malware programs, leading to the development of a new methodology and detection technique that provides an additional layer of defense in the fight against malware programs – even against previously unknown ones.

We first focused on identifying common features and capabilities among malware programs, such as the type of information that is requested by malware programs on endpoints and the common data they require in order to continue their malicious activities. To do so, predefined scripts were injected into endpoint information to constantly look for malware vulnerabilities. We also focused on identifying the goals and objectives of malware programs, so that we could provide a comprehensive solution that would combat a wide range of malwares, including previously unknown ones. Malware goals include the exfiltration of sensitive data such as people's credentials, passwords, documents and credit card details, and the carrying out of a denial of service (DOS) attack on a specific target. Unlike other techniques, our new generation detection technique is not based on the specific attributes or behaviors of the malwares, but rather on our taking advantage of their malicious achievements at the endpoints.

We also examined in detail the various stages of the attack lifecycle [3], to fully understand why classic solutions are often unable to detect new malware attacks, and to develop a solution that would provide an additional layer of detection for existing solutions [1, 2]. Moreover, analysis was conducted on the attacker's superiority factors over the defender. We found that numerous security holes exist, and the attacker only has to find and exploit one of them to achieve malicious goals. We took this finding and turned it around, using it against the attacker. Together along with C&C stage examination, this was adopted to constantly attack C&C web panels (remote administration panel of the bots) using an XSS cross site scripting attack technique (a type of injection in which malicious scripts are injected into websites [4]). Hence, a new methodology and detection technique was created.

In other words, our novel approach issues a counter-attack on malware based on the information that the latter derives from endpoints, and utilizing the trusted communications between the malware and its C&C. First, this technique detects malware infection by modifying endpoint information to consist of an XSS string. Once this information is queried by the malware and is presented on a web page, the technique then alerts an isolated server regarding the infection. In this manner, our method – combined with classis threat-detection techniques – provides an extra layer of detection to deal with previously unknown malware.

The contributions of this technique are numerous. First, it expands on and enhances existing threat detection techniques by exploiting malware vulnerabilities. Second, it reduces the serious gap between cyber defenders and cyber attackers. Third, this novel technique challenges malware developers in the defense aspect. Finally, it paves the way for a variety of generic offensive detection techniques in the future.

2   Related Work

There has been an extensive amount of research on malware detection techniques in general and on the vulnerabilities of malware C&C panels in particular. In the closest work to ours, Sood et al. [5] used the properties and features of remote panels to detect other C&C panels and in turn more infections. Sood collected the unique property values of the malware programs' C&C panels (Zeus, ICE IX and Citadel) using malware analysis and network forensic techniques. These values were then used to detect new C&C panels of the above-mentioned malware programs via Google dorks, C&C trackers, and network traffic analysis. However, these unique values were only collected after dynamic analysis of the malwares and a leak of their C&C panel. In contrast, our detection technique, that exploits vulnerabilities in malware C&C panels, is generic and can detect any insecure C&C panel, regardless of its unique properties or values. In addition, a dynamic analysis of the malware isn't a prerequisite in our detection technique.

Additional studies [6, 7] found vulnerabilities in the Phase bot and ICE IX C&C panels, and showed that malware C&C panels can be taken advantage of in the war against hackers. Sood, for example, discovered an XSS vulnerability in the main login page of the ICE botnet C&C panel. In a different study [8], SQL injection (SQLI) vulnerability was detected in Dexter C&C panels. Exploiting this vulnerability can grant the attacker control of the remote administration panel of the bot. In another study, researchers at Prolexic (Akamai) [9, 10] identified vulnerable code in the Dirt Jumper Family's (DJF) C&C that enables the configuration file to be take advantage of and downloaded using the SQLI technique. This file consists of all the account credentials needed to take over the attacker's C&C panel. Moreover, even the notorious Zeus malware was found vulnerable [11]. A security researcher found a vulnerability, and developed a simple program to take advantage of it. His script allows users to directly upload and execute the code of their choice on the server that is running the Zeus C&C.

In another work similar to ours, Grange [12] disclosed several exploits on the C&C servers of Gh0st Remote Access Trojan (Rat), PlugX and XtremeRat malware. These exploits enable the defender to take full control of malware C&C. Our work differs from Grange's in a number of aspects. First, although our work can also be used for counter-hacking, its main purpose is to detect malware attack. Furthermore, our work uses malware vulnerabilities in order to detect their C&C panel, whereas Grange's work uses them after malware C&C is already known. Second, these vulnerabilities are only designed for the above malware, whereas our technique is not dependent on a specific malware and can detect new, previously unknown malware.

Leder et al. [13] demonstrated that proactive counter-measures are effective and necessary in the fight against real botnets. Focusing on Kraken, Storm worm and Waledac malware, their research found specific vulnerabilities that can be used to take control of malware. Dereszowski discovered a buffer overflow vulnerability and developed an exploit against the C&C server of Poison Ivy RAT that provides remote code execution [14]. Researchers at malware.lu developed a scanner tool for identifying new C&C of Poison Ivy and Terminator RAT that were used to compromise their C2 servers [15]. Shawn Denbow and Jesse Hertz found an SQL injection vulnerability and an arbitrary file download vulnerability on Dark Comet RAT. Exploiting these vulnerabilities enable defenders to counter-attack the RAT C&C server [16].

The following two works present a similar approach. Eisenbarth et al. [17] in 2013 comes up with a technique to detect C&C servers of botnets capable of distributed denial-of-service (DDoS) through the arbitrarily sending out of valid bot requests and verifying the proper response signature. Gundert et al. [18] developed a new approach to detect RAT controllers. This method is primarily based on the fact that RATs return a unique response when a proper request is presented on its controller's listener port. This unique response is a signature indicating that a RAT control panel is running on the responding endpoint. This technique relies on large-scale Internet scanning combined with sending predefined requests to detect live instances of RAT control panels. However, in order to detect DDOS/RAT controllers with the above methods, samples and/or full packet captures (PCAP) must be obtained. In other words, it can only detect DDOS/RAT controllers for known and previously analyzed malware programs. Our detection technique, on the other hand, does not demand Internet scanning or traffic capture of the malicious file, but executes each file on the host and pops up an alert on a malicious remote controller without prior knowledge of the executed file.

The first to be published work on exploiting vulnerabilities of remote administration panels was conducted by a security guru named Eriksson [19]. The innovative idea was malware authors are not too careful when it comes to security, rendering malware programs and their remote panels are exposed to security holes. Eriksson discovered how to crash the graphical user interface (GUI) software by sending it random commands, in addition he found a heap overflow vulnerability that allowed him to take control on the attacker's machine. Eriksson later utilized the above techniques on PCShare RAT. However, Eriksson didn't use the vulnerabilities he found to detect these malware programs, but rather to crash or control them. Moreover, the exploits in his work are specific only for the above malware programs.

Watkins et al. [9, 20] proposed fighting back against known malware programs using fuzzing techniques to find new vulnerabilities in their command and control servers. The idea is to analyze a malware program's behavior and communication to fuzz the malware program's traffic values sent to a remote administration panel and examine the impact on the C&C server. These findings are partially related to our work, one of the main differences in our innovative technique is that conducting malware analysis is not an imperative part in our work. Moreover, we utilized vulnerabilities to detect infections and not to retaliate or take control of the malware.

3   Methodology

This section defines the novel approach of our study and presents the offensive detection technique in detail.

3.1 Main Concept

The Cyber Kill Chain, a computer network intrusion modeling method developed by Lockheed Martin (2011), consists of seven steps: Reconnaissance, weaponization, delivery, exploitation, installation, command and control, and actions on objectives [3]. While classic detection techniques focus on detecting the intrusion in the delivery, exploitation and installation steps, our concept is to detect the intrusion in the command and control step. It is during the installation step that the malware is installed on the endpoint, providing it with constant access. Next, the malware communicates with the C&C server, sending it information through the infected host (such as the computer name, processor name and memory size). Table 1 presents a complete list of information that is collected by malware programs. This information is saved in the database (DB) of the C&C server and is displayed in the panel administration tool of the malware (web or application).

Table 1. Information collected by malware.

Information type	Details
System info	OS version, install date, keyboard language, time-zone, hostname, ...
Hardware info	Connected storage devices, Bios/CPU/RAM/Motherboard/ Network adapter info, installed video/audio devices, connected printers/ monitors
Local users info	Description, username, password expiry date
Running processes info	Creation-date, (parent) process ID, path
Installed services info	Description, path, status
Environment variables	
File associations	File extension, associated program name, associated command
Network info	Internal ip, configured DNS servers
Directory listing	Root C drive, Desktop, Documents, Temp folder, ...
Print screen	

In our work, communications between the infected endpoint and the malicious C&C panel is utilized to cause the latter to perform an involuntary action that will send an endpoint infection report. To do so, an XSS attack technique is used, injecting a XSS string into the queried information by the malware itself. A focus was placed on malware programs with a web-panel administration tool as web pages are a fundamental factor for exploiting XSS vulnerability. Editing was also only performed on information whose collection from endpoints (without the user's knowledge) and presentation on a web-remote panel is considered malicious.

In general, two entities are involved in a malware attacked: the infected endpoint host and the attacker's C&C server. The malware carries out its malicious activities on the endpoint by communicating with the C&C panel on a certain frequency. The panel is able to receive information and send commands to the infected endpoint. To construct our detection methodology, we added a new entity to the malware attack process – a monitoring server that we called Server C, in order to force the attacker's C&C panel to execute a predefined XSS script. Server C cunique role as it is not supposed to communicate with endpoints. Therefore, it must include a web remote administration panel in order to do so. This concept is implemented in the proof-of-concept (POC) tool that we developed, i.e., PhoeniXSS on endpoints, as described in detail in the following section.

3.2   Detection Methodology

In order to develop an effective tool for detecting malware, and without harming endpoints' applications in the organization on the infected endpoint, we set up a new environment that models both the organization gateway and the attackers' malicious C&C panel. As depicted in Fig. 1, this environment included the endpoint host with the PhoeniXSS tool, the C&C Server with a leaked version of a malware program's remote administration tool, and a monitoring server.

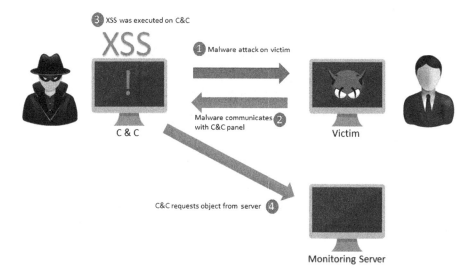

Fig. 1. Detection methodology.

The endpoint host was used to execute incoming binary files. Each file had to be executed by a PhoeniXSS tool for a 1-min period (maximum) before entering the organization. During file execution, PhoeniXSS hooks all predefined API functions and adds a predefined XSS string to their returned values. In this work, malware programs were tested with a specific XSS string using the PhoeniXSS tool. During infection, the above values are sent to the malware program's remote administration panel. Many malware web panels exhibit common values on the infected machines (e.g., computer name, user name and processor name). However, once one of these values is not validated and/or is treated incorrectly, the attacker's web panel becomes vulnerable to an XSS attack. We used this vulnerability to create an alert on our monitoring server. Finally, the monitoring server is programmed to wait for the http/s request. Although any internet device can access the server, no one should send a request to it. Therefore, a request is equal to an intrusion into the organization.

Every monitoring server can publish several image objects that are compatible with the injected XSS string to increase the intrusion detection rate. The PhoeniXSS environment is not limited in the number of hosts that run the PhoeniXSS tool and should run in parallel on a number of hosts. Each instance of the PhoeniXSS tool has to include a unique identifier to identify a compromised endpoint. Using the PhoeniXSS with an XSS cheat sheet on a large number of hosts improves the efficiency of the counter-attack.

4   Implementation

In this section, implementation details for the PhoeniXSS tool are provided.

4.1   PhoeniXSS

We implemented our approach in a POC tool called PhoeniXSS that is composed of two different projects: A. "The Application Verifier (AppVerif.exe) is a dynamic verification tool for user-mode applications. This tool monitors application actions while the application runs, subjects the application to a variety of stresses and tests, and generates a report about potential errors in application execution or design" [21]; B. MinHook – "API hooking library that provides basic hooking functionality for x64 and x86 environments" [22]. It replaces the prologue of the target API function with jump instructions to our hook function (i.e., our code). When the hooked function jumps to our code, we can call the API code which follows our jump instruction. The PhoeniXSS consists of a single DLL file (main DLL) that combines the above projects and is injected into the executed binary file at the beginning of the execution, using the Microsoft Application Verifier tool.

The PhoeniXSS tool executes binaries files with administrator credentials and hook (i.e., inline hooking) predefined API functions. As many malware programs use COM objects to query end-point hosts, our tool can also hook predefined COM queries. The DLL file of our tool is loaded into the memory before loading the imported DLLs of the execution file. However, hooking predefined API functions is executed after the imported DLLs are loaded into the memory. (i.e., our DLL is executed in the memory and hooks the API function "GetWindowTextA," which belongs to the User32.dll, only after the last one is loaded into the memory). After our main DLL is loaded into the memory, it hooks the original API function, overwrites the prologue of the function, and jumps to our new code segment. This segment calls the original API function, gets its returned values, and modifies them. This modification adds a specific predefined XSS string to the returned values by the original function.

The predefined XSS string consists of an image tag that loads the source of the image from server C. This server hosts a web application that publishes a number of GIF images. Each image is contained in a different XSS string. The length of the XSS string should be minimal to avoid length limitations in the C&C panel DB. In addition, the

XSS string should consist of a different characters to evade character limitations in the C2 panel code. The template of our XSS string is as follows:

<img/src=//j.en/X.gif>

when: - 'X' is a number and the name of the GIF image. The GIF image is a transparent image so it can remain stealthy after XSS execution in the browser. Furthermore, the image size is 1 × 1 pixel, in order to avoid significant overhead requests in the C&C panel and remain stealthy. After the modifications of the PhoeniXSS, the returned values of the API functions appear in the following format: Original Value + Image Tag. For instance, if the original value of the API function "GetWindowTextA" was "explorer.exe", then after PhoeniXSS modifications, the new value will be "explorer.exe<img/src=//j.en/1.gif>".

"GetWindowTextA" function is an example of one function from the list of predefined functions in our tool. The predefined hooked API functions in the PhoeniXSS tool are "GetComputerNameW"; "GetComputerNameA"; "GetWindowTextW"; and "GetWindowTextA". The GetComputerName function retrieves the NetBIOS name of the local computer, while the GetWindowText function retrieves the text of the specified window's title bar.

As mentioned earlier, the PhoeniXSS tool also intercepts calls to COM interfaces. It implements "vtable patching" hooking method for the function "ExecQuery" of "IWbemServices" interface. After our main DLL is loaded into the memory, it waits for the "wbemsvc.dll" file to be loaded, to create an object (i.e., an instance) of the above interface and hook the above function. The PhoeniXSS modifies the object's virtual methods table that contains pointers to all public methods of a COM object, so they can be replaced with the pointers to hook functions. This hook function calls the original function, receives the returned values from it, and adds the above XSS string to them.

5   Experimental Evaluation

We assembled an isolated environment to host the C&C server, the victim's machine, and our monitoring server in order to evaluate our detection technique on malware programs that exist in the wild. We obtained the complete toolkits for a number of malware programs that were leaked. The toolkits consisted of bot builder software and the C&C web application. The toolkits were for the following malware programs: MegalodonHTTP version 1, Dexter POS version 1, and DiamondFox version 4.2.650. We assumed that the leaked toolkits are essentially identical to the sources that were found in the wild. We did not perform a test on "live" malwares because counter-attacks (i.e., hacking back) is illegal in our country. We investigated the functionality of the above-mentioned malware programs and their back-end C&C code to determine if our detection technique can be applicable against them. To test if the PhoeniXSS is capable of detecting the above malware programs, we set up malwares program C&C web panels on our servers and configured our PhoeniXSS tool.

Our malware experiments were conducted as follows: The first tested malware communicated with its remote administration panel in clear text and without input validation on the victim machine. The second tested malware communicated with its remote administration panel using encryption and without input validation. The last tested malware communicated with its remote administration panel using an encryption and with input validation, yet was still discovered as being vulnerable to our XSS detection technique. The different phases of our experiment are described in the following section.

5.1 Experimental Setup

Our test environment included three virtual machines (VM) that ran on Intel Core i7-4720HQ CPU with 16 GB of RAM and 230 GB of disk space. The first machine is the victim machine, running on a Windows 7 VM, 32-bit, Service Pack 1. This machine was responsible for executing the payload file using the PhoeniXSS tool. We set up another machine to host malware programs' C&C panels. This host machine ran on a Windows 7 VM, 32-bit, Service Pack 1. Three different browsers were installed on the C&C server with default configuration. Using each of the following browsers – Firefox browser version 53.0.3, Chrome 51.0.2704.103 and Microsoft Edge 38.14393.1066.0 – our detection technique was tested on the collected malware program web panels. The C&C panels of the collected malware programs were hosted on a Windows machine with an Apache web server and MySQL DB. The last machine, our monitoring server, ran on a Windows Server 2008 VM, 32bit, Service Pack 1.

5.2 Experimental Procedure

After assembling our test environment, we used our tool to execute the payload of the above collected malware programs. The testing of each malware was performed in a clean and separate image. Moreover, each panel was installed separately from other panels.

Our first test was performed on a MegalodonHTTP botnet. MegalodonHTTP, is a piece of malware designed to power distributed denial-of-service (DDoS) botnets. Such botnets can launch seven types of DDoS attacks, open a remote shell on the infected system, mine crypto currency, and kill antiviruses. In order to run, it requires a .NET framework to be installed on the victim's machine [23]. The MegalodonHTTP communicates with its C&C panel over HTTP in clear text.

After building its DB and setting up the web panel of the botnet, we investigated the server-side code of the malware to evaluate XSS vulnerabilities. In addition, we inspected the bots' information that MegalodonHTTP exposes on its panel, e.g., country name, hardware ID, IP, computer name, operating system, CPU, GPU, RAM, AV type and more. This information is sent to the C&C panel and saved in the DB, yet without validating the input. Therefore, the MegalodonHTTP panel is probably vulnerable to a stored XSS attack. We provided a visual of the MegalodonHTTP panel in Fig. 2. Furthermore, another check is required to enable the XSS attack. The stolen information from the bot is sent to the server and is saved in its DB tables. Each table has different fields. These fields have a different length. If the length of these fields is too short for our XSS string, the XSS attack isn't feasible. There were several fields where their length wasn't restricted enough. For example, the maximum length of a Windows computer machine is 15 characters and the length of this field in MegalodonHTTP DB is 255 characters. We chose to modify the CPU field value of the infected machine to exploit XSS vulnerability. MegalodonHTTP receives the above value using the following WQL query:

SELECT Name FROM Win32_Processor

The Win32_Processor WMI class retrieves the processor status and information for both single and multiprocessor machines. The above query returns the processor name value of the infected endpoint. The PhoeniXSS tool is programmed to modify the CPU field value using its ability to hook COM interfaces.

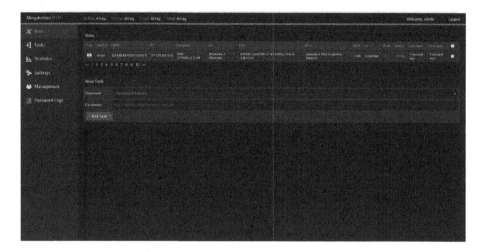

Fig. 2. MegalodonHTTP web administration panel.

Using the MegalodonHTTP bot builder, we built the payload file with a default configuration and executed it using the PhoeniXSS tool on our victim's machine. Following execution, information from the victim's machine was presented on the botnet panel. The injected XSS string of our tool was not displayed on the panel because the browser treated the injected string as the html source of the page. Therefore, from the attacker's point-of-view, it seemed like the original information was being presented. At the same time, however, the defender (i.e., the victim) detected that its machine had been infected and that its monitoring server was providing information to the attacker's machine. A visual of MegalodonHTTP panel after our XSS attack is provided in Fig. 3.

Another test was conducted on Dexter malware. "Dexter is a computer virus or point of sale (POS) malware which infects computers running Microsoft Windows. It infects POS systems worldwide and steals sensitive information such as credit card and debit card information and sends it to a C&C server" [24]. The harmful executable installed on the POS endpoint is known as POSGrabber. It is programmed so that it can run as an executable or be injected into another process. The RAM examining process is very straightforward. It initially identifies every single running process, then opens a handle

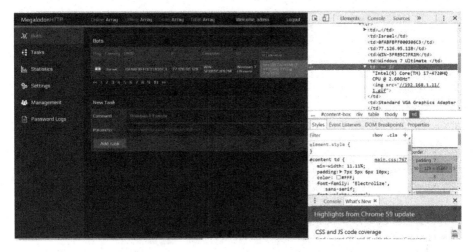

Fig. 3. MegalodonHTTP web administration panel after PhoeniXSS attack.

to selected processes and reads chunks of RAM to search for strings of credit card track information. The POSGrabber communicates with the C&C panel is over HTTP. This communication is encoded, and much of it is "encrypted" [8].

One of the main differences between MegalodonHTTP and Dexter malware is the traffic that is sent to their C&C panels. The former's traffic is sent in clear text, while the latter's traffic is sent encrypted using Base64 encoding and XOR cipher. Using the PhoeniXSS tool on Dexter malware, we substantiated that our detection technique works on malware programs that use encryption to communicate with their C&C.

We followed the same installation steps as with the MegalodonHTTP setup. The panel has three main pages: Dump Viewer, Bot Control, and File Uploader. The first presents the credit card tracks, the second enables the sending of commands to the bots, and shows stolen information from the infected endpoints, while the third page allows the loading of additional files which can then be used by the POSGrabbers.

We inspected the server-side code and observed the bot information that was displayed on the Dexter panel and found the following stolen information on bots: UID, Version, IP, user name, computer name, user-agent, operating system (OS), OS architecture, and more. Each field of stolen information in the DB and the methods that were used to retrieve it from the infected machine were inspected, revealing that there was no input sanitization on the sent information to the Dexter C&C panel. We chose to modify the value of the Computer Name field that is sent to the Dexter panel. The payload file with default configuration was built using the source code of the malware, and then executed in our isolated environment.

After execution, the stolen information was presented on the Bot Control page on the panel. The server-side code decrypted the information that was sent to the remote administration panel of the malware, including the XSS string that was injected by the PhoeniXSS tool. The attacker's browser treated the injected string as the HTML source of the page, thus an infection alert was raised on the defender's monitoring server. A visual of the Dexter panel after infection is provided in Fig. 4.

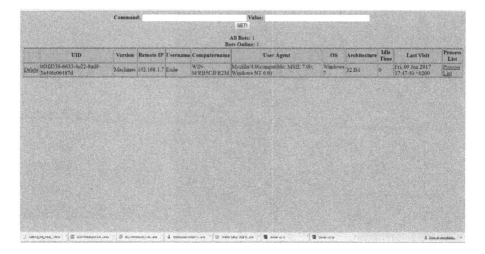

Fig. 4. Dexter web administration panel after PhoeniXSS attack.

The final test was conducted on DiamondFox malware, which is multiple purpose, including theft of credit card information as well as credentials at POS systems. It is highly accessible to even the most limited hackers, as it is distributed in many hacker forums. DiamondFox communicates over encrypted HTTP with a key statically built into the C&C and the bots [25].

DiamondFox's web installation page was used to set up the remote administration panel automatically and configure its bot builder to export a payload file with its key logger module. To substantiate that the DiamondFox panel is vulnerable, we investigated the server-side code of the panel and the received information by the botnet from the victim's machine.

The following information was displayed on the "client's" page on the DiamondFox panel: ID (based on HID), country, IP, computer name, operating system, and additional information about the victim's machine (HDD size, RAM size and more). The DiamondFox panel uses several functions to handle data input from its bots before inserting it into its DB.

The two most important functions are: A. "mysql_real_escape_string" – "escapes special characters in a string for use in an SQL statement" [26], B. "htmlentities" – "convert all applicable characters to HTML entities" [27]. This input escaping check prevented our tool from exploiting this part of the botnet panel. However, DiamondFox is a multipurpose botnet with a variety of modules, with one of them being its key logging capability. This module also includes an input escaping check on data input, however not on the entire input. A visual of DiamondFox's key logging module is provided in Fig. 5.

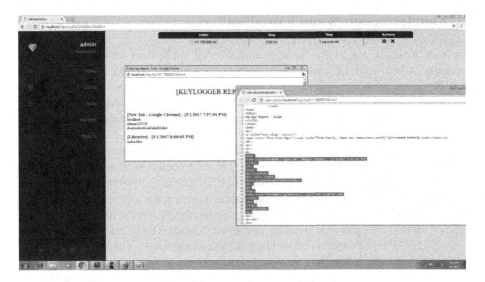

Fig. 5. DiamondFox's key logging module.

The payload used the API function, GetWindowTextA, to track user keyboard typing between different windows. However, the botnet panel failed to validate the data input in function GetWindowTextA, thereby rendering the botnet panel vulnerable to an XSS attack and being detected by our technique. A visual of DiamondFox's key logging module after using the PhoeniXSS tool is provided in Fig. 6.

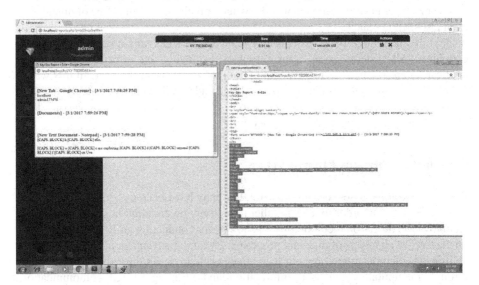

Fig. 6. DiamondFox's key logging module after using PhoeniXSS tool.

6   Countermeasures and False-Positive

An XSS attack is one of the most prominent web application security flaw and is currently ranked on the OWASP (Open Web Application Security Project) top 10 chart [28]. However, there are some methods that can prevent such attacks, such as validating and escaping user input. The most efficient method to prevent XSS attack is by escaping user input. Escaping data means taking the data an application has received and rendering it as text by the browser, ensuring it is secure before presenting it to the end user. By escaping user input, key characters in the data received by a web page will be prevented from being interpreted in any malicious way. Key characters such as '<' and '>' are disallowed from being rendered. Validating user input means ensuring an application is rendering the correct data and preventing malicious characters from doing harm to the site and user. However, input validation is not a primary prevention method for vulnerabilities such as XSS and should not be used alone to battle XSS attacks. Although these methods could be efficient in preventing XSS attacks, web applications today still struggle to cover all XSS attack vectors. Furthermore, our use of the XSS attack technique for detecting malware, as described in this paper, is just the first stage of possible future developments for detecting malware.

Wide usage of our PhoeniXSS tool, on the other hand, could alert C&C administrators of suspicious scripts in their panels. This result is divided into two: displayed scripts with detection in the C&C panel and displayed scripts without detection. In the first case, we succeed in detecting the attack, with the probable result of the attackers quickly ending their attack campaign, with a probability for mistakes. In the second case, displayed scripts on C&C panels could affect the attackers, taking them out of their comfort zone and leaving them to wonder whether their campaign is free of mistakes or even has been revealed.

Finally, many legitimate systems and applications access and query information of endpoints. All systems using, transmitting or displaying this information on a web application are required to receive explicit authorization from the user. In this way, a list of all the legitimate systems can be collected and identified in a defined "white-list" in advance to avoid false-positives indications. Any other indication will be considered malicious and should be treated with caution.

7   Conclusion and Future Work

Our work explores a new area of research that may have been overlooked by security research experts. In order to detect malicious activity on endpoint hosts, we developed an innovative detection methodology and a POC for conducting an XSS counter-attack on the C&C web panels of malware programs. By taking advantage of malware vulnerabilities, we demonstrated how our technique is able to detect three different malwares through manipulating endpoint information and utilizing trusted communications between the malware and its C&C panel.

This research promotes an innovative solution that focuses on the offense, that will enable malware victims to hack back (i.e., conduct a counter-attack) in a legitimate manner. Furthermore, our methodology can be expanded in various directions in the future, such as exploiting SQLI on application panels or using crafted documents to trigger an alert when opened, and more. In future work, we hope to be able to detect new families and variants of malware programs by using all of the above-mentioned techniques in the wild. Such a notion has the potential to change cyber-war rules and create a balanced power relationship between defense and attack forces within the cyber dimension.

Adversaries will continue to develop new technologies in order to avoid detection and achieve their purposes. We must not lag behind but must turn the tables on them, using offensive techniques that incorporate new, high-fidelity detection methods.

References

1. Saeed, I., Selamat, A., Abuagoub, A., Abdulaziz, S.: A survey on malware and malware detection systems. Int. J. Comput. Appl. **67**, 25–32 (2013). https://doi.org/10.5120/11480-7108
2. Feily, M., Shahrestani, A., Ramadass, S.: A survey of botnet and botnet detection. In: Proceedings of 2009 3rd International Conference on Emerging Security Information, Systems and Technologies, SECURWARE 2009, pp. 268–273 (2009). https://doi.org/10.1109/securware.2009.48
3. Cyber Kill Chain®. https://www.lockheedmartin.com/en-us/capabilities/cyber/cyber-kill-chain.html
4. Cross-site Scripting (XSS) – OWASP. https://www.owasp.org/index.php/Cross-site_Scripting_(XSS)
5. Sood, A.K.: Exploiting fundamental weaknesses in botnet Command and Control (C & C) panels. Presented at the 2014 (2014)
6. Sood, A.K.: Malware at Stake: For Fun - XSS in ICE IX C&C Panel. https://secniche.blogspot.co.il/2012/06/for-fun-xss-in-ice-ix-bot-admin-panel.html
7. Phase Bot – Exploiting C&C Panel | MalwareTech. https://www.malwaretech.com/2014/12/phase-bot-exploiting-c-pane.html
8. Wallace, B.: A Study in Bots: Dexter. https://blog.cylance.com/a-study-in-bots-dexter-pos-botnet-malware
9. Watkins, L., Silberberg, K., Morales, J.A., Robinson, W.H.: Using inherent command and control vulnerabilities to halt DDoS attacks. In: 2015 10th International Conference on Malicious Unwanted Software, MALWARE 2015, pp. 3–10 (2016). https://doi.org/10.1109/malware.2015.7413679
10. Goodin, D.: White hats publish DDoS hijacking manual, turn tables on attackers | Ars Technica. https://arstechnica.com/information-technology/2012/08/ddos-take-down-manual/
11. Goodin, D.: Zeus botnets' Achilles' Heel makes infiltration easy • The Register. http://www.theregister.co.uk/2010/09/27/zeus_botnet_hijacking
12. Grange, W.: Digital Vengeance: Exploiting the Most Notorious C & C Toolkits Ethics of Hacking back (2017)
13. Geers, K., Czosseck, C.: The Virtual Battlefield: Perspectives on Cyber Warfare. Network Security. IOS Press, Amsterdam (2009). 305 pages
14. Dereszowski, A.: Targeted attacks: from being a victim to counter attacking, pp. 1–28 (2010)
15. Rascagnères, P.: Public document APT1: technical backstage malware analysis. General Information History, pp. 1–48 (2013)
16. Denbow, S., Hertz, J.: Pest control: taming the rats (2012)
17. Eisenbarth, M., Jones, J.: BladeRunner: adventures in tracking botnets. In: Botconf (2013)

18. Gundert, L.: Proactive Threat Identification Neutralizes Remote Access Trojan Efficacy (2015)

19. Singel, R.: Security Guru Gives Hackers a Taste of Their Own Medicine | WIRED. https://www.wired.com/2008/04/researcher-demo/

20. Watkins, L., Kawka, C., Corbett, C., Robinson, W.H.: Fighting banking botnets by exploiting inherent command and control vulnerabilities. In: Proceedings of the 9th IEEE International Conference on Malicious Unwanted Software, MALCON 2014, pp. 93–100 (2014). https://doi.org/10.1109/malware.2014.6999411

21. Application Verifier | Microsoft Docs. https://docs.microsoft.com/en-us/windows-hardware/drivers/debugger/application-verifier

22. Kageyu, T.: MinHook - The Minimalistic x86/x64 API Hooking Library. https://www.codeproject.com/Articles/44326/MinHook-The-Minimalistic-x-x-API-Hooking-Libra

23. Kovacs, E.: Alleged Author of MegalodonHTTP Malware Arrested | SecurityWeek.Com. https://www.securityweek.com/alleged-author-megalodonhttp-malware-arrested

24. Dexter (malware). https://en.wikipedia.org/wiki/Dexter_(malware)

25. Wallace, B.: A Study in Bots: DiamondFox. https://www.cylance.com/a-study-in-bots-diamondfox

26. PHP: mysql_real_escape_string – Manual. http://php.net/manual/en/function.mysql-real-escape-string.php

27. PHP: htmlentities – Manual. http://php.net/manual/en/function.htmlentities.php

28. Top 10-2017 Top 10 – OWASP. https://www.owasp.org/index.php/Top_10-2017_Top_10

29. Agmon, O., Posener, B.E., Schuster, A., Mu, A.: Ginseng: Market-Driven Memory Allocation

30. Sharfman, I., Schuster, A., Keren, D.: Shape sensitive geometric monitoring categories and subject descriptors. In: PODS (2008). https://doi.org/10.1145/1376916.1376958

31. Friedman, A., Keren, D.: Privacy-preserving distributed stream monitoring. In: NDSS, pp. 23–26 (2014)

32. Ben-Yehuda, O.A., Ben-Yehuda, M., Schuster, A., Tsafrir, D.: The Resource-as-a-Service (RaaS) cloud. Commun. ACM **57**, 76–84. https://doi.org/10.1145/2627422

33. Gilburd, B., Schuster, A., Wolff, R.: k-TTP: a new privacy model for large-scale distributed environments. In: Proceedings of the Tenth ACM SIGKDD International Conference on Knowledge Discovery and Data Mining, pp. 563–568 (2004). https://doi.org/10.1145/1014052.1014120

34. Schuster, A., Wolff, R., Gilburd, B.: Privacy-preserving association rule mining in large-scale distributed systems. In: Proceedings of Cluster Computing and Grid, pp. 1–8 (2004)

35. Verner, U., Schuster, A., Silberstein, M., Mendelson, A.: Scheduling processing of real-time data streams on heterogeneous multi-GPU systems. In: Proceedings of the 5th Annual International Systems and Storage Conference - SYSTOR 2012, pp. 1–12 (2012). https://doi.org/10.1145/2367589.2367596

A Planning Approach to Monitoring Computer Programs' Behavior

Alexandre Cukier[1(✉)], Ronen I. Brafman[1], Yotam Perkal[2], and David Tolpin[2]

[1] Computer Science Department, Ben Gurion University, Beersheba, Israel
alexandre.cukier@gmail.com, brafman@cs.bgu.ac.il
[2] PayPal, Tel Aviv, Israel
yperkal@paypal.co, dvd@offtopia.net

Abstract. We describe a novel approach to monitoring high level behaviors using concepts from AI planning. Our goal is to understand what a program is doing based on its system call trace. This ability is particularly important for detecting malware. We approach this problem by building an abstract model of the operating system using the STRIPS planning language, casting system calls as planning operators. Given a system call trace, we simulate the corresponding operators on our model and by observing the properties of the state reached, we learn about the nature of the original program and its behavior. Thus, unlike most statistical detection methods that focus on syntactic features, our approach is semantic in nature. Therefore, it is more robust against obfuscation techniques used by malware that change the outward appearance of the trace but not its effect. We demonstrate the efficacy of our approach by evaluating it on actual system call traces.

1 Introduction

Malware is a serious threat for computer and Internet security for both individuals and entities. 430 million new unique pieces of malware were detected by Symantec in 2015, and 94.1 millions of malware variants during only the month of February 2017. Not surprisingly, to counter this threat, many techniques for malware detection have been proposed.

In this paper we are interested in the more general problem of understanding the behaviors taking place in the system. Given this information, one can determine whether they are malicious or not, and if malicious, provide an informed response.

The standard approach to this problem is to use pattern-recognition methods, which are syntactic in nature. Roughly speaking, they view the input, whether code or events, as a long string of symbols, and seek properties of these strings that help classify them. To fool these methods, malware attempts to obfuscate its behavior by changing the sequence's properties [27]. Semantic methods, instead, try to model the underlying system, seeking to understand the input's meaning,

© Springer International Publishing AG, part of Springer Nature 2018
I. Dinur et al. (Eds.): CSCML 2018, LNCS 10879, pp. 243–254, 2018.
https://doi.org/10.1007/978-3-319-94147-9_19

where in this paper, the input used is the system-call trace.[1] Therefore, they have the potential to be more robust to obfuscation attempts.

The most extreme and most accurate semantic approach is a faithful simulation of every trace followed by careful analysis of the resulting system state.

This is impractical: the analysis of the state of a computer following each trace is a non-trivial time consuming task that requires deducing high-level insights from the low level state and can only be conducted by experts.

Instead, we propose a methodology that uses an abstract system model based on AI-planning languages and models. It requires a one-time, off-line effort by an expert, and can be used automatically to analyze each trace: An expert that understands the semantics of system calls generates a planning operator for every system call. Each operator describes how the state of the system changes in response to the application of some system call. Each operator is an abstraction that attempts to capture the system call's relevant effects. The abstraction process also involves the generation of a set of propositions describing the system state. Now, given a system call trace, instead of simulating it on the real system, we simulate the corresponding planning operators on the abstract state. The propositions true in the resulting state give us the needed information about what behaviors were carried out by this code. This approach is fast and difficult to fool: obfuscation techniques that do not impact the actual behavior will not impact relevant aspects of the state.

In what follows we describe this methodology using examples, and demonstrate its advantages by comparing it to statistical methods on actual system calls related to a mail application.

2   Related Work

[9] is considered the seminal work which pushed forward research on methods and representations of operating system process monitoring based on system calls.

[24] provides an early comparison of machine learning methods for modeling process behavior. [10] introduces the model of execution graph, and behavior similarity-measure based on the execution graph. [17] combines multiple models into an ensemble to improve anomaly detection. [26] applies continuous time Bayesian network (CTBN) to system call processes to account for time-dependent features and address high variability of system call streams over time. [13] applies a deep LSTM-based architecture to sequences of individual system calls, treating system calls as a language model.

Initially, only system call indices were used as features [9,24]. [14] compares three different representations of system calls: n-grams of system call names, histograms of system call names, and individual system calls with associated parameters. [18] proposes the use of system call sequences of varying length

---

[1] A system call is a mechanism used by a program to request from the operating system services it cannot perform directly, such access to hardware, files, network or memory.

as features. [14, 22] investigate extracting features for machine learning from arguments of system calls. [25] studies novel techniques of anomaly detection and classification using n-grams of system calls. [4] conducts a case study of n-gram based feature selection for system-call based monitoring, and analyses the influence of the size of the n-gram set and the maximum n-gram length on detection accuracy.

Other work attempted to detect behaviors in a semantic way, using abstract representations of behaviors based on low level events and various techniques for detection. They all carry the notion of state, keeping track of effects of previous events. [5] is the first to introduce semantics to characterize malicious behaviors. It builds behavior templates from binaries using formal semantics, which is used through a semantics-aware algorithm for detection. [16] builds multi-layered behavior graphs from low level events used through a behavior matcher. [12] uses attribute grammars for abstraction and specification, using parallel automata for parsing and detection. [23] specifies behaviors through UML activity diagrams from which one generates colored Petri Nets for detection. [2] uses first-order linear temporal logic to specify behaviors and model checking techniques for detection. [7] offers an advanced state-full approach where behaviors are specified as finite state machines. Our approach is more fine-grained and general. We model the actual operators, not the target behaviors, although the model is informed by the behaviors. We illustrate this using the *reverse shell* example in the next section.

Behavior recognition is closely related to plan and goal recognition [21]. Given a sequence of observed actions, the goal is to try to infer the actor's intentions. Typically, the output is a ranked list of hypothesized goals. Most work assumes a library of possible behavior instances, i.e., plans, an approach limited in its ability to go beyond known instances. Probabilistic techniques, such as [1] use Bayesian methods to assess the probability of various goals based on the actions involved. An influential recent approach is plan-recognition as planning [19], where the authors do away with the assumption of an explicit plan library. The plan library is replaced by a model of the domain (which implicitly defines the set of possible plans), and the goal is to compute a good plan that is closest to the observed behavior. This line of work is appropriate when the observations are a subset of the actual actions taken, or when we attempt to recognize the goal before plan completion. We attempt to recognize malicious behavior off-line given a complete trace, although extensions for the online setting are natural.

3 Background

3.1 AI Planning

AI Planning is a decision making technique used to find sequences of actions that can transform a system from some initial state into a goal state. Formally, a classical planning problem is a tuple: $\pi = \langle P, A, I, G \rangle$, where: P is a set of primitive propositions describing properties of interest of the system; A is the action set. Each action transforms the state of the system in some way; I is the start state; and G is the goal condition — usually a conjunction of primitive propositions. A state of the world, s, assigns truth values to all $p \in P$. Recall that a literal is simply a primitive proposition or its negation.

An action $a \in A$ is a pair, $\{pre(a), effects(a)\}$, where $pre(a)$ is a conjunction of literals, and $effects(a)$ is a set of pairs (c, e) denoting conditional effects. We use $a(s)$ to denote the state that is obtained when a is executed in state s. If s does not satisfy all literals in $pre(a)$, then $a(s)$ is undefined. Otherwise, $a(s)$ assigns to each proposition p the same value as s, unless there exists a pair $(c, e) \in effects(a)$ such that $s \models c$ and e assigns p a different value than s. We assume that a is well defined, that is, if $(c, e) \in effects(a)$ then $c \wedge pre(a)$ is consistent, and that if both $(c, e), (c', e') \in effects(a)$ and $s \models c \wedge c'$ for some state s, then $e \wedge e'$ is consistent.

The classical planning problem is defined as follows: given a planning problem π, find a sequence of actions $\{a_1, \ldots, a_k\}$ (a.k.a. a plan) such that $a_k(\cdots (a_1(I)) \cdots) \models G$.

To illustrate this model, consider a simplified domain with three action types: *socket*, *listen*, and *accept*. These actions model the effect of system calls that create a socket, listen for an incoming connection, and accept a connection. For the sake of this example, we ignore various parameters of these system calls, and assume that system calls do not fail.

The set P contains: { *(opened socket-descriptor)*, *(listening socket-descriptor)*, *(connected socket-descriptor)*}, where *socket-descriptor* is a parameter that we abbreviate as *sd*. The set of actions is:

- *socket(returned-sd)* with precondition: ¬*(opened returned-sd)*, and the effect: *(opened returned-sd)*[2]
- *listen(sd)* has no precondition and the conditional effect: *(listening sd)* when *(opened sd)* ∧¬*(listening sd)*
- *accept(sd, returned-sd)* has the preconditions: *(listening sd)* and the effects: *(opened returned-sd)* and *(connected returned-sd)*

The plan *socket(sd1)*, *listen(sd1)*, *accept(sd1, sd2)* is a legal plan. Initially, all propositions are false. Because *(opened sd1)* is false, we can apply *socket(sd1)*. Once applied, *(opened sd1)*, the precondition of *listen*, becomes true. This results in *(listening sd1)* becoming true. Finally, *accept* needs a socket descriptor in the state *listening (sd1)* and another having *opened* false *(sd2)*.

It now sets *sd2* to *opened* and *connected*. Given the resulting state, we recognize that a host connected itself to our local server.

On the other hand, the plan: *socket(sd1)*, *accept(sd1, sd2)* is invalid because the preconditions of accept are not all satisfied: *(listening sd2)* is not set to true.

Typically, planning models are used for generating plans. Thus, in the above example, a planning algorithm could find the (abstracted) sequence of system calls required to achieve various goals. Our focus in this paper is on the planning

---

[2] A more faithful model will use conditional effects instead, and will also consider their return value.

model itself — the propositions and the operators, as an abstraction of the operating system. The acting agent is a running process, the OS is the environment in which it is acting, and its system-call trace defines the plan, via our mapping. To determine what the process is doing, we simply observe the abstract state of the OS. For the purpose of this paper, we consider that the OS abstraction has a unique running and single thread process.

4   Our Approach

We propose to build an abstract system model and simulate an abstraction of the system call trace on it.

The manual part of our approach is the construction of the abstraction. We associate an action with each system call, with preconditions (typically empty) and effects (typically conditional). The set of propositions that we use to describe the system is informed by the type of behaviors we want to capture. For example, whether channels were opened, files accessed, information transmitted over a channel, etc. An action describes what new facts will become true following the execution of the system call it models, possibly conditional on other facts being true prior to its execution.

We illustrate this using the example of a remote shell: a command line interface controlled by a remote host often used by attackers to execute system commands. We focus on the *reverse shell*, where a host connects itself to a remote server. Starting a reverse shell requires a few steps: (1) Create a socket. (2) Independently connect the socket to an endpoint and duplicate the socket descriptor to the standard input and output (so that the input and output streams go through the socket). (3) Execute a shell.

We use system calls *socket, connect, dup, fcntl, close* and *execve* that, respectively, create a socket, connect a socket to a remote host, duplicate a socket, set properties to a socket, close a socket, and execute a program.

Propositions. The propositions are: *(opened fd), (is-socket fd), (equal-fds fd1 fd2), (close-on-exec fd), (connected sd), (is-shell path), (remote-shell-started)*

Initial State. The initial state initiates the resources used by a process when it starts, and taints the ones that have targeted properties:

- Propositions *(opened fd0), (opened fd1), (opened fd2)* are set to true, as fd0/1/2 denote standard input/output/error, respectively, and these files are open.
- Shell executable paths are tainted. We assume that we know all of those presents on the operating system. We handle two of them in this example: */bin/sh* and */bin/bash* that we name respectively *sh* and *bash*. Thus, *(is-shell sh)* and *(is-shell bash)* are set to true.

Actions. Planning operators are a simplified abstraction of the system calls. Since system calls called with wrong arguments do not make programs crash, and have no effect, the corresponding actions use conditional effects only – i.e., they are always executable but change the state only if their conditions are met.

- *socket(returned-sd, cloexec)* has the effects:
  The flag *FD_CLOEXEC* is represented by the boolean *cloexec.*
  - *(opened returned-sd)∧(is-socket returned-sd)* if *¬(opened returned-sd)*
  - *(close-on-exec returned-sd)* if *¬(opened returned-sd)∧(= cloexec True)*
- *connect(sd)* has the effects:
  - *(connected sd)* if *(opened sd)∧(is-socket sd)∧¬(connected sd)*
  - *∀fd, (connected fd)* if *(equal-fds sd fd)∧(opened sd)∧(is-socket sd)∧¬(connected sd)*
- *dup(sd, returned-sd)* has the effects:
  - *(opened returned-sd)∧(equal-fds sd returned-sd)∧(equal-fds returned-sd sd)* if *(opened sd)∧¬(opened returned-sd)*
  - *(is-socket returned-sd)* if *(is-socket sd)∧(opened sd)∧¬(opened returned-sd)*
  - *(connected returned-sd)* if *(connected sd)∧(opened sd)∧¬(opened returned-sd)*
  - *∀fd, (equal-fds fd returned-sd)∧(equal-fds returned-sd fd)* if *(equal-fds fd sd)∧¬(opened returned-sd)*
- *fcntl(sd, command, returned-sd, cloexec)* has the effects:
  returned-sd is the argument of the command *F_DUPFD* and *cloexec* is the argument of the command *F_SETFD. F_DUPFD_CLOEXEC* uses both.
  The flag *FD_CLOEXEC* is represented by the boolean *cloexec.*
  - same effects as *dup(sd, returned-sd)* if *(= command F_DUPFD)∨(= command F_DUPFD_CLOEXEC)*
  - *(close-on-exec sd)* if *[[(= command F_SETFD)∧(= cloexec True)]∨(= command F_DUPFD_CLOEXEC)]∧(opened sd)∧¬(opened returned-sd)*
  - *¬(close-on-exec sd)* if *(= command F_SETFD)∧(= cloexec False)∧(opened sd)∧¬(opened returned-sd)*
- *close(sd)* has the effects:
  - *¬(opened sd)∧¬(is-socket sd)∧ ¬(connected sd)∧¬(close-on-exec sd)*
  - *∀fd, ¬(equal-fds sd fd)∧¬(equal-fds fd sd)*
- *execve(path)* has the effect:
  - *(remote-shell-started)* if *(is-shell path) ∧∃fd, (connected fd)∧¬(close-on-exec fd)∧[(= fd fd0)∨(equal-fds fd fd0)]∧[(= fd fd1)∨(equal-fds fd fd1)]*

Valid Plans The five different valid plans shown in Fig. 1 show how diverse the plans are even for such a simple example.

Plan 1 is the standard sequence performed to establish a reverse shell, which appears in most shellcode databases. Plan 2 uses the fact that we know that system call *socket* allocates the lowest file descriptor available. Calling *close(fd0)* before *socket* avoids the duplication of the socket on the file descriptor 0. Plan 3 replaces one system call by an equivalent one: *dup* is replaced by *fcntl* called with the command *F_DUPFD*. Plan 4 demonstrates that planning captures and updates correctly properties set by flags and through different system calls. The flag *FD_CLOEXEC* is first set through system call *socket*, and reset later by *fcntl* called on *F_SETFD*. Plan 5 shows that planning is able to follow complex flow

```
1. (socket fd3 False)   3. (socket fd3 False)        5. (socket fd3 False)
   (connect fd3)           (connect fd3)                (dup fd3 fd4)
   (close fd0)             (close fd0)                  (connect fd4)
   (dup fd3 fd0)           (dup fd3 fd0)                (close fd4)
   (close fd1)             (close fd1)                  (close fd0)
   (dup fd3 fd1)           (fcntl fd3 F_DUPFD fd1 False) (dup fd3 fd0)
   (execve sh)             (execve sh)                  (close fd3)
                                                        (close fd1)
2. (close fd0)          4. (close fd0)                  (dup fd0 fd1)
   (socket fd0 False)      (socket fd0 True)            (execve sh)
   (connect fd0)           (connect fd0)
   (close fd1)             (close fd1)
   (dup fd0 fd1)           (dup fd0 fd1)
   (execve sh)             (fcntl fd0 F_SETFD NULL False)
                           (execve sh)
```

Fig. 1. Valid plans for the reverse shell domain

of operation on file descriptors. The key point is that, despite major differences in appearance, which are likely to fool syntactic methods (certainly, if some of the plans were not available previously), our semantic approach recognizes the behavior they implement.

The main effort required by our approach is building an appropriate model for each system call. This model is informed by the basic set of low-level behaviors one would like to model. Once completed, we can simulate any sequence of system calls by applying them to an initial state of the abstract system using any planning simulator/validator. By examining the final state of the system, we can recognize which behaviors took place. Thus, the off-line modeling task is done once, and the resulting model can be used repeatedly, automatically, and very cheaply, to analyze programs.

The (manual) abstraction process is flexible. We can use it to identify simple behaviors, such as create a socket, connect to a remote host, read data from socket, open file for writing, write into file, etc. And we can also recognize complex behaviors by detecting combinations of simple behaviors. For example, downloading a file requires reading data from a connected socket and writing it to an opened file. Thus, once we have the low-level behaviors, it is easy to capture the higher level ones. We can do this by either modifying the action model or by adding axioms, which are a method of adding a simple form of inference to planning.

With such a layered approach, basic behaviors can be reused to identify multiple high level behaviors.

As this model is an abstraction, some information is lost in this model, and the method cannot be 100% correct and capture every nuance. Much can be captured by building a more elaborate model, but some aspects, such as accurate modeling of system resources, are not likely to be practical.

5 Empirical Evaluation

In the previous section we demonstrated the capabilities of our approach to recognize behaviors on the reverse shell domain, where our planning model is able to recognize the same behavior generated in different ways. We now want to highlight our ability to recognize complex, higher level behaviors that are built from lower level behaviors, compared to statistical methods that are quite popular in this area. To do this, we consider the behavior of real processes involved in a mail service. Given the system call trace logs of several processes, we attempt to recognize which behavior is realized by each of the processes, such as sending an email via SMTP, collecting an email from a remote server via IMAP, and so on. The code and data set used for the empirical evaluation can be obtained from a Git repository at https://github.com/alexEnsimag/planning-for-syscall-monitoring.

5.1 Data Collection

We generate system call traces of processes running in a mail service (Fig. 2). The setup consists of two hosts: the client and the server, and involves a number of processes, denoted in what follows in italic. The hosts collect emails from an external server. In order to provide sufficient volume and diversity of the data processed, we opened a dedicated email account with a web-based email service, and subscribed to multiple promotion and notification mailing lists. On the client, *fetchmail* is used to retrieve emails from a web-based email provider via the IMAP protocol. Then, *procmail* dispatches received emails, which are then sent by *postfix* to the server via SMTP protocol. The server's *postfix* process receives the emails, passes them through the *amavis* antivirus and stores in the local filesystem. The *dovecot* process serves emails via the IMAP protocol. The emails are retrieved by the client's *fetchmail*, and stored in the filesystem. We use *Docker* [11] to run containers encapsulating the mail server and mail client hosts, and *sysdig* [6] to record the system calls.

We analyze system call traces of the following processes: *smtpd*, *fetchmail* on the client, *fetchmail* on the server, *imap-login*, all other processes.

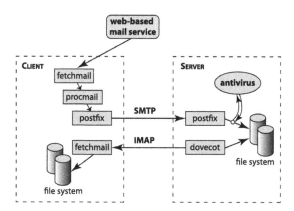

Fig. 2. A mail service setup for evaluation of planning approach on mail delivery activities

These processes realize the following behaviors:

- receiving an email over the SMTP protocol;
- receiving an email over IMAP protocol;
- forwarding an email from the client to the server;
- IMAP connection setup and authentication;
- other behaviours not tracked by the system.

We use 440 samples for each process (including *other*). Data is split into training and test sets as 66%/33%.

Statistical Classification. We compare our goal-tracking approach to a baseline, commonly used statistical classification of processes based on system call sequences. We train a statistical classifier (random forest) on the collected system call traces. This classification approach is similar to approaches used in system call monitoring literature [3,8]. We use two types of n-gram vectorization of the system call names: bi-grams and 1-skip-2-grams. In each case, the vocabulary contains 100 most recurrent n-grams in the corpus. Thus, each sample is represented by a vector of 100 elements, where each element in the vector represents one of the n-grams in the vocabulary and the values represent the *number of times* a specific n-gram appeared in that sample. For example, if the first 4 elements in the vector correspond to bi-grams (`open, read`), (`read, write`), (`write, read`), and (`write, close`), a system call trace

```
open, read, write, read, write, close
```

will produce a bi-gram vector: $1, 2, 1, 1, \ldots$

We proceed with the empirical evaluation as follows, for both bi-grams and 1-skip-2-grams:

1. We train a classifier that classifies each of the behaviors based on non-obfuscated system call sequences. The classifier achieves 98% accuracy for both models.
2. We create obfuscated samples in a way that 'breaks' the bi-grams by inserting a system call that has no effect on the process behavior (for example, `sleep` with a sufficiently small argument) in between each couple of system calls in the sequence. This method is called *adding semantic no-ops* and is the focus of [20]. When testing the statistical model on the obfuscated data we get 0% accuracy with both models (all samples in the test set are classified as 'other').
3. We retrain our model on both obfuscated and non-obfuscated data. It now achieves $\approx 66\%$ accuracy with bi-grams, and $\approx 98\%$ with 1-skip-2-grams.

Planning-Based Classification Information contained in the system logs and manual inspection of system call traces in the training set are used to specify the planning domain and the goal for each of the behaviors. Then, the VAL plan validation system [15] is used to classify system call traces in the test set.

The planning-based classifier based on the domain built for the original, non-obfuscated system call logs is applied to both non-obfuscated and obfuscated system call traces. In both cases, the planning based classifier reaches **over 98% accuracy**. Manual inspection of the misclassified samples suggests that the samples correspond to failed communication between components of the mail service.

Table 1. Classification accuracy on real system logs

	Random forest bi-grams	Random forest skip-grams	Planning-based
Original	98%	98%	**98%**
Obfuscated	0%	0%	**98%**
Re-trained	66%	98%	**98%**

Table 1 summarizes the results. We chose a 1-skip-2-gram model because it counters the noise inserted. The trace given previously, obfuscated as:

```
open, sleep, read, sleep, write, sleep, read, sleep, write, sleep,
close
```

and vectorized using a 1-skip-2-grams, will give the unique bi-grams (`open`, `read`), (`read, write`), (`write, read`), (`write, close`), and (`sleep, sleep`). The obfuscation technique used is clearly ineffective, as the model reveals the original bi-grams.

Our planning approach obtains the same accuracy as the 1-skip-2-gram model trained with obfuscated samples, that is, the same as a statistical model built to handle a specific obfuscation technique. However, statistical methods might be challenged by more complex obfuscations. One can overcome this to some extent using feature extraction methods that cover a wide range of obfuscations. But in an adversarial setting, which is often the case in system security, there is always a new obfuscation technique which defeats feature extraction unless known to the model builder. On the other hand, planning-based (or, more generally, semantic) classification will stay robust to obfuscation, whether known or newly invented, as long as the actual behavior of the program is preserved.

6   Discussion and Future Work

We presented an approach for monitoring computer programs using an abstract model of the system state and the basic "actions" that operate on this state — system calls, in our case. The method is semantic in nature, and hence not prone to the weaknesses of syntactic methods that consider the command sequences form rather than their meaning. Unlike statistical methods that, in principle, can be fully automated, our approach has a non-trivial, one-time manual modeling step. But once the model is constructed, it can be used automatically and with little cost.

We demonstrated the effectiveness of our method by first showing how we capture a simple low level behavior that has diverse implementations using a simple model. Syntactically, each implementation is quite different, yet the common semantics can be captured by modeling just a few system calls. Then, we showed how we detect more complex, higher level behavior with almost perfect accuracy, without being affected by obfuscation techniques that easily fool state-of-the-art statistical methods.

The approach used here can be used for other applications beyond system-call logs, such as analysis of transactions, HTTP logs, and more. Moreover, we believe that it could complement statistical methods by allowing us to run statistical analysis on the higher level features generated by our abstract state.

References

1. Baker, C.L., Tenenbaum, J.B., Saxe, R.R.: Bayesian models of human action understanding. In: Proceedings of the 18th International Conference on Neural Information Processing Systems, NIPS 2005, pp. 99–106. MIT Press, Cambridge (2005). http://dl.acm.org/citation.cfm?id=2976248.2976261
2. Beaucamps, P., Gnaedig, I., Marion, J.-Y.: Abstraction-based malware analysis using rewriting and model checking. In: Foresti, S., Yung, M., Martinelli, F. (eds.) ESORICS 2012. LNCS, vol. 7459, pp. 806–823. Springer, Heidelberg (2012). https://doi.org/10.1007/978-3-642-33167-1_46
3. Canali, D., Lanzi, A., Balzarotti, D., Kruegel, C., Christodorescu, M., Kirda, E.: A quantitative study of accuracy in system call-based malware detection. In: ISSTA 2012, New York, NY, USA, pp. 122–132 (2012). https://doi.org/10.1145/2338965.2336768, http://doi.acm.org/10.1145/2338965.2336768
4. Canzanese, R., Mancoridis, S., Kam, M.: System call-based detection of malicious processes. In: International Conference on Software Quality, Reliability and Security, QRS 2015, pp. 119–124 (2015)
5. Christodorescu, M., Jha, S., Seshia, S.A., Song, D., Bryant, R.E.: Semantics-aware malware detection. In: Proceedings of the 2005 IEEE Symposium on Security and Privacy, SP 2005, pp. 32–46. IEEE Computer Society, Washington (2005). https://doi.org/10.1109/SP.2005.20
6. Draios Inc: Sysdig (2012–2016). http://sysdig.com/
7. Ezzati-Jivan, N., Dagenais, M.R.: A stateful approach to generate synthetic events from kernel traces. Adv. Soft. Eng. **2012**, 6:6–6:6 (2012). https://doi.org/10.1155/2012/140368
8. Firdausi, I., lim, C., Erwin, A., Nugroho, A.S.: Analysis of machine learning techniques used in behavior-based malware detection. In: ACT 2010, pp. 201–203 (2010). https://doi.org/10.1109/ACT.2010.33
9. Forrest, S., Hofmeyr, S.A., Somayaji, A., Longstaff, T.A.: A sense of self for unix processes. In: IEEE Symposium on Security and Privacy, pp. 120–128, May 1996. https://doi.org/10.1109/SECPRI.1996.502675
10. Gao, D., Reiter, M.K., Song, D.: Gray-box extraction of execution graphs for anomaly detection. In: Proceedings of the 11th ACM Conference on Computer and Communications Security, CCS 2004, pp. 318–329. ACM, New York (2004). http://doi.acm.org/10.1145/1030083.1030126
11. Hykes, S.: Docker (2013–2017). http://docker.com/

12. Jacob, G., Debar, H., Filiol, E.: Malware behavioral detection by attribute-automata using abstraction from platform and language. In: Kirda, E., Jha, S., Balzarotti, D. (eds.) RAID 2009. LNCS, vol. 5758, pp. 81–100. Springer, Heidelberg (2009). https://doi.org/10.1007/978-3-642-04342-0_5

13. Kim, G., Yi, H., Lee, J., Paek, Y., Yoon, S.: Lstm-based system-call language modeling and robust ensemble method for designing host-based intrusion detection systems. arXiv preprint arXiv:1611.01726 (2016)

14. Liu, A., Martin, C., Hetherington, T., Matzner, S.: A comparison of system call feature representations for insider threat detection. In: Proceedings from the Sixth Annual IEEE SMC Information Assurance Workshop, pp. 340–347, June 2005. https://doi.org/10.1109/IAW.2005.1495972

15. Long, D.: VAL: The plan validation system (2014). https://github.com/KCL-Planning/VAL

16. Martignoni, L., Stinson, E., Fredrikson, M., Jha, S., Mitchell, J.C.: A layered architecture for detecting malicious behaviors. In: Lippmann, R., Kirda, E., Trachtenberg, A. (eds.) RAID 2008. LNCS, vol. 5230, pp. 78–97. Springer, Heidelberg (2008). https://doi.org/10.1007/978-3-540-87403-4_5

17. Mutz, D., Valeur, F., Vigna, G., Kruegel, C.: Anomalous system call detection. ACM Trans. Inf. Syst. Secur. 9(1), 61–93 (2006). https://doi.org/10.1145/1127345.1127348

18. Poulose Jacob, K., Surekha, M.V.: Anomaly detection using system call sequence sets. J. Software 2(6) (2007)

19. Ramírez, M., Geffner, H.: Plan recognition as planning. In: IJCAI 2009, pp. 1778–1783 (2009). http://ijcai.org/Proceedings/09/Papers/296.pdf

20. Rosenberg, I., Gudes, E.: Bypassing system calls-based intrusion detection systems. Concurrency Comput. Pract. Experience 29(16) (2017). https://doi.org/10.1002/cpe.4023

21. Sukthankar, G., Geib, C., Bui, H., Pynadath, D., Goldman, R.P. (eds.): Plan, Activity, and Intent Recognition. Elsevier (2014)

22. Tandon, G., Chan, P.K.: On the learning of system call attributes for host-based anomaly detection. Int. J. AI Tools 15(06), 875–892 (2006). https://doi.org/10.1142/S0218213006003028

23. Tokhtabayev, A., Skormin, V., Dolgikh, A.: Dynamic, resilient detection of complex malicious functionalities in the system call domain. In: MILCOM 2010, pp. 1349–1356, October 2010. https://doi.org/10.1109/MILCOM.2010.5680136

24. Warrender, C., Forrest, S., Pearlmutter, B.: Detecting intrusions using system calls: alternative data models. In: Proceedings of the 1999 IEEE Symposium on Security and Privacy (Cat. No.99CB36344), pp. 133–145 (1999). https://doi.org/10.1109/SECPRI.1999.766910

25. Wressnegger, C., Schwenk, G., Arp, D., Rieck, K.: A close look on n-grams in intrusion detection: Anomaly detection vs. classification. In: Proceedings of the 2013 ACM Workshop on Artificial Intelligence and Security, AISec 2013, pp. 67–76. ACM, New York (2013). https://doi.org/10.1145/2517312.2517316, http://doi.acm.org/10.1145/2517312.2517316

26. Xu, J., Shelton, C.R.: Intrusion detection using continuous time Bayesian networks. JAIR 39, 745–774 (2010)

27. You, I., Yim, K.: Malware obfuscation techniques: a brief survey. In: BWCCA 2010, pp. 297–300 (2010). https://doi.org/10.1109/BWCCA.2010.85

One-Round Secure Multiparty Computation of Arithmetic Streams and Functions
(Extended Abstract)

Dor Bitan[1(✉)] and Shlomi Dolev[2]

[1] Department of Mathematics, Ben-Gurion University of the Negev,
Beer-Sheva, Israel
dorbi@post.bgu.ac.il
[2] Department of Computer Science, Ben-Gurion University of the Negev,
Beer-Sheva, Israel
dolev@cs.bgu.ac.il

Abstract. Efficient secure multiparty computation (SMPC) schemes over secret shares are presented. We consider scenarios in which the secrets are elements of a finite field, $\mathbb{F}_p$, and are held and shared by a single participant, *the user*. Evaluation of any function $f : \mathbb{F}_p^n \to \mathbb{F}_p$ is implemented in one round of communication by representing f as a multivariate polynomial. Our schemes are based on partitioning secrets to sums or products of random elements of the field. Secrets are shared using either (multiplicative) shares whose product is the secret or (additive) shares that sum up to the secret. Sequences of additions of secrets are implemented locally by addition of local shares, requiring no communication among participants, and so does sequences of multiplications of secrets. The shift to handle a sequence of additions from the execution of multiplications or vice versa is efficiently handled as well with no need to decrypt the secrets in the course of the computation. On each shift from multiplications to additions or vice versa, the current set of participants is eliminated, and a new set of participants becomes active. Assuming no coalitions among the active participants and the previously eliminated participants are possible, our schemes are information-theoretically secure with a threshold of all active participants. Our schemes can also be used to support SMPC of boolean circuits.

1 Introduction

Cloud services have become very popular in recent years. Companies like Amazon, Google, Microsoft, IBM, etc., are offering storage devices and computing engines to both private users and organizations. The usage of clouds for storage and computing has significant benefits in price, speed, and manageability. Nonetheless, it requires users to send their information to an untrusted third party. In some cases, the information held by a user is confidential, and hence

© Springer International Publishing AG, part of Springer Nature 2018
I. Dinur et al. (Eds.): CSCML 2018, LNCS 10879, pp. 255–273, 2018.
https://doi.org/10.1007/978-3-319-94147-9_20

the distribution of the information to untrusted parties cannot be considered. One solution to this problem may be a cryptographic scheme that enables a user to upload encrypted data to the cloud, perform computations in the cloud over the encrypted data and retrieve an encrypted output of the desired result. Such an encryption scheme enables the user to take advantage of the storage and computing services provided by the cloud without compromising the confidentiality of the data.

SMPC schemes over a distributed system suggest partial solutions to this problem [BMR90, BOGW88, CCD88, DFK+06, DIK10, GIKR02]. These schemes, which are best in their security level being information-theoretically secure, support such computations at the cost of communication between participants. At each round of communication, each participant sends at most one message to each of the other participants, performs arbitrary computations and/or receives at most one message from each of the other participants (not necessarily in this order) [KN06]. Typically, communication between participants is used for reducing the degree of the polynomial that encrypts the data after each multiplication during the computation [BOGW88, CCD88, DGL15, DLY07, DL16].

Existing fully homomorphic encryption schemes suggest a centralized, rather than distributed, computationally secure solutions to the above mentioned problem [BP16, Gen09, GHS12, GHS16, SV10, VDGHV10, XWZ+18]. Unfortunately, beyond being only computationally secure (rather than information-theoretically secure), they are currently too slow to be used in practice.

Related Work. In their seminal work from 1988, Ben-Or, Goldwasser and Wigderson [BOGW88] showed that every function of N inputs can be efficiently computed by N participants with a threshold of $N/2$ in case of honest-bcut-curious participants, or $N/3$ in case of malicious participants. Their methods are based on Shamir's secret sharing scheme [Sha79] and their protocols require rounds of communication proportional to the depth of the arithmetic circuit. Substantial efforts have been spent to achieve a better communication complexity in such tasks. Bar-Ilan and Beaver [BIB89] were the first to suggest a way to evaluate functions in a constant number of rounds of communication, followed by further works that attempt to minimize communication complexity of SMPC protocols. In 2002, Gennaro et al. [GIKR02] proved that, in the presence of malicious participants, some functions do not admit SMPC protocols with less than three rounds of communication. Specifically, they have shown that the functions XOR_2^n and AND_2^n do not admit protocols of SMPC with only two rounds of communication, assuming malicious participants are present. Nonetheless, they have shown that functions that depend only on the inputs of a single participant can be securely computed in two rounds of communication. When relaxing the assumptions and considering honest-but-curious (rather than malicious) participants the round complexity of general SMPC protocols is reduced to two rounds of communication [BOGW88, IK02].

Our Contribution. We suggest a different approach to tackle the problem described above concerning outsourcing of computation. The main novelty of our schemes is allowing evaluation of arithmetic functions in one round of communication. Our approach is based on representing each secret either as a sum of secret shares or as a product of secret shares, shifting between representations when necessary. The schemes we present use two sets of participants. At each stage, only one of the sets holds shares of the secrets. The participants in the first set handle multiplications and are called *the multiplication participants*, or *the M.parties*, in short. The participants in the second set handle additions and are called *the addition participants*, or *the A.parties*. We detail the operations of the participants in a sequence of the same operation and the communication between them when there is a switch in operations and the immediate elimination of the previous participants (virtual machines, containers, etc.). Our schemes require communication among participants only when switching between operations, and support communicationless sequences of multiplications followed by communicationless sequences of additions and vice versa. Our schemes are information-theoretically secure against attacks of coalitions consisting of all-but-one of the active participants.

Paper Organization. In Sect. 2 we suggest information-theoretically SMPC schemes that support arithmetic streams. These schemes give rise to information-theoretically SMPC schemes that enable evaluation of any arithmetic function in one round of communication, presented in Sect. 3. The schemes presented in Sects. 2 and 3 use two participants in each set of currently-active parties and support arithmetic circuits. Several extensions of our schemes appear in Sect. 4. Amongst them, working with a larger number of participants, supporting boolean (rather than arithmetic) circuits and evaluating functions over inputs held by more than one participant. Section 5 discusses security issues. It contains ways to mask the circuit itself, ways to handle malicious participants, and an analysis of the threshold of our schemes. Lastly, conclusions appear in Sect. 6. Formal proofs and some of the details are omitted from this extended abstract.

2   Stream Computation

Consider the following scenario. A user receives a stream of values and arithmetic operations produced by some source. The user wishes to perform an arithmetic computation over the values received according to the operations received on the fly. The stream begins with an initial value, denoted by m_0. At this stage the user sets a value, referred to as *the computation state* and denote by st and initializes st as m_0. Afterward, at each stage a pair of value and arithmetic operation are produced by the source and received by the user, who in turn updates the computation state st accordingly. Explicitly, at stage i (for $i \geq 1$) a pair, consisting of a value m_i and arithmetic operation op_i are produced, where op_i is either addition '+' or multiplication '·'. The user updates the state either by multiplying st by m_i or by adding m_i to st, according to op_i. Namely, $st := st \; op_i \; m_i$. We call this kind of scenario *a stream computation*.

We assume that the values received from the source are confidential, and so is the computation state that they yield in each stage. The user cannot keep (and update) st on her hardware, since it might be hacked by an adversary.

We seek for a cryptographic scheme that will allow the user to outsource the aforementioned computation, while keeping the values m_i and st information-theoretically secure at all stages, without keeping st as plaintext at any stage. The value m_i should also be eliminated at the end of each stage. The user should be able to retrieve st at any time she so wishes.

At this point we would like to make an observation concerning outsourcing of computation in general. Usually, when talking about outsourcing of computation, we consider a user that wishes to use a remote strong computer to run a computation over private data, where the main reason for outsourcing the computation is computing power. One of the main interests in such scenarios is to involve the user in the computation as less as possible and to shift most of the computational tasks to the cloud. In our scenario, since the values are produced by the source and received by the user on-the-fly, the user must be on-line during the computation and take an active part in the computation. We assume that the user does have enough computing power to run the computation herself, but since the values and the computation state produced during computation are confidential, she cannot save them as plaintext anywhere.

We now present schemes for outsourcing stream computations in two different cases – the *non-vanishing* case and the *p-bounded* case.

Non-vanishing Stream. We suggest a scheme for performing secure outsourcing of stream computation under the following assumptions. For the particular example we detail in the sequel, we assume that the values m_i are elements of the field $\mathbb{F}_p$ of prime order p (the arithmetic operations are carried out in the field), and that the values and operations produced never yield $st = 0$. Such a stream is *non-vanishing*. We assume that the user has a secure connection channel with (at least) four honest-but-curious servers denoted $\mathcal{P}^{(j)}$ (for $1 \leq j \leq 4$). The basic four participants scheme can be generalized to one with a larger number of participants as explained in Sect. 4. For ease of presentation, we first present the scheme for the case of four participants. In the scheme we suggest, some of the participants hold shares of the computation state st, denoted $st^{(j)}$. The shares do not reveal any information about st and enable extrapolation of st by the user at any stage.

The scheme we suggest has two modes: *multiplication mode* and *addition mode*. $\mathcal{P}^{(1)}$ and $\mathcal{P}^{(2)}$ are the *M.parties* and the rest of the servers are the *A.parties*. The scheme is composed of five procedures as follows:

- *Init* – Initializing.
- *MinM* – Multiplication in multiplication mode.
- *M* → *A* – Switching mode from multiplication to addition.
- *AinA* – Addition in addition mode.
- *A* → *M* – Switching mode from addition to multiplication.

The general idea behind the scheme is that multiplications are handled by the *M.parties* and additions are handled by the *A.parties*. At stage i, some of the procedures are invoked to update st according to op_i and shift (if necessary) the shares from one set of participants to another, eliminating the previous set of participants. When the shares of st are being held by the $M.parties$ (respectively, $A.parties$), we say that the scheme is in multiplication (respectively, addition) mode.

The scheme stages are as follows:

- Run $Init$ – distributing (multiplicative) shares of m_0 to $\mathcal{P}^{(1)}$ and $\mathcal{P}^{(2)}$.
- At stage i, upon receiving (m_i, op_i):
  - If the received operation op_i does not match the current mode (i.e., receiving $op_i = $ '·' in addition mode or $op_i = $ '+' in multiplication mode), then run $M \rightarrow A$ or $A \rightarrow M$ to switch mode and eliminate the previous set of participants.
  - Run $AinA$ or $MinM$ to update the shares of st according to (m_i, op_i).

We now describe each procedure. All operations are carried out in $\mathbb{F}_p$. We begin with the initializing procedure.

Procedure 1: $Init$ – *Initializing.* This procedure is invoked by the user and the $M.parties$ only at stage zero, when the initial value m_0 is produced and received by the user. First, the user picks a non-zero element x_0 of $\mathbb{F}_p$ uniformly at random and computes y_0 such that $x_0 \cdot y_0 = m_0$. We call this procedure *mult.-random-split* of m_0 into two multiplicative shares. Then, the user sends x_0 to $\mathcal{P}^{(1)}$ and y_0 to $\mathcal{P}^{(2)}$, who in turn set $st^{(1)}$ to x_0 and $st^{(2)}$ to y_0, respectively. The values $st^{(1)}$ and $st^{(2)}$, kept by the $M.parties$ after the execution of this protocol, are their shares of st. Since x_0 is picked randomly, y_0 is also random. Hence, no information concerning m_0 is revealed to the $M.parties$.

Procedure 2: $MinM$ – *Multiplication in multiplication mode.* This procedure is invoked by the user and the $M.parties$ at stages i such that op_i is multiplication (after switching to multiplication mode if necessary, using $A \rightarrow M$). Similarly to $Init$, first the user mult.-random-splits m_i to $x \cdot y$. Then the user sends x to $\mathcal{P}^{(1)}$ and y to $\mathcal{P}^{(2)}$. The $M.parties$ in turn update the shares of st they hold. $\mathcal{P}^{(1)}$ sets $st^{(1)}$ to $st^{(1)} \cdot x$ and $\mathcal{P}^{(2)}$ sets $st^{(2)}$ to $st^{(2)} \cdot y$. Now, the shares of the $M.parties$ are updated according to the current computation state. The fact that x and y are random implies that no information is revealed to the participants neither about m_i nor about st.

Procedure 3: $M \rightarrow A$ – *Switching mode from multiplication to addition.* This procedure is invoked by all the participants at stages i such that op_i is addition and the current mode is multiplication: First, $\mathcal{P}^{(1)}$ picks an element a of $\mathbb{F}_p$ and computes b such that $a + b = st^{(1)}$. We call this procedure *add.-random-split* of $st^{(1)}$ into two additive shares, a and b. Then, $\mathcal{P}^{(1)}$ sends a to $\mathcal{P}^{(3)}$ and b to $\mathcal{P}^{(4)}$. $\mathcal{P}^{(2)}$ sends $st^{(2)}$ to both $\mathcal{P}^{(3)}$ and $\mathcal{P}^{(4)}$. At this stage the $M.parties$ are eliminated. Then, the $A.parties$ multiply the values they received and set $st^{(j)}$ to the product ($j = 3, 4$). Observe that:

$$st = st^{(1)} \cdot st^{(2)} = (a + b) \cdot st^{(2)} = a \cdot st^{(2)} + b \cdot st^{(2)} = st^{(3)} + st^{(4)}.$$

Namely, from the two mult.-random-split shares of st that were held by the $M.parties$, the $A.parties$ receive add.-random-split shares of st. Since a and b

are random elements of the field, the *A.parties* gain no information about *st* and the *M.parties* are eliminated.

Procedure 4: *AinA – Adding in addition mode.* This procedure is invoked by the user and the *A.parties* at stages i such that op_i is addition (after switching to addition mode if necessary, using $M \to A$). First, the user add.random-splits m_i to $x + y$ and sends x to $\mathcal{P}^{(3)}$ and y to $\mathcal{P}^{(4)}$. Then, in order to update its share of the computation state, each *A.party* adds the value it received from the user to $st^{(j)}$, $(j = 3, 4)$. Since x and y are random elements of the field, neither of the *A.parties* gain any information about m_i or about the current state.

Procedure 5: $A \to M$ – *Switching mode from addition to multiplication.* This procedure is invoked by the user and all the participants at stages i such that op_i is multiplication and the current mode is addition. The user mult.-random-splits $1 \in \mathbb{F}_p$ to $r \cdot r^{-1}$, and sends r^{-1} to $\mathcal{P}^{(1)}$ and r to the *A.parties*. Then, $\mathcal{P}^{(1)}$ sets $st^{(1)}$ to r^{-1}. Each of the *A.parties*, $\mathcal{P}^{(j)}$, $(j = 3, 4)$, multiplies r by $st^{(j)}$ and sends the product to $\mathcal{P}^{(2)}$. At this stage the *A.parties* are eliminated. Then $\mathcal{P}^{(2)}$ adds the values received and sets $st^{(2)}$ to the sum.

Observe that:

$$st = st^{(3)} + st^{(4)} = r^{-1} \cdot \left(r \cdot (st^{(3)} + st^{(4)})\right) = r^{-1} \cdot \left(r \cdot st^{(3)} + r \cdot st^{(4)}\right) = st^{(1)} \cdot st^{(2)}.$$

Thus, from the two add.-random-split shares of st that were held by the *A.parties*, the *M.parties* receive mult.-random-split shares of st. At this stage, $\mathcal{P}^{(1)}$ obviously has no information about st. Since $st \neq 0$ and r is random, $\mathcal{P}^{(2)}$ also has no information about the current state.

Observe that at any stage of the scheme:

- The computation state is not saved as plaintext anywhere.
- The values m_i received by the user are immediately random-split into shares, the shares are distributed and m_i is eliminated.
- None of the participants gains any information about the values m_i or about st.
- At each stage the user can retrieve the shares of st from the participants and efficiently compute st.

Hence, this scheme enables a user to perform information-theoretically secure outsourcing of any non-vanishing stream using four participants.

Bounded Stream. In the scheme suggested above, the depth and length of the arithmetic circuit are practically unbounded. One can use it to outsource arbitrarily long computation streams containing any number of multiplications and additions in $\mathbb{F}_p$. There is a limitation though, on the possible result of each stage of the computation. Namely, none of them may be zero. In some cases, one has a computation stream that does not meet this limitation. How can we outsource stream computations that are not non-vanishing? We now suggest an answer to this question assuming that the depth and length of the stream are bounded. The scheme we suggest is based on that suggested above for the non-vanishing case. Similarly to the assumptions of the non-vanishing scheme, we assume that the values m_i are elements of a finite field $\mathbb{F}_q$ (q is prime), and that the arithmetic operations are multiplication and addition in $\mathbb{F}_q$.

We begin with an observation. Assume $M = (m_0, m_1, \ldots, m_n) \in \mathbb{F}_q^{n+1}$ is a sequence of values produced by a source in some stream computation, and that $OP = (op_1, \ldots, op_n) \in \{+, \cdot\}^n$ is the sequence of operations produced by it corresponding to M. At each stage of the computation, the computation state st is the result of applying the operations in OP to the corresponding values in M, where operations are carried out in $\mathbb{F}_q$. One gets the exact same result by performing the computation over the positive integers and taking the result modulo q. Formally, for each entry m_i of M, let $a_i \in \{1, 2, \ldots, q\} \subseteq \mathbb{N}$ denote the minimal (strictly) positive integer such that $a_i \equiv m_i \pmod q$. The a_i's are *the integer correspondents* of the m_i's. Then, performing the stream computation over the a_i's (while using the same operations over the integers), we obtain an integer result $st_\mathbb{N}$ such that $st_\mathbb{N} \equiv st \pmod q$. Assume a computation stream over elements in $\mathbb{F}_q$ is such that, when performing the corresponding stream computation over the integers, we obtain an integer-computation state, $st_\mathbb{N}$, that never exceeds a large prime p. We will call such a computation stream *p-bounded*.

We now suggest a scheme to perform information-theoretically secure outsourcing of a bounded computation stream. As in the non-vanishing scheme, we assume that the user has a secure connection channel with four honest-but-curious participants $\mathcal{P}^{(j)}$ ($1 \leq j \leq 4$). The general idea behind the scheme is to run at each stage the procedures described in Sect. 2 over the integer correspondents of the m_i's, modulo p, where operations are carried in $\mathbb{F}_p$.

The scheme stages are as follows:

- Upon receiving the initial value $m_0 \in \mathbb{F}_q$, run $Init$ to distribute multiplicative shares of $a_0 \pmod p$ to $\mathcal{P}^{(1)}$ and $\mathcal{P}^{(2)}$, where a_0 is the integer correspondent of m_0.
- At stage i, upon receiving $m_i \in \mathbb{F}_q$ and an operation op_i
  - If op_i does not match the current mode then run $M \to A$ or $A \to M$ to switch mode eliminating the *M.parties* or the *A.parties*.
  - Run $MinM$ or $AinA$ to update the computation state shares according to $a_i \pmod p$ and op_i.

The user can extrapolate $st \in \mathbb{F}_q$ at any stage by retrieving the shares from the participants, compute the computation state modulo p, and then take the integer correspondent to the result modulo q. The correctness of the scheme is derived from the fact that the stream is p-bounded. The security of this scheme for p-bounded streams is derived from the security of the non-vanishing stream scheme, since from the participants perspective, there is no difference between the cases.

3   SMPC of Arithmetic Functions in One Round of Communication

The ideas from the previous section, used to outsource stream computations, give rise to SMPC schemes that allow evaluation of arithmetic functions in one round of communication. In this section we suggest schemes that support this task in two different cases: the *non-vanishing* case and the *p-bounded* case. In the schemes we suggest, the set of variables over which the function is evaluated may be dynamic, and so may be the function itself.

One-Round SMPC of Arithmetic Functions Over Non-Zero Elements. We suggest a SMPC scheme that enables a user to securely outsource storage and computations of data under the following assumptions.

– The user holds a sequence $m = (m_1, \ldots, m_n) \in \mathbb{F}_p^n$.
– The user has a private connection channel with four participants $\mathcal{P}^{(k)}$, $(1 \leq k \leq 4)$. As in the arithmetic streams scenario, this scheme can be generalized to one with a larger number of participants, as detailed in Sect. 4.
– The participants are honest-but-curious.

At each stage of the scheme, the participants hold shares of m. The scheme we suggest now enables a user to secret share $m = (m_1, \ldots, m_n)$ amongst honest-but-curious servers in a way that allows the user to evaluate $f(m)$ using computing engines provided by the servers, where $f : \mathbb{F}_p^n \to \mathbb{F}_p$.

We begin with an observation concerning functions from $\mathbb{F}_p^n$ to $\mathbb{F}_p$. Since $\mathbb{F}_p$ is a finite field, any function $f : \mathbb{F}_p^n \to \mathbb{F}_p$ can be represented as a multivariate polynomial. Since $x^p \equiv x \pmod{p}$, this representation is not unique. Given a function f, we would like to assign f with a *minimal-multivariate-polynomial-representation* of it. To this end, we consider the representation of f as a multivariate polynomial such that the degree of each variable is at most $p - 1$. For any given f there is exactly one such multivariate polynomial. We denote this polynomial by Q_f and assign f with Q_f as its minimal-multivariate-polynomial-representation. We occasionally abuse notation and write f instead of Q_f. We note that the total degree[3] of Q_f is at most $n(p-1)$ and write

$$f(m) = \sum_{i=(i_1,\ldots,i_n)\in\mathcal{I}} a_i \cdot m_1^{i_1} \ldots m_n^{i_n}$$

where $\mathcal{I} = \{0, \ldots, p-1\}^n$ and $a_i \in \mathbb{F}_p$. There are p^{p^n} such functions. For example, if $n = 6, p = 11$, then one of them is: $3m_1^3 m_2^3 m_5 + 6m_3^4 m_1 + 2m_3 m_6$. The fact that each variable in each monomial can appear with any exponent between zero and $p - 1$ implies that there are p^n different monomials (for most functions f used in practice, most of the monomials are irrelevant since they have leading coefficient zero. Nevertheless, for some functions f, taking a representation of f

---

[3] The total degree of a multivariate polynomial is the maximal sum of exponents in a single monomial of it.

as a multivariate polynomial may imply exponential growth of the representation of f). For $i = (i_1, \ldots, i_n) \in \mathcal{I}$ we denote $m_1^{i_1} \ldots m_n^{i_n}$ by A_i. We refer to A_i as *the i'th monomial.*

The scheme we suggest is composed of two protocols:

- *The Distribution protocol* – invoked by the user to secret share m amongst the participants.
- *The Evaluation protocol* – invoked by the user to perform SMPC of a function f over m using the participants.

The general idea behind the scheme is as follows. As in the arithmetic stream schemes presented in Sect. 2, $\mathcal{P}^{(1)}$ and $\mathcal{P}^{(2)}$ are the *M.parties* and $\mathcal{P}^{(3)}$ and $\mathcal{P}^{(4)}$ are the *A.parties*. In the Distribution protocol, the user secret shares m amongst the *M.parties*. The Evaluation protocol is composed of four stages. At the first stage, the user sends information regarding f to the participants, and the *M.parties* perform operations over their shares of m that correspond to SMPC of each of the (non-zero leading coefficient) monomials A_i of f. At the second stage, the *M.parties* send to the *A.parties* information that allows the *A.parties* to achieve additive shares of each of the monomials of f. At this point the *M.parties* are eliminated. At the third stage, the *A.parties* use the information they received from the *M.parties* to achieve shares of $f(m)$. At the fourth stage, the user can choose between either retrieve the shares of $f(m)$ from the *A.parties* and compute $f(m)$, or shift the information from the *A.parties* to a new set of *M.parties* (as in $A \to M$) to allow further computations over $(m, f(m))$ without decrypting $f(m)$. We now describe each protocol. We begin with the Distribution protocol.

The Distribution Protocol.
This protocol is invoked by the user to secret share $m = (m_1, \ldots, m_n) \in \mathbb{F}_p^n$ amongst the *M.parties*. For each m_j, $1 \leq j \leq n$, the user mult.-random-splits m_j to two multiplicative shares x_j and y_j. Then, the user distributes $(x_1, \ldots, x_n)$ to $\mathcal{P}^{(1)}$ and $(y_1, \ldots, y_n)$ to $\mathcal{P}^{(2)}$.

The Evaluation Protocol.
This protocol is invoked by the user to perform SMPC of a function f over m using the participants. The protocol has four stages. We now describe each of them.

- *Stage 1 – MonEv – Monomial evaluation.* At this stage the user sends information about f to the participants. The *M.parties* compute multiplicative shares of the monomials of f. As mentioned, above we write f in the form

$$f(m_1, \ldots, m_n) = \sum_{i \in \mathcal{I}} a_i \cdot A_i,$$

where A_i is the i'th monomial and is determined by $i = (i_1, \ldots, i_n)$. At this stage, for each monomial A_i with non-zero leading coefficient, the user sends $i \in \mathcal{I}$ to the *M.parties* and $a_i \in \mathbb{F}_p$ to the *A.parties*. Each of the *M.parties* evaluates each monomial A_i over his shares. $\mathcal{P}^{(1)}$ sets $A_{x_i} := \Pi_{j=1}^n x_j^{i_j}$ and $\mathcal{P}^{(2)}$ sets $A_{y_i} := \Pi_{j=1}^n y_j^{i_j}$. Observe that A_{x_i} and A_{y_i} are multiplicative shares of A_i evaluated at m:

$$A_{x_i} \cdot A_{y_i} = \Pi_{j=1}^n x_j^{i_j} \cdot \Pi_{j=1}^n y_j^{i_j} = \Pi_{j=1}^n (x_j y_j)^{i_j} = \Pi_{j=1}^n m_j^{i_j} = A_i.$$

- *Stage 2 – SMA – Shift from M.parties to A.parties.* At this stage, for each i received from the user, the *M.parties* manipulate the multiplicative shares of A_{x_i} and A_{y_i} and send information to the *A.parties* that enables the *A.parties* to achieve additive shares of A_i. For each i received, $\mathcal{P}^{(1)}$ add.-random-splits A_{x_i} into two additive shares $b_i + c_i$ in $\mathbb{F}_p$. Then, $\mathcal{P}^{(1)}$ sends b_i to $\mathcal{P}^{(3)}$ and c_i to $\mathcal{P}^{(4)}$, while $\mathcal{P}^{(2)}$ sends A_{y_i} to the *A.parties*. The *M.parties* are now eliminated and the *A.parties* multiply the values received. Denote the products calculated by $\mathcal{P}^{(3)}$ and $\mathcal{P}^{(4)}$ by α_i and β_i, respectively.

  Observe that from the multiplicative shares of A_i that were held by the *M.parties*, the *A.parties* achieve additive shares of A_i:

$$A_i = A_{x_i} \cdot A_{y_i} = (b_i + c_i) \cdot A_{y_i} = b_i \cdot A_{y_i} + c_i \cdot A_{y_i} = \alpha_i + \beta_i.$$

  Since a_i and b_i are random, the *A.parties* gain no information about A_i.
- *Stage 3 – fEv – Evaluation of f.* At this stage the *A.parties* compute additive shares of $f(m)$ using the information received from the user at stage 1 and the additive shares of A_i achieved at stage 2.
  - $\mathcal{P}^{(3)}$ computes $u_3 := \sum_{i \in \mathcal{I}} a_i \cdot \alpha_i$.
  - $\mathcal{P}^{(4)}$ computes $u_4 := \sum_{i \in \mathcal{I}} a_i \cdot \beta_i$.

  Observe that u_3 and u_4 are additive shares of $f(m)$:

$$u_3 + u_4 = \sum_{i \in \mathcal{I}} a_i \cdot \alpha_i + \sum_{i \in \mathcal{I}} a_i \cdot \beta_i = \sum_{i \in \mathcal{I}} a_i \cdot (\alpha_i + \beta_i) = \sum_{i \in \mathcal{I}} a_i \cdot A_i = f(m).$$

- *Stage 4 – RetCas – Retrieving/Cascading.* At this stage the user has a choice between two options: *retrieving* and *cascading*. In the retrieving option, the user retrieves the additive shares of $f(m)$ from the *A.parties* and adds them to obtain $f(m)$. In the cascading option, the user has the *A.parties* manipulate the shares they hold and send information to a new set of *M.parties* (in the same fashion as in procedure $A \rightarrow M$ described in Sect. 2). Then the *A.parties* are eliminated and the user can begin a new computation over $(m_1, \ldots, m_n, f(m))$. We now describe the cascading option. The user performs mult.-random-split of $1 \in \mathbb{F}_p$ to two multiplicative shares x_{m+1} and r, and sends x_{m+1} to $\mathcal{P}^{(1)}$ and r to the *A.parties*. Each of the *A.parties* multiplies r by u_k, $k = 3, 4$, and sends the product to $\mathcal{P}^{(2)}$. At this stage the *A.parties* are eliminated. Now $\mathcal{P}^{(2)}$ adds the values received and sets y_{m+1} to the sum. Observe that:

$$f(m) = u_3 + u_4 = x_{m+1} \cdot \left(r \cdot (u_3 + u_4) \right) = x_{m+1} \cdot y_{m+1}.$$

Thus, from the additive shares of $f(m)$ that were held by the *A.parties*, the *M.parties* obtain multiplicative shares of $f(m)$, and further functions can be evaluated over $(m_1, \ldots, m_n, f(m))$ by the user and the participants using stages 1–3 of this protocol. We note that this option is secure only if $f(m) \neq 0$, since if $f(m) = 0$ then so is y_{m+1}.

Since each m_j is secret shared independently, the set of secrets over which any function can be evaluated is dynamic and further secrets can be shared on the fly. The fact that each monomial is evaluated over the secret shares independently implies that the function itself is dynamic in the sense that new monomials can be evaluated and added on the fly.

One-Round SMPC of p-Bounded Arithmetic Functions. In the scenario considered above, SMPC of arithmetic functions over non-zero elements, there is a limitation on the possible values that the m_j's may take. Namely, they cannot be zero. Moreover, if the user wishes to perform further computations over $(m, f(m))$ without first decrypting $f(m)$, then $f(m)$ too must be non-zero. Can we avoid these limitations over the m_j's and $f(m)$? We now consider a scenario in which some of the m_j's may be zero and f may vanish, and suggest a scheme that overcomes the limitations of the previous scenario assuming f is *p-bounded* for small enough p. The term p-bounded is defined below.

Similarly to the assumptions of the previous scenario, we assume that the values m_j are elements of a finite field of prime order q, denoted $\mathbb{F}_q$. We begin with an observation. Assume the user holds $m = (m_1, \ldots, m_n) \in \mathbb{F}_q^n$ and wishes to evaluate $f(m)$ for some f. One can compute $f(m)$ by performing operations in $\mathbb{F}_q$ on m according to a representation of f as a multivariate polynomial. The same result is obtained if one computes $f(m)$ over the positive integers and then takes the result modulo q. Formally, for each entry m_j of m let $a_j \in \{1, 2, \ldots, q\} \subseteq \mathbb{N}$ denote the minimal (strictly) positive integer such that $a_j \equiv m_j \pmod q$. Then, performing the computation over the a_j's using integer operations we obtain an integer result $f(m)_{\mathbb{N}}$, such that $f(m)_{\mathbb{N}} \equiv f(m) \pmod q$. Assume a function $f : \mathbb{F}_q^n \to \mathbb{F}_q$ is such that for every $m \in \mathbb{F}_q^n$, computation of $f(m)$ over the integers yields an integer-result, $f(m)_{\mathbb{N}}$, which is strictly smaller than a large prime p. Such a function is *p-bounded.*[4]

The scheme we suggest now is based on that suggested in the previous case for non-zero elements and enables SMPC of p-bounded functions over elements, some of which may be zero. As in the non-zero scheme, we assume that the user has a secure connection channel with four honest-but-curious servers, $\mathcal{P}^{(k)}$, $1 \leq k \leq 4$. The general idea behind the scheme is to run at each stage the same procedures as in the scheme suggested in the previous case, over the integer correspondents of the m_j's, modulo p.

The scheme stages are as follows. For $m = (m_1, \ldots, m_n) \in \mathbb{F}_q^n$, let $\tilde{m} = (\tilde{m}_1, \ldots, \tilde{m}_n) \in \mathbb{F}_p^n$ denote the $\mathbb{F}_p^n$ element corresponding to m. That is, $\tilde{m}_j := a_j \pmod q$ for $1 \leq j \leq n$. Similarly, for $f : \mathbb{F}_q^n \to \mathbb{F}_q$ let $\tilde{f} : \mathbb{F}_p^n \to \mathbb{F}_p$ denote the

---

[4] Actually, all functions $f : \mathbb{F}_q^n \to \mathbb{F}_q$ are p-bounded for $p \geq q^{nq+1}$ (considering the minimal-multivariate-polynomial-representation of f). This fact is not useful for large p.

function corresponding to f in the p-world. The Distribution and Evaluation protocols are as follows.

Distribution.
For $m \in \mathbb{F}_q^n$ use the Distribution protocol of the non-zero scheme to secret share $\tilde{m} \in \mathbb{F}_p^n$ among the *M.parties*.

Evaluation.
For $f : \mathbb{F}_q^n \to \mathbb{F}_q$ use the Evaluation protocol of the non-zero scheme to evaluate $\tilde{f}$ over $\tilde{m}$:

- The first three stages are the same as in the non-zero protocol.
- At the fourth stage, $RetCas$,:
  - Decryption is done by retrieving $\tilde{f}(\tilde{m})$ by the dealer and taking the integer corresponding to $\tilde{f}(\tilde{m})$ modulo q.
  - Cascading (performing further computations over $(m_1, \ldots, m_n, f(m))$ without first decrypting $f(m)$) can be done under the following assumptions. Assume the user wishes to perform SMPC of $g : \mathbb{F}_q^{n+1} \to \mathbb{F}_q$ over $(m_1, \ldots, m_n, f(m))$. Use Q_f to write g as a function from $\mathbb{F}_q^m$ to $\mathbb{F}_q$. If g is p-bounded considering its representation as a multivariate polynomial obtained by using Q_f to write g as a function from $\mathbb{F}_q^m$ to $\mathbb{F}_q$, then SMPC of g over $(m_1, \ldots, m_n, f(m))$ can be done with no need to first decrypt $f(m)$ by the user.

We note that this protocol has the same dynamic attributes as those suggested in the previous scenario and it requires one round of communication.

4   Extensions

The Case of More Than Four Participants. The schemes described above employ four participants. However, the ideas behind the procedures from which the schemes are composed generalize to a larger number of participants. Assume one wishes to run the schemes using $n_1 \geq 2$ *M.parties* and $n_2 \geq 2$ *A.parties*. We now suggest ways to generalize the procedures described above to suit $n_1 + n_2$ participants.

Random-Split.
The procedure *mult.-random-split* described above can be generalized to n_1 *M.parties* by taking $n_1 - 1$ random non-zero elements of the field, $x_1, \ldots, x_{n_1-1}$, and computing the x_{n_1} that yields $\pi_{i=1}^{n_1} x_i = m$. The generalization of *add.-random-split.* to n_2 participants is analogous.

Additive Shares from Multiplicative Shares.
Procedure $M \to A$ of the arithmetic streams scenario and procedure SMA of the Evaluation protocol in the arithmetic functions scenario demonstrate shifting of information from two *M.parties*, $\mathcal{P}^{(1)}$ and $\mathcal{P}^{(2)}$, to two *A.parties*, $\mathcal{P}^{(3)}$ and $\mathcal{P}^{(4)}$. These procedures are used to create additive shares of the secret shared data

from multiplicative shares of it. These procedures generalize to procedures by which information is shifted from n_1 *M.parties* to n_2 *A.parties* in the following way. Assume n_1 *M.parties*, $\mathcal{P}^{(i)}$, $1 \leq i \leq n_1$, hold n_1 multiplicative shares, x_i, of an element m. To achieve n_2 additive shares of m held by n_2 *A.parties*, $\mathcal{P}^{(1)}$ add.-random-splits x_1 to n_2 additive shares b_j, $1 \leq j \leq n_2$, and sends each b_j to the j'th *A.party*. The rest of the *M.parties*, $\mathcal{P}^{(i)}$, $2 \leq i \leq n_1$, send x_i to each of the *A.parties*. At this stage the *M.parties* are eliminated and the *A.parties* multiply the values received to obtain additive shares of m. Observe:

$$m = \prod_{i=1}^{n_1} x_i = x_1 \cdot \prod_{i=2}^{n_1} x_i = \left(\sum_{j=1}^{n_2} b_j \right) \cdot \prod_{i=2}^{n_1} x_i = \sum_{j=1}^{n_2} \left(b_j \cdot \prod_{i=2}^{n_1} x_i \right).$$

Multiplicative Shares from Additive Shares.
Procedure $A \rightarrow M$ of the arithmetic streams scenario and procedure *RetCas* of the Evaluation protocol (the cascading options of it) in the arithmetic functions scenario demonstrate shifting of information from two *A.parties* to two *M.parties*. These procedures are used to create multiplicative shares of the secret shared data from additive shares of it. These procedures generalize to procedures by which information is shifted from n_2 *A.parties* to n_1 *M.parties* in the following way. Assume n_2 *A.parties*, $\mathcal{P}^{(i)}$, $1 \leq i \leq n_2$, hold n_2 additive shares, x_i, of an element m. To achieve n_1 multiplicative shares of m held by n_1 *M.parties*, the user performs mult.-random-split of 1 to n_1 multiplicative shares. The user sends to each of $n_1 - 1$ *M.parties* one (distinct) multiplicative share of 1, and sends the last share of 1 to all of the *A.parties*. Each of the *A.parties* then multiplies the multiplicative share of 1 received by its additive share of m and send the product to the last *M.party*. At this stage the *A.parties* are eliminated and the last *M.party* adds the values received. Now the *M.parties* hold multiplicative shares of m.

Evaluation of Boolean Circuits. The schemes suggested in Sects. 2 and 3 can be used to perform computations of boolean streams and SMPC of boolean circuits by working in $\mathbb{F}_2$. A *True* boolean value is $1 \in \mathbb{F}_2$ and a *False* boolean value is $0 \in \mathbb{F}_2$. Boolean operations may be identified with field operations in the following way. The $\wedge$ operation is identified with $\mathbb{F}_2$ multiplication, the $\oplus$ operation is identified with $\mathbb{F}_2$ addition, and the $\neg$ operation is identified with adding 1 in $\mathbb{F}_2$. The $\vee$ operation of two literals is identified with $x + y + x \cdot y$, where x and y are the elements of $\mathbb{F}_2$ corresponding to the literals. Then, given a boolean circuit C over boolean literals $b = (b_1, \ldots, b_n) \in \{True, False\}^n$, one can use the schemes suggested above for p-bounded functions to perform boolean streams computation and SMPC of boolean circuits by taking $m = (m_1, \ldots, m_n) \in \mathbb{F}_2^n$, where the m_i's are the $\mathbb{F}_2$ correspondents of the b_i's. The boolean circuit $C : \{True, False\}^n \rightarrow \{True, False\}$ will be taken as a function $\tilde{C} : \mathbb{F}_2^n \rightarrow \mathbb{F}_2$.

Evaluating Functions Over Inputs Held by More Than One Participant. The scheme suggested in Sect. 3 can be used to perform SMPC of arithmetic functions over inputs held by several participants. Instead of having only one participant holding inputs, assume $\mathcal{D}^{(1)}, \ldots, \mathcal{D}^{(k)}$ are k users, each of them is holding a set of secret values in $\mathbb{F}_p$, and that the users wish to privately evaluate a function f over the entire set of variables. Let each of the users distribute shares of her secrets independently to the $M.parties$ using the distribution protocol described in Sect. 3. Let one of the users invoke the evaluation protocol described above sending the relevant information about f to the $M.parties$ and the $A.parties$. In the final stage of the evaluation protocol, let the $A.parties$ send their outputs to all of the users. Adding these outputs, each of the users obtains the result of evaluating f over the entire set of secrets. This way we extend our scheme to one that supports SMPC of functions over inputs held by several participants. We note that, in addition to their original role in the scheme, the $M.parties$ and $A.parties$ can also take the role of being users, holding secret values and following the scheme described above as users. This will not affect the security nor the correctness of the schemes.

Handling Additions and Multiplications by the Same Participants. The schemes described above use a different set of participant for additions and for multiplications. We can use one set of participants for both operations by using, on each shift from multiplications to additions or vice versa, (at least) two temporary auxiliary participants. To switch from multiplication to addition mode, the participants that are holding multiplicative shares of the secrets use $M \to A$ to allow the auxiliary parties receive additive shares of the secrets. The auxiliary parties use $AinA$ to add zero to the secret shared message and shift the information back to the permanent set of participants. Similarly, to switch from additions to multiplications, the participants that are holding additive shares of the secrets use $A \to M$ to allow the auxiliary parties receive multiplicative shares of the secrets. The auxiliary parties use $MinM$ to multiply the secret shared message by 1 and shift the information back to the permanent set of participants. This adjustment costs in communication complexity.

5   Security

Keeping the Circuits Secure. In the schemes suggested in Sects. 2 and 3, some information about the circuit itself is revealed to the participants. In the arithmetic streams schemes the $M.parties$ (respectively, $A.parties$) know exactly how many consecutive multiplications (respectively, additions) are computed in a specific part of the circuit. In the SMPC schemes some information about f itself is revealed to the participants, as according to the Evaluation protocol, the user sends the relevant elements $i \in \mathcal{I}$ to the $M.parties$ and the corresponding a_i's to the $A.parties$. We now suggest ways to prevent that leakage of information by adding noise to the procedure in cost of communication complexity.

Securing Arithmetic Streams.
We can adjust the arithmetic streams schemes to prevent leakage of information about the computation circuit itself by performing, at each stage of the computation, both addition and multiplication operations that yield the same result

that would have been obtained normally. If at stage i one has $op_i = +$ (meaning that the user needs to multiply st by m_i), then

- use $MinM$ to multiply st by m_i,
- use $M \to A$ to switch from multiplication mode to addition mode and eliminate the $M.parties$,
- use $AinA$ to add zero to st,
- use $A \to M$ to switch back from addition mode to multiplication mode using a new set of $M.parties$ and eliminate the $A.parties$.

If at stage i the user needs to add m_i to st, then

- use $MinM$ to multiply st by 1,
- use $M \to A$ to switch from multiplication mode to addition mode and eliminate the $M.parties$,
- use $AinA$ to add m_i to st,
- use $A \to M$ to switch back from addition mode to multiplication mode using a new set of $M.parties$ and eliminate the $A.parties$.

This adjustment costs in communication complexity, but it keeps the arithmetic circuit secure in a way that neither of the participants can tell what are the arithmetic operations that are actually being performed.

Securing Arithmetic Functions.
Recall that the information held by the user is $m = (m_1, \ldots, m_n) \in \mathbb{F}_p^n$. We can take redundant copies of each (or some) of the m_i's, take redundant variables that equal $1 \in \mathbb{F}_p$, take redundant variables that equal $0 \in \mathbb{F}_p$ and permute them all to obtain $m' = (m'_1, \ldots, m'_r)$ which contains the information we began with along the added redundancy. This expansion of m costs in communication complexity but now we can hide f in several ways.
    Recall that
$$f(m) = \sum_{i \in \mathcal{I}} a_i \cdot A_i, \quad a_i \in \mathbb{F}_p,$$

where A_i is the i'th monomial. In most applications, most of the a_i's are zero and we will call their corresponding monomials *the zero monomials*. We will call the other monomials *the non-zero monomials*. Now, one can mask f by the following procedures. To evaluate $f : \mathbb{F}_p^n \to \mathbb{F}_p$ over m, take some suitable $f' : \mathbb{F}_p^r \to \mathbb{F}_p$ and evaluate it over m' in such a way that $f(m) = f'(m')$. f' can mask f in the following ways.

- The non-zero monomials of f can be represented in various forms. Since m' contains redundant copies of the variables of m and redundant 1-variables, one can compute monomials of f by various choices of monomials of f'. For example, if one of the monomials of f is x_1^8, and m' contains redundant copies of m_1, $m'_2 = m'_3 = m'_4 = m_1$ and $m'_5 = 1$, then the corresponding monomial of f' may be $x_2^2 x_3^3 x_4^3 x_5^3$.

– Since m' contains redundant 0-variables, one can take f' which contains
redundant monomials that contain a redundant 0-variable. For example, if
$f(m) = m_1^2$, then one can take $f'(m') = {m_1'}^2 + 4{m_6'}^3 m_8'$, where $m_1' = m_1$
and m_6' or m_8' equal zero. The user should keep in mind the indices of the
redundant variables.

These procedures add noise to the computation circuit and cost in an expansion of m and communication complexity.

Malicious Participants and Threshold Analysis. The correction and security of our schemes are based on the assumption that the participants are honest-but-curious, and that they do not form coalitions. That is, we assume that each of the participants follows the exact directions of each procedure of the scheme and is not sending to any of the other participants information not supposed to be sent. Nevertheless, we assume that the participants are trying to learn information about the secret shared inputs and about the computation circuits through the data received during the execution of the scheme. In case of deviation of a participant from the directions of the scheme either the scheme might yield an incorrect solution or the security of the secret shared data may be compromised. We now discuss ways to detect incorrect outputs caused by malicious participants and analyze the threshold for ensuring the security of the schemes against coalitions of participants.

Output Verification.
Detection of incorrect output caused by malicious participants is achieved either by repeating the same computations while using different sets of participants, or by computing different representations of the same function. Assume one runs our scheme using a total of n participants. For a positive integer, s, one can use $s \cdot n$ participants where each n participants run the same protocol independently. As s is taken to be larger, the correction of the output can be verified with higher certainty. Another approach to detect incorrect output is computing the same circuit several times using the same n participants with different randomization in each computation and different representations of the same circuit in each iteration. In this case, one may use schemes for masking the computation as described above, thus ensuring that the participants cannot force repeated incorrect output in successive computations of the same circuit. One exception to this end is a repeated incorrect zero output that can be forced by a malicious $M.party$ by outputting zero regardless of the inputs received. These two approaches can be combined to reveal malicious participants in the following way. The user can use more than n participants and repeat the same computations (independently) using different n participants on each iteration. Assuming the user receives different outcomes, she can eliminate both sets of participants and repeat the process until identical results are obtained.

Security.
We now discuss the security of our schemes against attacks of coalitions of participants. That is, participants that join their shares of m to learn information

about the secret shared inputs. Assume a user runs a scheme as suggested above using n_1 *M.parties* and n_2 *A.parties*. In case of an attack of a coalition of *M.parties*, if the size of the coalition is up to $n_1 - 1$, no information about the secret shared input can be gained by the coalition, since for each product of $n_1 - 1$ non-zero elements of a finite field, x_i, $1 \leq i \leq n_1 - 1$, and for each non-zero element m of the field, there exists exactly one element x_{n_1} in the field such that the product of all the n_1 elements x_i yields m. Similarly, in case of an attack of a coalition of *A.parties*, if the size of the coalition is up to $n_2 - 1$, no information about the secret shared input can be gained by the coalition, since for each sum of $n_2 - 1$ elements of a finite field, x_i, $1 \leq i \leq n_2 - 1$, and for each element m of the field, there exists exactly one element x_{n_2} in the field such that the sum of all the n_2 elements x_i yields m. Hence, the threshold of the schemes is the number of currently-active participants.

6   Conclusions

We have suggested schemes for information-theoretically SMPC of arithmetic streams and evaluation of arithmetic functions in one round of communication. The schemes presented above consider two sets of participants. The first set consists of n_1 *M.parties* that handle sequences of multiplications, and the second set consists of n_2 *A.parties* that handle sequences of additions. Such sequences are handled locally and require no communication between parties. We switch from sequences of multiplications to sequences of additions and vice versa without decrypting the information. We have suggested ways to secure the arithmetic circuit being computed in cost of communication complexity. We have suggested ways to detect incorrect outputs caused by malicious parties. Our schemes are secure against attacks of coalitions of participants that are smaller than the number of currently active participants. Our schemes can be used to perform computations of boolean circuits among many other scopes.

Acknowledgments. We thank Dani Berend for being involved during the entire research providing original ideas throughout, in particular suggesting to use polynomial representation instead of circuits.

The research was partially supported by the Rita Altura Trust Chair in Computer Sciences; the Lynne and William Frankel Center for Computer Science; the Ministry of Foreign Affairs, Italy; the grant from the Ministry of Science, Technology and Space, Israel, and the National Science Council (NSC) of Taiwan; the Ministry of Science, Technology and Space, Infrastructure Research in the Field of Advanced Computing and Cyber Security; and the Israel National Cyber Bureau.

References

[BIB89] Bar-Ilan, J., Beaver, D.: Non-cryptographic fault-tolerant computing in constant number of rounds of interaction. In: Proceedings of the Eighth Annual ACM Symposium on Principles of Distributed Computing, pp. 201–209. ACM (1989)

[BMR90] Beaver, D., Micali, S., Rogaway, P.: The round complexity of secure protocols. In: Proceedings of the Twenty-Second Annual ACM Symposium on Theory of Computing, pp. 503–513. ACM (1990)

[BOGW88] Ben-Or, M., Goldwasser, S., Wigderson, A.: Completeness theorems for non-cryptographic fault-tolerant distributed computation. In: Proceedings of the Twentieth Annual ACM Symposium on Theory of Computing, pp. 1–10. ACM (1988)

[BP16] Brakerski, Z., Perlman, R.: Lattice-based fully dynamic multi-key FHE with short ciphertexts. In: Robshaw, M., Katz, J. (eds.) CRYPTO 2016. LNCS, vol. 9814, pp. 190–213. Springer, Heidelberg (2016). https://doi.org/10.1007/978-3-662-53018-4_8

[CCD88] Chaum, D., Crépeau, C., Damgard, I.: Multiparty unconditionally secure protocols. In: Proceedings of the Twentieth Annual ACM Symposium on Theory of Computing, pp. 11–19. ACM (1988)

[DFK+06] Damgård, I., Fitzi, M., Kiltz, E., Nielsen, J.B., Toft, T.: Unconditionally secure constant-rounds multi-party computation for equality, comparison, bits and exponentiation. In: Halevi, S., Rabin, T. (eds.) TCC 2006. LNCS, vol. 3876, pp. 285–304. Springer, Heidelberg (2006). https://doi.org/10.1007/11681878_15

[DGL15] Dolev, S., Gilboa, N., Li, X.: Accumulating automata and cascaded equations automata for communicationless information theoretically secure multi-party computation. In: Proceedings of the 3rd International Workshop on Security in Cloud Computing, pp. 21–29. ACM (2015)

[DIK10] Damgård, I., Ishai, Y., Krøigaard, M.: Perfectly Secure multiparty computation and the computational overhead of cryptography. In: Gilbert, H. (ed.) EUROCRYPT 2010. LNCS, vol. 6110, pp. 445–465. Springer, Heidelberg (2010). https://doi.org/10.1007/978-3-642-13190-5_23

[DL16] Dolev, S., Li, Y.: Secret shared random access machine. In: Karydis, I., Sioutas, S., Triantafillou, P., Tsoumakos, D. (eds.) ALGOCLOUD 2015. LNCS, vol. 9511, pp. 19–34. Springer, Cham (2016). https://doi.org/10.1007/978-3-319-29919-8_2

[DLY07] Dolev, S., Lahiani, L., Yung, M.: SECRET SWARM UNIT reactive k-secret sharing. In: Srinathan, K., Rangan, C.P., Yung, M. (eds.) INDOCRYPT 2007. LNCS, vol. 4859, pp. 123–137. Springer, Heidelberg (2007). https://doi.org/10.1007/978-3-540-77026-8_10

[Gen09] Gentry, C.: A fully homomorphic encryption scheme. Stanford University, Stanford (2009)

[GHS12] Gentry, C., Halevi, S., Smart, N.P.: Fully homomorphic encryption with polylog overhead. In: Pointcheval, D., Johansson, T. (eds.) EUROCRYPT 2012. LNCS, vol. 7237, pp. 465–482. Springer, Heidelberg (2012). https://doi.org/10.1007/978-3-642-29011-4_28

[GHS16] Gentry, C.B., Halevi, S., Smart, N.P.: Homomorphic evaluation including key switching, modulus switching, and dynamic noise management, 8 March 2016. US Patent 9,281,941

[GIKR02] Gennaro, R., Ishai, Y., Kushilevitz, E., Rabin, T.: On 2-round secure multiparty computation. In: Yung, M. (ed.) CRYPTO 2002. LNCS, vol. 2442, pp. 178–193. Springer, Heidelberg (2002). https://doi.org/10.1007/3-540-45708-9_12

[IK02] Ishai, Y., Kushilevitz, E.: Perfect constant-round secure computation via perfect randomizing polynomials. In: Widmayer, P., Eidenbenz, S., Triguero, F., Morales, R., Conejo, R., Hennessy, M. (eds.) ICALP 2002. LNCS, vol. 2380, pp. 244–256. Springer, Heidelberg (2002). https://doi.org/10.1007/3-540-45465-9_22

[KN06] Kushilevitz, E., Nissan, N.: Communication Complexity. Cambridge University Press, United Kingdom (2006)

[Sha79] Shamir, A.: How to share a secret. Commun. ACM **22**(11), 612–613 (1979)

[SV10] Smart, N.P., Vercauteren, F.: Fully homomorphic encryption with relatively small key and ciphertext sizes. In: Nguyen, P.Q., Pointcheval, D. (eds.) PKC 2010. LNCS, vol. 6056, pp. 420–443. Springer, Heidelberg (2010). https://doi.org/10.1007/978-3-642-13013-7_25

[VDGHV10] van Dijk, M., Gentry, C., Halevi, S., Vaikuntanathan, V.: Fully homomorphic encryption over the integers. In: Gilbert, H. (ed.) EUROCRYPT 2010. LNCS, vol. 6110, pp. 24–43. Springer, Heidelberg (2010). https://doi.org/10.1007/978-3-642-13190-5_2

[XWZ+18] Jian, X., Wei, L., Zhang, Y., Wang, A., Zhou, F., Gao, C.: Dynamic fully homomorphic encryption-based merkle tree for lightweight streaming authenticated data structures. J. Netw. Comput. Appl. **107**, 113–124 (2018)

Brief Announcement: Gradual Learning of Deep Recurrent Neural Network

Ziv Aharoni[⊠], Gal Rattner, and Haim Permuter

Ben-Gurion University, 8410501 Beer-Sheva, Israel
{zivah,rattner}@post.bgu.ac.il, haimp@bgu.ac.il

Abstract. Deep Recurrent Neural Networks (RNNs) achieve state-of-the-art results in many sequence-to-sequence modeling tasks. However, deep RNNs are difficult to train and tend to suffer from overfitting. Motivated by the Data Processing Inequality (DPI) we formulate the multi-layered network as a Markov chain, introducing a training method that comprises training the network gradually and using layer-wise gradient clipping. In total, we have found that applying our methods combined with previously introduced regularization and optimization methods resulted in improvement to the state-of-the-art architectures operating in language modeling tasks.

Keywords: Data-processing-inequality · Machine-learning
Recurrent-neural-networks · Regularization · Training-methods

1 Introduction

Several forms of deep Recurrent Neural Network (RNN) architectures, such as LSTM [7] and GRU [2], have achieved state-of-the-art results in many sequential classification tasks [3,5,6,14,15,17] during the past few years. The number of stacked RNN layers, i.e. the network depth, has key importance in extending the ability of the architecture to express more complex dynamic systems [1,12]. However, training deeper networks poses problems that are yet to be solved.

   In this paper, we suggest an approach that breaks the optimization process into several learning phases. Each learning phase includes training an increasingly deeper architecture than the previous ones. In this way, we gradually train and extend the network depth, reducing the deleterious effects of degradation and backpropagation problems. Additionally, by adjusting the appropriate training scheme (mainly the regularization) at every learning phase, we are able to maximize the network performance even further.

2 Gradual Learning

2.1 Notation

Let us represent a network with l layers as a mapping from an input sequence $X \in \mathcal{X}$ to an output sequence $\hat{Y}_l \in \mathcal{Y}$ by $\hat{Y}_l = S_l \circ f_l \circ f_{l-1} \circ \cdots \circ f_1(X; \Theta_l)$,

© Springer International Publishing AG, part of Springer Nature 2018
I. Dinur et al. (Eds.): CSCML 2018, LNCS 10879, pp. 274–277, 2018.
https://doi.org/10.1007/978-3-319-94147-9_21

where the term $\Theta_l = \{\theta_1, \ldots, \theta_l, \theta_{S_l}\}$ denotes the network parameters, such that θ_k are the parameters of the kth layer. We also define the lth layer cost function by $J(\Theta_l) = \text{cost}(\hat{Y}_l, Y)$, where $\theta^l = \{\theta_1, \ldots, \theta_l\}$. Next, we define the gradient vector with respect to $J(\Theta)$ by $\mathbf{g} = \frac{\partial}{\partial \Theta} J(\Theta)$, and the gradient vector of the kth layer parameters with respect to $J(\Theta)$ by $\mathbf{g}_k = \frac{\partial}{\partial \theta_k} J(\Theta)$.

2.2 Theoretical Motivation

The structure of a neural network comprises a sequential processing scheme of its input. This structure constitutes the Markov chain $Y - X - T_1 - T_2 - \cdots - T_L$. The goal is to estimate $P_{Y|T_L}(y|t)$ by $Q^{\Theta}_{Y|T_L}(y|t)$. Driven by the Markov relation we state two theorems (without proofs due to space constraints).

Theorem 1 (Maximum Likelihood Estimator (MLE) and minimal negative log-likelihood). *Given a training set of N examples $S = \{(x_i, y_i)\}_{i=1}^{N}$ drawn i.i.d from an unknown distribution $P_{X,Y} = P_X P_{Y|X}$, the MLE of $Q^{\Theta}_{Y|T_L}$ is given by $P_{Y|X}$ and the optimal value of the criteria is $H(Y|X)$.*

Theorem 2. *If $Q^{\Theta}_{Y|T_L}$ satisfies the optimality conditions of Theorem 1, then $I(X;Y) = I(T_l; Y)$ $\forall l = 1, \ldots, L$.*

We show that by satisfying the optimality criteria of Theorem 1 we necessarily did not lose relevant information of Y by processing X to T_L. In particular, we show that a necessary condition to achieve the MLE is that the network states, namely $\{T_l\}_{l=1}^{L}$, will satisfy $I(Y; X) = I(Y; T_l)$.

2.3 Implementation

Due to the fact that shallow networks are easier to train, we propose a greedy training scheme, where we break the optimization process into L phases (as the number of layers), optimizing $J(\Theta_l)$ sequentially as l increases from 1 to L. The training scheme is depicted in Fig. 1.

3 Layer-Wise Gradient Clipping (LWGC)

Previous studies [4,8,13] have shown that *covariate shift* has a negative effect on the training process among deep neural architectures. *Covariate shift* is the change in a layer's input distribution during training, also manifested as *internal covariate shift*. We suggest that treating each layer weights' gradient vector individually and clipping the gradients vector layer-wise can reduce *internal covariate shift* significantly. LWGC for a network with L different layers is formulated as

$$[\hat{\mathbf{g}}_1^T, \ldots, \hat{\mathbf{g}}_L^T]^T := \left[\frac{\mu_1}{\max(\mu_1, \|\mathbf{g}_1\|)} \mathbf{g}_1^T, \ldots, \frac{\mu_N}{\max(\mu_N, \|\mathbf{g}_N\|)} \mathbf{g}_N^T \right]^T. \quad (1)$$

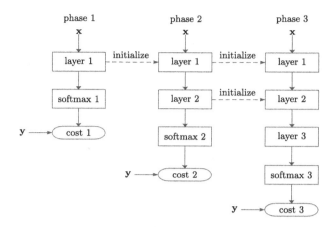

Fig. 1. Depiction of our training scheme for a 3 layered network. At phase 1 we optimize the parameters of layer 1 according to cost 1. At phase 2, we add layer 2 to the network, and then we optimize the parameters of layers 1, 2, when layer 1 is copied from phase 1 and layer 2 is initialized randomly. At phase 3, we add layer 3 to the network, and then we optimize all of the network's parameters, when layers 1, 2 are copied from phase 2 and layer 3 is initialized randomly.

4   Experiments

We present results on a dataset from the field of natural language processing, the PTB, conducted as a word-level dataset.

We conducted two models in our experiments, a *reference* model and a *GL-LWGC LSTM* model that was used to check the performance of our methods. Our GL-LWGC LSTM model compared the state-of-the-art results with only two layers and 19M parameters, and achieved state-of-the-art results with the third layer phase. Results of the *reference* model and *GL-LWGC LSTM* model are shown in Table 1.

Table 1. Single model validation and test perplexity of the PTB dataset

Model	Size	Valid	Test
Zoph and Le [18] - NAS	25M	-	64.0
Melis et al. [10] - 2-layer skip connection LSTM	24M	60.9	58.3
Merity et al. [11] - AWD-LSTM	24M	60.0	57.3
Yang et al. [16] - AWD-LSTM-MoS + finetune	22M	56.54	54.44
Ours - 2-layers GL-LWGC-AWD-MoS-LSTM + finetune	19M	55.18	53.54
Ours - GL-LWGC-AWD-MoS-LSTM + finetune	26M	**54.24**	**52.57**
Krause et al. [9] AWD-LSTM + dynamic evaluation	24M	51.6	51.1
Yang et al. [16] AWD-LSTM-MoS + dynamic evaluation	22M	48.33	47.69
Ours - GL-LWGC-AWD-MoS-LSTM + dynamic evaluation	26M	**46.64**	**46.34**

References

1. Bianchini, M., Scarselli, F.: On the complexity of neural network classifiers: a comparison between shallow and deep architectures. IEEE Trans. Neural Netw. Learn. Syst. (2014)
2. Cho, K., Van Merriënboer, B., Bahdanau, D., Bengio, Y.: On the properties of neural machine translation: encoder-decoder approaches (2014). arXiv preprint arXiv:1409.1259
3. Cho, K., van Merrienboer, B., Gulcehre, C., Bahdanau, D., Bougares, F., Schwenk, H., Bengio, Y.: Learning phrase representations using RNN Encoder-Decoder for statistical machine translation (2014). arXiv preprint arXiv:1406.107
4. Cooijmans, T., Ballas, N., Laurent, C., Gülçehre, Ç., Courville, A.: Recurrent batch normalization (2016). arXiv preprint arXiv:1603.09025
5. Ha, D., Dai, A., Le, Q.V.: Hypernetworks (2016). arXiv preprint arXiv:1609.09106
6. He, K., Zhang, X., Ren, S., Sun, J.: Deep residual learning for image recognition (2015). arXiv preprint arXiv:1512.03385
7. Hochreiter, S., Schmidhuber, J.: Long short-term memory. Neural Comput. 9(8), 1735–1780 (1997)
8. Ioffe, S., Szegedy, C.: Batch normalization: Accelerating deep network training by reducing internal covariate shift (2015). arXiv preprint arXiv:1502.03167
9. Krause, B., Kahembwe, E., Murray, I., Renals, S.: Dynamic evaluation of neural sequence models (2017). arXiv preprint arXiv:1709.07432
10. Melis, G., Dyer, C., Blunsom, P.: On the State of the Art of Evaluation in Neural Language Models. ArXiv e-prints, July 2017
11. Merity, S., Shirish Keskar, N., Socher, R.: Regularizing and Optimizing LSTM Language Models. ArXiv e-prints, August 2017
12. Montufar, G., Pascanu, R., Cho, K., Bengio, Y.: On the number of linear regions of deep neural networks (2014). arXiv preprint arXiv:1402.1869
13. Shimodaira, H.: Improving predictive inference under covariate shift by weighting the log-likelihood function. J. Stat. Plann. Infer. 90(2), 227–244 (2000)
14. Smith, L.N., Hand, E.M., Doster, T.: Gradual dropin of layers to train very deep neural networks (2015). arXiv preprint arXiv:1511.06951
15. Sutskever, I., Martens, J., Dahl, G., Hinton, G.: On the importance of initialization and momentum in deep learning. In: Proceedings of the 30th International Conference on International Conference on Machine Learning, ICML 2013, vol. 28, pp. III-1139–III-1147 (2013). JMLR.org
16. Yang, Z., Dai, Z., Salakhutdinov, R., Cohen, W.W.: Breaking the softmax bottleneck: a high-rank RNN language model (2017). arXiv preprint arXiv:1711.03953
17. Zilly, J.G., Srivastava, R.K., Koutník, J., Schmidhuber, J.: Recurrent highway networks (2016). arXiv preprint arXiv:1607.03474
18. Zoph, B., Le, Q.V.: Neural architecture search with reinforcement learning (2016). arXiv preprint arXiv:1611.01578

Brief Announcement: Adversarial Evasion of an Adaptive Version of Western Electric Rules

Oded Margalit[1,2(✉)]

[1] IBM Cybersecurity Center of Excellence, Beersheba, Israel
odedm@il.ibm.com
[2] CS Department, BGU, Beersheba, Israel
https://researcher.watson.ibm.com/researcher/view.php?person=il-ODEDM

Abstract. Western-Electric are one of the earliest, and widely used, anomaly detection rules. In this paper we describe an adaptive scenario using these rules and show how a malicious player can optimally fabricate data to deceive the algorithm to enlarge the standard deviation of the data while avoiding being detected.

1 Introduction

Western Electric created rules to detect anomalies in time series. The rules were first published on 1956 [1] and since then became a well accepted standard in the industry. Examples of recent usage of these rules can be found, for example, at [3,4]. These rules can be used, for example, for anomaly detection in Domain Name System (DNS) requests; in HTTP traffic; for Data Leakage Prevention (DLP); etc.

In Sect. 2 we define the rules; in Sect. 3 we define our own version (the adaptive version); in Sect. 4 we show how an adversary can attack the model; and in Sect. 5 we summarize the work and give ideas for future work.

2 Western Electric Rules

Western Electric has four statistical process control rules that are aimed to detect anomalies in normally distributed $N(\mu, \sigma)$ time series data:

(1) Any point outside of the $[\mu - 3\sigma, \mu + 3\sigma]$ range.
(2) Two out of three consecutive points in the range $[\mu - 3\sigma, \mu - 2\sigma)$ or two out of three consecutive points in the range $(\mu + 2\sigma, \mu + 3\sigma]$.
(3) Four out of five consecutive points in the range $[\mu - 2\sigma, \mu - \sigma)$ or four out of five consecutive points in the range $(\mu + \sigma, \mu + 2\sigma]$.
(4) Nine consecutive points in the range (μ, ∞) (above the average) or nine consecutive points in the range $(-\infty, -\mu)$ (below the average).

This work was not supported by any organization.

These rules assume that the distribution of the measured values is normal (otherwise, the Chebyshev's inequality will give us weaker bounds on the probability of values far away from the mean) which might not be the case; but by the central limit theory, when the data is aggregated, i.e. it is the average of many independent random variables, it converges to a normal distribution.

3 Adaptive Model

Western Electric rules are designed to detect anomalies in data which has normal distribution with known mean μ and standard deviation σ. So the usual use-case is when you know the behavior of your data beforehand.

Sometimes the mean and the standard deviation is not known in advance. A common practice in such cases, that we chose to investigate in this paper, is a slightly modified model, where we estimate the average (μ) and the standard deviation (σ) from the data itself. Let $\{x_i\}_{i=1}^{\infty}$ be the values of the data series and define expected average μ for $t \geq 1$ to be

$$\mu_t = \frac{1}{t}\sum_{i=1}^{t} x_i$$

and define the standard deviation σ for $t > 1$ to be

$$\sigma_t = \sqrt{\frac{\sum_{i=1}^{t}(x_i - \mu_t)^2}{t}}.$$

For the $(n+1)^{th}$ data-point we use μ_n and σ_n.

4 Adversarial Attack

Let's assume that we are working in DLP domain (we can get similar results in other domains). The attacker is trying to exfiltrate (leak) data out of a system which is protected by Western-Electric rules which implement egress filtering by monitoring the rate of data going out of the protected system. The attacker can decide how much data (how many packets of information) she sends at any given point in time. Her goal is to send as much data as possible, but without alerting the defender by violating the Western-Electric rules. So her problem is to find x_τ, the number of packets sent at time τ such that the sequence $\{x_\tau\}_{\tau=1}^{t}$ will pass the rules while maximizing σ_t. The amount of data leaked is proportional to μ_t but increasing σ_t is more important to the attacker since it will allow her more degrees of freedom to manipulate data in the future.

Using the modified model of Sect. 3 we can fabricate data with extreme values while avoiding detection. Before getting into the specific rules, we formulate and prove a theorem:

Theorem 1. If the sequence $\{x_i\}_{i=1}^t$ is defined as $x_{n+1} = \mu_n \pm 3\sigma_n$ then σ_t is the same regardless of the signs chosen in the process; where σ_τ is the standard deviation of the first τ values in the x_i sequence.

Proof. If $x_{n+1} = \mu_n + \alpha\sigma_n$ then

$$\mu_{n+1} = \frac{1}{n+1} \sum_{i=1}^{n+1} x_i$$

$$= \frac{1}{n+1} \left(\sum_{i=1}^{n} x_i + x_{n+1} \right)$$

$$= \frac{1}{n+1} \left(n\mu_n + (\mu_n + \alpha\sigma_n) \right)$$

$$= \mu_n + \frac{\alpha\sigma_n}{n+1}$$

So we can compute the new variance:

$$\sigma_{n+1}^2 = \frac{1}{n+1} \sum_{i=1}^{n+1} (x_i - \mu_{n+1})^2$$

$$= \frac{1}{n+1} \left(\sum_{i=1}^{n} (x_i - (\mu_n + \frac{\alpha\sigma_n}{n+1}))^2 \right.$$

$$\left. + ((\mu_n + \alpha\sigma_n) - (\mu_n + \frac{\alpha\sigma_n}{n+1}))^2 \right)$$

$$= \frac{1}{n+1} \left(\sum_{i=1}^{n} \left((x_i - \mu_n)^2 \right. \right.$$

$$\left. -2(x_i - \mu_n)\frac{\alpha\sigma_n}{n+1} + \left(\frac{\alpha\sigma_n}{n+1}\right)^2 \right)$$

$$\left. + \left(\frac{n\alpha\sigma_n}{n+1}\right)^2 \right)$$

$$= \frac{1}{n+1} \left(n\sigma_n^2 - 2 \cdot 0 \cdot \frac{\alpha\sigma_n}{n+1} + \frac{n\alpha^2\sigma_n^2}{(n+1)^2} \right.$$

$$\left. + \frac{n^2\alpha^2\sigma_n^2}{(n+1)^2} \right)$$

$$= \frac{n\sigma_n^2}{n+1} \left(1 + \frac{\alpha^2}{(n+1)^2} + \frac{n\alpha^2}{(n+1)^2} \right)$$

$$= \frac{n\sigma_n^2}{n+1} \left(1 + \frac{\alpha^2}{n+1} \right).$$

If, as in our case, the αs are ± 3, then the sign cancels out by the squaring and the resulted sequence $\{\sigma_i\}_{i=2}^n$ is the same regardless of the sign – so it does not matter if we make the data abnormal by letting rule number 1 *almost* fire upwards $(+3\sigma)$ or downwards (-3σ) – the drifted variance will be exactly the same. ∎

Let's start with the rules one by one.

(1) The simplest way to avoid the first rule is to keep the maximum possible value all the time: each value x_{n+1} will be $\mu_n + 3\sigma_n$. This way we can drift the standard deviation as fast as possible without letting rule number 1 fire.

This will make both μ_t and σ_t as high as possible. As we saw earlier, raising μ_t is nice but increasing σ_t is more important.

(2) The solution above will be caught by the second rule. To avoid this we must have at least a one of every consecutive three be no more than 2σ away from the average μ. Otherwise, using the pigeon-hole principle, at least two out of the three will be on the same side. This one third bound is tight, as the following sequence proves:

$$x_{n+1} = \begin{cases} \mu_n - 3\sigma_n, \, n = 0 \bmod (3) \\ \mu_n + 3\sigma_n, \, n = 1 \bmod (3) \\ \mu_n - 2\sigma_n, \, n = 2 \bmod (3) \end{cases}$$

Note that there *is* a small difference between $+3\sigma, -3\sigma, +2\sigma$ and $+2\sigma, +3\sigma, -3\sigma$ since careful examination of the proof of Theorem 1 shows that when α is changing, the effect on σ changes.

(3) The sequence above passes the first two rules, but fails the third. This time we can fix it without degrading the performance, by alternating the 2σ:

$$x_{n+1} = \begin{cases} \mu_n - 3\sigma_n, \, n = 0, 3 \bmod (6) \\ \mu_n + 3\sigma_n, \, n = 1, 4 \bmod (6) \\ \mu_n - 2\sigma_n, \, n = 2 \bmod (6) \\ \mu_n + 2\sigma_n, \, n = 5 \bmod (6) \end{cases}$$

(4) The forth rule is ok — no need to change anything to bypass it.

5   Summary and Future Work

In this paper we've demonstrated a simple way to bypass an adaptive version of half-a-century-old Western-Electric anomaly detection rules.

The original Western Electric rules were designed (and work great) under the assumption that monitored data has normal distribution with known parameters: expected value μ and standard deviation σ. Our revised model learns the parameters and the lesson to learn from it is that adaptive algorithms are prune to adversarial algorithms. New rules can be added, but it is a classic cat-and-mouse game of changing the anomaly detection to catch the adversary and the attacker changes his methods to avoid being caught.

As we saw, by the central limit theory, we can assume, in many realistic scenarios, that the distribution is normal; but what can we say about non-normal distribution?

Western-Electric rules are not the only set of anomaly detection rules. One can try to achieve similar results on other set of rules, like, for example, Nelson Rules [2].

Another research direction – find an elegant proof for Theorem 1.

Acknowledgment. I'd like to thank the anonymous referees for their helpful remarks, which helped me to improve this paper.

References

1. Western Electric Company: Statistical Quality Control Handbook. Western Electric Co, Indianapolis (1956)
2. Nelson, L.S.: The Shewhart control chart tests for special causes. J. Qual. Technol. **16**, 237–239 (1984)
3. Romano, M., Kapelan, Z., Savic, D.: Automated detection of pipe bursts and other events in water distribution systems. American Society of Civil Engineers (2012)
4. Lovell, D.P., Fellows, M., Marchetti, F., Christiansen, J., Elhajouji, A., Hashimoto, K., Kasamoto, S., Li, Y., Masayasu, O., Moore, M.M., Schuler, M., Smith, R., Stankowski, L.F., Tanaka, J., Tanir, J.Y., Thybaud, V., Van Goethem, F., Whitwell, J.: Analysis of negative historical control group data from the in vitro micronucleus assay using TK6 cells. Mutation Res./Genet. Toxicol. Environ. Mutagen. **825**, 40–50 (2018)

Brief Announcement: Deriving Context for Touch Events

Moran Azran, Niv Ben Shabat, Tal Shkolnik, and Yossi Oren[✉]

Department of Software and Information Systems Engineering, Ben Gurion
University, Beer Sheva, Israel
{azranmo,nivb,talshko,yos}@post.bgu.ac.il

Abstract. To quantify the amount of high-level context information
which can be derived by observing only a user's touchscreen interactions,
we performed a user study, in which we recorded 160 touch interaction
sessions from users running different applications, and then applied both
classical machine learning methods and deep learning methods to the
results. Our results show that it is possible to derive higher-level user
context information based on touch events alone, validating the efficacy
of touch injection attacks.

Keywords: Machine learning · Malicious hardware · Smart phone

1 Introduction

Smart phone touchscreens are often produced by third-party manufacturers and
not by the phone vendors themselves. According to a 2015 study, more than 50%
of global smartphone owners have damaged their phone screen at least once, and
21% of global smartphone owners are currently using a phone with a cracked or
shattered screen [1]. These shattered screens are often replaced with aftermarket
components of questionable origin. In [2,3], Shwartz et al. showed how malicious
touchscreen hardware can launch a **touch injection attack** that allows the
touchscreen to impersonate the user and exfiltrate data. One limitation of this
attack approach is that the attacker knows the position and timing of touches on
the victim's screen, but does not have any higher-level **contextual** information
such as the user's current activity or current running application.

The main objective of our research is to quantify the amount of high-level
context information the attacker can derive by observing only the user's touchscreen interactions. If an attacker can understand the context of certain events,
he can use this information to create a customized attack which will be more
effective. For example, the attacker can know when he should steal information from the user or to insert malicious touches. To quantify the amount of
high-level context information which can be derived by observing only a user's
touchscreen interactions, we performed a user study, in which we recorded 160
touch interaction sessions from users running different applications.

This research was supported by Israel Science Foundation grants 702/16 and 703/16.

ⓒ Springer International Publishing AG, part of Springer Nature 2018
I. Dinur et al. (Eds.): CSCML 2018, LNCS 10879, pp. 283–286, 2018.
https://doi.org/10.1007/978-3-319-94147-9_23

2   Method and Results

2.1   Experiment Setup

The experiment was conducted on a group of third year university students. In the first part of the experiment, the subjects were required to fill in a personal survey which included questions such as: age, gender, which hand do you usually hold the cell phone? Do you usually hold the cell phone with both hands? when did you last play on your cell phone? When was the last time you drank coffee? In addition, subjects were asked whether they knew certain games. In the second stage of the experiment, each subject was asked to record four different touch interaction sessions on the test phone. First, the subjects were asked to play the game "Color Infinity". The objective of this game is to pass the ball through various obstacles. The game included fast and short touches around the screen. Next, subjects were required to play a game called "Bricks". The objective of this game is to move the bricks to the appropriate color when at some point the color of the brick changes. This game includes continuous screen touches. When the subjects finished playing games, the subjects were asked to launch the phone's web browser and perform a web search by typing the word "Facebook" in the browser search bar. Finally, the subjects were asked to enter an e-mail application on the cell phone and send an e-mail containing a subject and content line.

2.2   Machine Learning Methods

Feature Selection: The features we selected for classical machine learning were derived from the features described in [4], including median velocity of the five last points of the trajectory, mean resultant length, largest absolute perpendicular distance between the end-to-end connection, stroke duration and inter stroke time. We augmented the feature set of [4] with three additional features suggested by Meng et al. in [5]: average touch movement speed per direction, average single-touch/multi-touch time and number of touch movements per stroke (NTM).

Classifier Selection: We evaluated multiple ML models, including Logistic Regression, Linear Discriminant Analysis, K Nearest Neighbors, Decision Tree, Gaussian Naive Bayes, Random Forest and Quadratic Discriminant Analysis. All models were instantiated using their default parameters. To compare the performance of classical ML algorithms with deep learning algorithms, we also analyzed the raw touch information using a deep learning convolutional neural network (CNN). Our CNN had three Conv1D layers, three MaxPool1D layers and a final softmax activation layer.

2.3   Data Collection and Initial Processing

Data collection was conducted on 01/03/2018 during a university hackathon event. We collected 153 touch recordings from 40 different subjects. The experiment took 4 h in total. To record the touches, we used a specially modified LG

Table 1. Performance of classical machine learning classifiers

Algorithm	Prediction rate
Logistic Regression	0.8954
Linear Discriminant Analysis	0.9215
KNeighbors	0.8039
Decision Tree	0.8692
GaussianNB	0.9281
Random Forest	0.9019
Quadratic Discriminant Analysis	0.8954

Table 2. Predictor ranks for context recognition (as output by the relieff algorithm)

Predictor rank	Feature	Predictor importance weight
1	Stop_Y	0.1938
2	Y_Avg	0.1913
3	Stop_X	0.1790
4	Stroke_Duration	0.1275
5	Start_X	0.0978
6	Start_Y	0.0895
7	Pressure_Avg	0.0634
8	X_Avg	0.0575

Nexus 5X Android phone. The phone was modified at the root-kit level with a touch recording functionality, which runs in the background and outputs a CSV file with the touch screen locations, pressure and timestamp. Data from the phone was downloaded to a workstation running Matlab and Python for further analysis. For classical machine learning we used Matlab's Classification Learner tool any Python's scikit-learn toolkit. For deep learning we used the TensorFlow framework running on Python.

2.4 Machine Learning Results

The performance of the classical machine learning classifiers is summarized in Table 1. The classical machine learning classifiers were highly effective in determining the activity context of the user from the supplied touch data, with the best-performing classifier (Linear Discriminant Analysis) providing a prediction rate of over 92% over the 4 activity contexts evaluated. The relative ranking of the different features, as output by the relieff algorithm, is summarized in Table 2, and shows that the most significant features are the final Y coordinate and the average Y coordinate of each stroke.

We ran our deep learning classifier on the raw data with 30 epochs and a 90–10 validation_split. The deep learning classifier was able to detect the correct activity 87.5% of the time on the validation set, a level of performance similar to that of the classical methods.

3 Conclusion

Our results show that it is possible to derive higher-level user context information based on touch events alone, validating the efficacy of touch injection attacks. Applying touch context analysis on the defensive side can also have a benefit, since it can prevent attacks by identifying anomalous interaction and therefore protect against abnormal use of the phone.

References

1. Motorola Mobility. Cracked screens and broken hearts - the 2015 motorola global shattered screen survey. https://community.motorola.com/blog/cracked-screens-and-broken-hearts
2. Shwartz, O., Shitrit, G., Shabtai, A., Oren, Y.: From smashed screens to smashed stacks: attacking mobile phones using malicious aftermarket parts. In: 2017 IEEE European Symposium on Security and Privacy Workshops, EuroS&P Workshops 2017, Paris, France, 26–28 April 2017, pp. 94–98. IEEE (2017)
3. Shwartz, O., Cohen, A., Shabtai, A., Oren, Y.: Shattered trust: when replacement smartphone components attack. In: Enck, W., Mulliner, C. (eds.) 11th USENIX Workshop on Offensive Technologies, WOOT 2017, Vancouver, BC, Canada, 14–15 August 2017. USENIX Association (2017)
4. Frank, M., Biedert, R., Ma, E., Martinovic, I., Song, D.: Touchalytics: on the applicability of touchscreen input as a behavioral biometric for continuous authentication. IEEE Trans. Inf. Forensics Secur. 8(1), 136–148 (2013)
5. Meng, Y., Wong, D.S., Schlegel, R., Kwok, L.: Touch gestures based biometric authentication scheme for touchscreen mobile phones. In: Kutyłowski, M., Yung, M. (eds.) Inscrypt 2012. LNCS, vol. 7763, pp. 331–350. Springer, Heidelberg (2013). https://doi.org/10.1007/978-3-642-38519-3_21

Author Index

Printed in the United States
By Bookmasters